TECHNOLOGY AND SOCIETY

TECHNOLOGY AND SOCIETY

Advisory Editor
DANIEL J. BOORSTIN, author of
The Americans and Director of
The National Museum of History
and Technology, Smithsonian Institution

THE BUILDERS OF THE BRIDGE

*The Story of
John Roebling and His Son*

D[avid] B[arnard] Steinman

ARNO PRESS
A NEW YORK TIMES COMPANY
New York • 1972

Reprint Edition 1972 by Arno Press Inc.

Copyright © 1945, 1950 by Harcourt, Brace
and Company, Inc.
Reprinted by permission of Harcourt, Brace
Jovanovich, Inc.

Reprinted from a copy in The Pennsylvania
State Library

Technology and Society
ISBN for complete set: 0-405-04680-4
See last pages of this volume for titles.

Manufactured in the United States of America.

Library of Congress Cataloging in Publication Data

Steinman, David Barnard, 1886-1960.
   The builders of the bridge.

   (Technology and society)
   Bibliography: p.
   1. Roebling, John Augustus, 1806-1869.
2. Roebling, Washington Augustus, 1837-1926. I. Title. II. Series.
TA140.R7S8 1972   624.2'092'2 [B]   72-5074
ISBN 0-405-04724-X

THE BUILDERS OF THE BRIDGE

D. B. STEINMAN

THE BUILDERS OF THE BRIDGE

*The Story of
John Roebling and His Son*

New York
HARCOURT, BRACE AND COMPANY

COPYRIGHT, 1945, 1950, BY
HARCOURT, BRACE AND COMPANY, INC.

All rights reserved, including the right to reproduce this book or portions thereof in any form.

PRINTED IN THE UNITED STATES OF AMERICA

DEDICATED

TO THE MEMORY OF

THE BUILDERS OF THE BRIDGE

THAT THEIR

UNCONQUERABLE SPIRIT

MAY LIVE ON AS AN

ENDURING INSPIRATION TO

THE BUILDERS OF TOMORROW

PREFACE

A boy grew up in the shadows of the Bridge. He loved to walk over the span and to explore its marvels. He was awed by its vastness, by the majesty of the towers and by the power of the cables; and he was fascinated by all the details of the construction—the anchorages and the cables, the trussing and the beams, the slip-joint at mid-span, the machinery of the cable railway, the stone work of the towers, and the magic of the radiating stays. When he returned from these pilgrimages he would recount to his playmates and to his elders the wonders he had seen. To him it was truly a "miracle bridge"; and, as he wondered how so marvelous a work could have been created, he was fired with the ambition to become a builder of suspension bridges. In a background of poverty, this far-flung ambition seemed beyond the boy's reach; but the spirit of the Bridge, and later the story of its builders, had entered his heart—and his dream came true.

In partial discharge of that debt of inspiration, the writing of this book has been undertaken. It has been a labor of love, in the truest sense, with no counting of the cost. The amount of research required was enormous; the material had to be gleaned from literally thousands of sources—original manuscripts, family letters, diaries, memoirs, notes, reports, periodicals, newspaper files, biographical works, scrapbooks, technical literature, records of historical societies, and correspondence. The Roebling family, through the courtesy of Joseph M. Roebling, a great-grandson of the founder, generously entrusted to the writer a vast treasure of source material—a trunk-load of original papers and manuscript letters, many of them more than a century old.

An enumeration of the sources for reference and for credit had to be abandoned; it would be prohibitively voluminous. An existing bibliography on suspension bridges has forty printed pages listing technical references on the Brooklyn Bridge alone; practically all of these were explored by the writer in his hopeful quest for human-interest leads. Sometimes valuable parts of the story were discovered in the most unexpected places—buried in printed documents and official reports that seemed dry, dusty, and formidable.

Although some gaps still remain in the Roebling story despite the most exhaustive research, the exploration has, on the whole, been fruitful. Important sections of the narrative—such as the epic of the Niagara bridges, or the drama of the Brooklyn Bridge caissons—are here assembled for the first time, each gathered from a hundred scattered sources.

Only in minor features has the writer drawn upon his imagination or taken slight liberties with strict chronology for the sake of the story; these deviations are limited to the early portion of the biography. Every endeavor has been applied to make the main part of the narrative—particularly the bridgebuilding sections—faithfully and historically accurate.

D. B. STEINMAN.

New York, 1944.

CONTENTS

PART ONE—THE DESIGN

I.	THE WALLED TOWN	3
II.	THE EMIGRANT	15
III.	AMERICA, 1831	29
IV.	THE SURVEYOR	44
V.	THE INVENTOR	61
VI.	THE BRIDGEBUILDER	78
VII.	A LONELY BOYHOOD	107
VIII.	SO LITTLE TIME	123

PART TWO—THE FOUNDATION

IX.	THE INDUSTRIALIST	135
X.	SPANNING NIAGARA	157
XI.	THE MIND AND THE HEART	194
XII.	"THE BRIDGE WILL BE BEAUTIFUL"	205
XIII.	A RIVER DIVIDES	217
XIV.	THE SOLDIER	247
XV.	THE OHIO IS SPANNED	267

PART THREE—THE STRUCTURE

XVI.	A CITY'S DREAM	295
XVII.	THE GOAL IN SIGHT	306
XVIII.	DOWN IN THE CAISSONS	323
XIX.	OF GRANITE AND STEEL	368
XX.	THE BRIDGE IS DEDICATED	409
	EPILOGUE	419
	BIBLIOGRAPHY	421
	INDEX	447

LIST OF ILLUSTRATIONS

INCLINED PLANE UP MOUNT PISGAH	100
INCLINED PLANES ON THE MORRIS CANAL	100
THE DELAWARE RIVER AQUEDUCT AT LACKAWAXEN	101
JOHN A. ROEBLING	132
ROEBLING'S DRAWING OF THE MONONGAHELA BRIDGE	133
PITTSBURGH IN 1854; SHOWING THE MONONGAHELA BRIDGE	133
THE NIAGARA RAILWAY SUSPENSION BRIDGE	196
ROEBLING'S PITTSBURGH-ALLEGHENY BRIDGE	197
THE OHIO RIVER BRIDGE AT CINCINNATI	260
ORIGINAL DESIGN OF THE BROOKLYN BRIDGE	261
WASHINGTON A. ROEBLING	292
THE BROOKLYN CAISSON	293
DOWN IN THE CAISSON	324
THE NEW YORK ANCHORAGE	325
CABLE-MAKING ON THE BRIDGE	356
THE FOOTBRIDGE USED DURING CONSTRUCTION	357
ARCHES IN THE APPROACH TO THE BRIDGE	388
THE BROOKLYN BRIDGE	389

PART ONE
THE DESIGN

CHAPTER I

THE WALLED TOWN

JOHN AUGUST ROEBLING—destined to become the master pioneer of modern bridgebuilding, to be the founder of a great American industry, and to play a dramatic role in the cavalcade of progress in the New World—was born June 12, 1806, in the ancient, slumbering, old-world town of Mühlhausen in Thuringia.

> A sleepy town, where under the same wheel,
> The same old rut is deepened year by year.

The land of Thuringia is high and stony and cold, yielding small return to the farmer. Only by hard work and the utmost frugality were the people able to procure a bare subsistence. Families were large: there were many mouths to feed, and every pair of hands had to be usefully employed.

Mühlhausen is one of the oldest cities in Thuringia. For a thousand years it had been a walled town, both in fact and in spirit. Its inhabitants were content to remember its historic past, to dwell on the evidences of their city's antiquity and one-time significance.

The medieval appearance of the town had changed but little through the centuries. The old churches and towers, the low, gabled town hall, with its ivied walls, its archways, and its cobbled courtyard, and the quaint architecture of the surrounding houses presented a picturesque, old-world pattern.

Leading an uneventful existence within its walls, the city, even in a later century, was little known to the outside world. Tourists and pilgrims passed it by as they flocked to neighboring shrines: Jena, with its ancient university; Gotha, with its medieval castles and grand-ducal palace; Erfurt, with its hallowed memories of Luther; and Weimar, with the fine house in which Goethe lived for fifty years and with the more humble lodging in which Schiller worked and died.

2

During its years of peaceful slumber there lived in Mühlhausen a well-liked, easy-going burgher named Christoph Polycarpus Roebling. His second name was given him in honor of a sainted bishop, the "Blessed Polycarp," who lived in the second century and who suffered martyrdom by burning in his old age. His eighteenth-century namesake was willing to smoke for his convictions, but not to burn for them. He was a tobacconist by trade, preparing his own stock with the help of his apprentices. Phlegmatic, unambitious, contented, he enjoyed the pleasant scene of his native town, the friendly greetings of his fellow townsmen, his tall stein of beer, and his long pipe of tobacco. It was said that he smoked more tobacco than he sold. To the old-time German, the pipe was more than a custom or habit—it was a mode of reverie and philosophical contemplation. Polycarpus would sit for hours in the doorway of his shop, peacefully smoking his pipe, finding quiet pleasure in the familiar scene, occasionally chatting with his cronies, and feeling annoyed when customers came and interrupted his leisure. Distant echoes of political upheavals and military turmoil left him unmoved. Change, he piously thanked God, never upset the ways of Mühlhausen.

Contentedly his gaze would travel over the peaceful scene: the cobbled pavements, the arched doorways, the dormer windows, the steeply gabled roofs, and the ancient church spires reaching skyward above the housetops. All this was pleasant to look upon as Christoph Polycarpus puffed placidly at his pipe, and no man of sense would ask for anything else.

But his wife, Friederike, five years his senior, was of a different mold. Alert, ambitious, grimly determined, she rebelled against the dull, unprogressive life that her lethargic husband found so profoundly satisfying. To her Mühlhausen was a prison. Walled in by custom and complacency, provincialism and poverty, she rebelled against fate. Chained to a treadmill existence, and facing the blankness of drudgery and despair, she beat in vain against the walls.

3

There were five children of this disparate union. The youngest was baptized Johann August, in the beautiful old twelfth-century church of St. Blasius, where, in 1707, Johann Sebastian Bach had married and where he had composed, for uncomprehending burghers, his first great church cantatas.

The house in which Johann August was born was a modest dwelling, with the tobacco shop below and the living quarters upstairs. The fragrance of the father's wares filled the humble home and blended with the atmosphere of peace and contentment.

But peace and contentment were not for Friederike Roebling. With the birth of her children, she had ceased longing for some richer life for herself, only to transfer her dreams to her sons. There she was—a middle-class *hausfrau*, occupied with humble domestic duties, serving and caring for a household of a dozen persons—thinking day and night of the future of her sons and trying to plan careers for them, unwilling that they should merely follow in the footsteps of their father as small-town tradesmen. She tried to infuse her own vital energy into her family, as she secretly nourished ambitions which became tragedies through their very hopelessness.

Only in her youngest son, Johann, did she discover endowments and qualities like her own—quick intelligence, nervous energy, and an active brain. At last her ambition had a definite goal, her life had a single object: the education of her boy.

Thenceforth everything she did was for that son. With new hope in her heart, she worked and slaved from dawn to night. She was never done with the added labors she assumed, and still she never wearied. By scrimping and scraping, patching and mending, fasting and self-denial, every possible pfennig was saved. She grew thrifty to the point of miserliness, self-sacrificing to the verge of physical breakdown, bending everything to her own driving urge: that her youngest son should have his opportunity. In this singleness of purpose her will became as iron. All softer feelings—the weakness of womanhood, the tenderness of motherhood—were grimly crushed. In her fierce drive she gave her last resources of strength and energy and back-breaking toil; to give, in addition, warmth and light to gladden the hearts and brighten the lives of her growing children was beyond her capacity.

Her love, her pride, her hopes for her son took the form of a fanatic fervor, a self-immolation, a drying up of the wellsprings of emotion. As she toiled without wearying, she accepted pain and suffering without a quiver of the lips or the shedding of a tear. She grasped the thorns, but not the beauty or the fragrance of life; all the cares and martyrdom of motherhood, without its smiles or its tenderness. What a tragedy is the cradle without a lullaby, childhood without the warmth of affection, life without laughter or tears!

4

In this strange domestic setting John Roebling passed his early years, his growing, sensitive nature exposed to a contrast of life patterns, a conflict of personalities. Rare were the smiles of understanding or signs of affection, while the silent clashes of opposed temperaments were recurrent. When a letter arrived by post, a rare event in those days, Polycarpus would let it lie a day or two before opening it, to show his unconcern (and perhaps a fleeting sense of marital independence), while his wife, consumed with curiosity, fairly trembled with uncontrolled impatience. In the growing child these experiences aroused wide-eyed wonder.

From this background of contrasts the boy acquired the complementary elements of his character: the caution and self-control, the philosophical composure and the contemplative spirit, of the father were fused with the passionate driving force of the mother.

There was much in the life of Friederike Roebling to leave a lasting impression upon a thoughtful child. The austere self-denial, the focused ambition, the intense practicality, the consecrated devotion to a lifework—traits clearly manifested in John Roebling's later life and personality—these he owed to the earnest woman who dominated his childhood. That restless, driving urge to strive ever further to the limit of one's strength, and even beyond it when necessary, he inherited from his mother. From her the youth received the endowments that most potently determined his personality and his life: an intense awareness of destiny, a deep-rooted sense of responsibility, and that unconquerable tenacity of purpose before which all obstacles melt. As Goethe has said, "Earnestness alone makes life a part of eternity."

5

Four months after John Roebling was born, war broke out between France and Prussia; and two months later the Prussian armies suffered a crushing defeat in the Battle of Jena. At Mühlhausen, only seventy miles away, after centuries of peaceful slumber, the smooth current of life was interrupted as Napoleon's victorious armies came sweeping over Germany. During the succeeding years, in the changing tides of battle, troops of various nationalities—French, Swedish, Russian, Austrian, Prussian, and Westphalian—marched through the town, each in turn levying tribute and taxes.

Thus the boy's earliest years received, at close range, the impact of world history in the making. After generations of isolated and cloistered existence the drowsy burghers of Mühlhausen were rudely awakened from their pleasant hibernation. Repeatedly, in the years that followed, Napoleon's troops moved in large numbers through the ancient town, advancing or retreating with the fortunes of war. One day, when John was seven years old, the city was encircled by the flaming campfires of the Cossacks, who swarmed through the streets. Soon they were gone, only to be followed by Swedish, Austrian, and Prussian soldiers.

The changing masters demanded crushing tributes; the defeated Grand Army brought in a terrifying typhus epidemic. But the most stirring experiences to excite the boy's heart were the bizarre and fantastic figures of the Kalmucks and the Bashkirs, and the whole shifting, motley, kaleidoscopic picture of fighting men, of strange garb and outlandish speech, camping in the streets and quartered in the homes of the village. As he moved, with wondering gaze and beating heart, among the colorful troops of varied races and nationalities, there dawned upon him the realization that the world was a bigger and more fascinating place than he had known; and with that knowledge there was born within him a hunger some day to seek strange lands and peoples beyond the seas.

In 1815 the nine-year-old schoolboy saw the older students put on uniforms, shoulder muskets, and march off to the front, as the people rose up once more to resist Napoleon's advance.

6

Physical environment played the part it always must in shaping the boy's character and molding his mind. To the ancient architecture of his native town can be traced the builder's innate feeling for the beautiful and the enduring in construction; and in the venerated churches of Mühlhausen can be seen the soaring Gothic lines of the masterworks of his maturity.

The traditions of the countryside became part of his heritage. The storied memories of Luther and Münzer imbued the boy with a spirit of intellectual revolt, a questioning of accepted custom and authority; the local echoes of Bach and Goethe filled his soul with a love of music and poetry.

The exciting times of his boyhood had a more direct influence on his life. It was not merely that the stirring military events and political upheavals of his early years contributed a quickened tempo, an adjustment to change, and a sense of participation in history. More than all that, for him and for his generation, the walls of Mühlhausen were down. Through the streets of the sleepy town had poured men of many countries, speaking many tongues, bringing new and challenging ideas. Even with peace restored, there was no going back to the past. Mühlhausen might sink back into its thousand years' apathy, but its young people were now sharply aware that outside the walls there was the undiscovered universe.

7

In the world of John Roebling's boyhood, education was a rare and costly privilege, beyond the aspirations or the means of middle-class families. The meager earnings of the tobacco shop barely sufficed to feed and clothe the four surviving children and the apprentices, and to keep a roof over their heads. Only by the untiring sacrifices and unwavering determination of his mother was the boy enabled to obtain an education.

Like other children of the town, John attended the Mühlhausen public schools. But, encouraged by his teachers, who were struck by his brilliance and his hunger for knowledge, he was advanced at a very early age to the town "Gymnasium," approximately the equivalent of our high schools. The course of study in those days included religion, mathematics, natural sciences, geography, world history, German, French, English, Latin, penmanship, drawing, and singing.

Pale, seemingly undernourished, and too serious for his age, but quick, alert, and confident in his studies, the tobacconist's son attracted the attention of his teachers and of his fellow pupils. Through common interests he formed a close friendship with an older schoolmate, Ludwig Lies, who, like himself, excelled in mathematics and whose ambition was engineering. Lies, who lived in Eschwege, invited the hungry-looking boy to spend a summer vacation there. There, with his friend's help, John ambitiously prepared himself for the "builder's examination"—a practical test in the principles and arts of carpentry, masonry, and general building construction. He passed the examination successfully and received an official certificate, qualifying him as a "Baumeister," or master builder. He was only fourteen years old.

John was artistically inclined; drawing, both architectural and classical, was one of his favorite pastimes. Beautiful examples of his freehand drawings, in pencil shading and in pen-and-ink rendering, have been preserved from his student days. During his visit to Eschwege John drew a skillful perspective sketch of the city, showing its topography and its architecture. This drawing, later published with other lithographic views in a volume entitled the *Chronicle of Eschwege*, depicted the thirteenth-century town, with its old castle and ancient Nikolai tower, and the arch bridge leading to Brückenhausen on a small island in the river.

Strange to relate, the "maturity examinations" for graduation from the Mühlhausen Gymnasium were never reached by its most famous pupil. While John Roebling demonstrated decided talent for geometry and technology, he did not show the same aptitude for the Latin and theology of the current classical education. As soon as he had finished the other parts of the curriculum, he left the Gymnasium to seek training more directly in line with his obvious talents. Following the advice and the precedent of a former schoolmate, Friederich Stueler, John went to the near-by city of Erfurt to study at the celebrated private Pedagogium of Dr. Ephraim Solomon Unger, a famous mathematician.

It was Dr. Unger who fully awakened and gave direction to the youth's latent talents. This great teacher not only trained his eager pupils in trigonometry, surveying, and science, he also taught them to use their minds in clear, original, analytical thinking. When John went to America, he took with him books on mathematics written by this inspired educator.

At the Pedagogium, too, a strong friendship grew up between the young scholar and his fellow student from Mühlhausen. Stueler—destined in later years to achieve all the honors that a bureaucratic government could confer upon him, including the post of court architect to the king of Prussia—was six years older than Roebling. Their political views were directly opposed, but their kindred interests and ambitions kept them warm friends all through their student years.

The friendships he formed were part of Roebling's education. As is usual with youths whose schooling is ahead of their years, he made friends, through common interests, with men older than himself. They were attracted first by his intense earnestness and obvious talents, and then by his enthusiastic personality. The interaction of minds was mutually stimulating, strengthening ambitions and widening horizons.

8

At seventeen, Roebling completed his studies at Erfurt and was ready for his major step: engineering. Financially this was the crisis of his life. But his mother's fierce saving of pfennigs had served its purpose. Thanks to her thrift, arrangements were made for John to go to Berlin, to enroll at the Royal Polytechnic Institute.

In those days the two hundred miles between Mühlhausen and Berlin meant an exhausting four days' journey by stagecoach and a long absence from home. John's mother helped him to pack his outfit in an old brass-bound trunk; and she took down, from its repository in an iron pot in the cupboard, the money she had saved to pay for his tuition and board. His sister

Amalia presented him with some blouses and kerchiefs she had sewed and embroidered for him; and his older brothers, "Christel" (Hermann Christian) and Carl, who had already begun to earn a little money at their trades, gave him a present of cash to purchase textbooks and drawing instruments. But the most welcome surprise of all was a gift from Polycarpus, a new suit of "city" clothes for the young scholar to wear at the university.

All through the day before John's departure the neighbors kept dropping in to leave small gifts and to bid the youth Godspeed. In the morning the whole family and many of the neighbors stood in front of the house, and there were mingled smiles and tears as the stagecoach drove away—rattling over the cobblestones, bearing the ambitious student to the distant capital.

9

With overnight stops and changes of horses at taverns along the post road John made the long journey from Mühlhausen through Leipzig to Berlin.

The large city, with its wide tree-lined avenues and impressive buildings, was a novel experience for the tall, awkward boy from the cloistered provincial town. The impact of crowds of people, of carriages and horses, of shops and public buildings and palaces, was overwhelming. In Mühlhausen he was regarded with friendly curiosity as the bright boy who had great ambition. He had set out full of confidence in himself and of pride in the letters which he was bringing from his enthusiastic teachers. But in the face of all this activity, these well-dressed, assured people, he felt strange and uncertain and lonely.

As soon as he had found a pension near the campus, providing lodging and meals at a price within his means, he hurried to the university to arrange for his matriculation. The ivy-covered buildings, the peaceful academic atmosphere, and the kindly interest of the professors soon restored his confidence and put him at ease.

The Royal Polytechnic Institute was then the foremost engineering school in the world, and it numbered the most famous teachers of the day among its faculty. Roebling studied engineering and architecture under Rabe and Sluter; foundation construction and bridgebuilding under Dietleyn; and hydraulics and dike construction under the celebrated Professor Eytelwein.

Johann Albert Eytelwein was a prolific writer of engineering treatises and the leading authority on river and harbor works. At that time he was making researches on the strength of iron, which were fascinating to the future bridgebuilder, as the tests and studies bore directly on the then exciting question of the relative merits of cast iron and wrought iron for bridges.

Most stimulating of all, however, were Dietleyn's lectures on bridge construction. To John Roebling's mind fascinating new horizons were opened as the enthusiastic lecturer graphically described five small suspension bridges that had but recently been constructed in England—bridges in which the roadway was hung from chains of linked iron bars. At the same time the attention of the future master builder was directed to the hazards of inadequate design, as Dietleyn reported that two of the English bridges had been blown down by the wind. He also told the class of a pioneer chain bridge that had been built in 1796 by James Finley in Pennsylvania; and then, most significant of all, he told about a *wire* suspension bridge, a flimsy span which had been constructed in 1816 over the Schuylkill Falls in Philadelphia and which had collapsed within the year.

Thus, in his impressionable student years, Roebling became acquainted with a new type of bridge construction, then in its pioneering infancy; and he learned of the groping attempts of men, in scattered parts of the world, to apply the idea and to make it practical.

10

While he was making strides in technical knowledge, John Roebling also realized the importance of other attainments. He was preparing himself for living, as well as for earning a living. His eager mind was busy absorbing new impressions; his imagination was stimulated by ideas from every source. Hungry for knowledge, he extended his studies in the humanities: history, modern languages, literature, music, esthetics, and philosophy. To all his studies, he brought much the same intensity of application as his mother had brought to making them possible for him. Knowing the price at which this opportunity had been bought for him, he realized its pricelessness.

As it happened, the world-famous philosopher, Georg Wilhelm Friedrich Hegel, was lecturing at the university. No young and receptive mind, coming in contact with that dominating personality, could help being influenced by him. John not only became Hegel's favorite pupil, but a friendship sprang up between the gray-haired professor and the shy youth from Mühlhausen. They spent hours together, at the teacher's home or strolling through country lanes. All his life, Roebling spoke with pride of this association of his student days. From Hegel, as initially from Unger, but now in a broader sense, he learned to think logically and clearly and to rely upon the validity of his own conclusions.

On one of those long walks the philosopher was stimulated by the eagerness of his young companion to a recollection of his own eager youth. The years of acquired conservatism rolled away as he relived the storm-and-stress period of his early days. In this confidential mood, he told Roebling that he had hotly defended the French Revolution, and that he and his friend Schelling, defying local Prussian authority, had gone out early one morning in 1790 to plant a "liberty tree" in the market place of their town. Continuing his reminiscences, he described his early years as a teacher at Jena and the uprooting of his life when Napoleon's victory over the Prussians threw the scholarly little city into confusion and terror; the French soldiers invaded Hegel's home and he had to take to his heels, carrying with him his only treasured possession, the manuscript of his first important book, the *Phenomenology of Spirit*. He had been so destitute for a time—he told his fascinated listener—that Goethe had lent him a few taler to keep him from starving. Finally, after years of poverty and struggle, he had received the coveted appointment at Berlin, establishing him in security and honor for the rest of his days.

It was through Hegel's lectures on the philosophy of history that Roebling's thoughts first turned toward America as the land of the future. "It is a land of hope for all who are wearied of the historic armory of old Europe," the aged professor declared, speaking with the vision of his youth and the wisdom of his years. "The deepest law of politics is freedom—the open avenue to change. History is the growth of freedom, and the state is, or should be, freedom organized."

II

At the Royal Polytechnic Institute, bridge construction was Roebling's favorite study. He had heard of suspension bridges, but he had never seen one. During his senior year he learned of one then under construction, the first of its kind in Germany, a small span suspended from four iron chains over the Regnitz at Bamberg. During a holiday vacation, he visited the beautiful old cathedral town to see, with his own eyes, the "miracle bridge."

It was a thrilling and unforgettable experience. He saw the bridge and, in the same flash, he saw his future lifework revealed to him. The suspension idea was beautiful in its logic and in its obvious efficiency. The small chain bridge before him was good —but he felt that he could improve upon its design and its construction. In that realization, his ambition was suddenly crystallized. He was going to be a builder of suspension bridges, and he was going to build them better and larger and stronger than any previously conceived.

He could hardly tear himself away from the small "miracle bridge" that had fascinated and thrilled him. He made a careful study of it, drew a plan of the structure, analyzed its design, and worked out a detailed estimate of the materials required. These studies he assembled into a report which he presented as his graduation thesis.

Passing his final examinations with high honors, Roebling graduated with the prized degree of Civil Engineer. He was highly trained in mathematics, engineering, and architecture. At the same time, he was a skilled linguist, a scholar in history and philosophy, and an accomplished musician. Whatever he had undertaken to do, he had done with thoroughness.

"One thing remember," Hegel had told him, "nothing great in the world has been accomplished without passion!"

CHAPTER II

THE EMIGRANT

JOHN ROEBLING was twenty when he began his professional career. He was equipped with as fine a cultural and technical education as was then available. He was sure of himself and confident of his own abilities. He was eager and ambitious. But he was not impulsive. He felt his way, step by step, planning surely for the next advance. In some odd way, he lacked youthful qualities.

He had ideas of his own. He wanted to be a builder of great bridges. The idea of suspending a span had inspired him, but he felt that bridgebuilders did not yet know the right way to do it. All he wanted was a chance to show what he could do.

But he was blocked at the very outset. Opportunities in Germany for a young engineer were limited. Under an autocratic regime, private enterprise was practically nonexistent, and the only chance for engineering employment was in the government service. There was no choice: Roebling made his application and secured his first job—as surveyor and assistant engineer in the employ of the Prussian Government on the building of roads and small bridges in Westphalia.

It did not take him long to discover that there was no place for new ideas in Germany. Government work was always the same, rigidly prescribed, strictly routinized, and restricted to the same old rules and traditional procedures. Originality was frowned upon; new ideas were regarded as presumptuous. If a thing had always been done in a certain way, it was heresy to ask questions or to suggest a change.

To Roebling the narrowness and tyranny of petty officialdom were galling; at every turn he found his abilities checked by red tape and official inertia. Nothing could be done in Germany, he recorded furiously in his diary, except with "an army of councilors, ministers, and other officials discussing the matter for ten years, making long journeys, and writing long reports, while the money spent in all these preliminaries comes to more than the actual accomplishment of the enterprise."

At the end of three years of this work—the prescribed period of apprenticeship—the young engineer was in search of some avenue of escape from a routine that had become oppressive. He found himself facing a critical decision: should he wait for his official government appointment, depressing and poorly paid, but secure; or should he strike out for himself on some new course that might lead to freedom and independence? In this rebellious mood he returned home to Mühlhausen in the winter of 1829-30.

If Polycarpus, smoking in the doorway, and his wife, busy with her housework, looked for indications that their son had justified his fine and expensive education by finding a place for himself, they were disappointed. John was smoldering with discontent. He brought with him no ambitious plans for an official career; he did not even talk about building bridges. He and his brother Carl talked only about the slow, smug, stodgy ways of their fatherland, and of the hopelessness of life under autocratic regimentation.

2

Unrest was abroad in the land, as it was in all Europe. There were new creative and revolutionary currents of thought in philosophy, in literature, and in government. True, it was the era of Napoleon and Metternich, seeking to rule through imperial conquest and subjection; but it was also the era of Kant and Hegel, of Goethe and Heine, of Beethoven and Schubert.

After the outbreak of the French Revolution the ideas of liberty, equality, and fraternity had spread into Germany and had taken deep root among writers and students, the thoughtful and the spiritually sensitive. The hearts of youth everywhere had turned toward the young republic of France and had lived on the light and hope it had kindled. The rise of Napoleon as a leader of the people and as a champion of human rights fired the imaginations of the liberal-minded in all lands. But when this idol of the people revealed his insatiable lust for conquest and power, there was bitter disillusionment. Beethoven, sincerely believing that the hero of France was motivated by humanistic ideals akin to his own, dedicated his majestic Third

Symphony, the *Eroica*, to Bonaparte. But when, on May 18, 1804, Napoleon assumed the title of Emperor, the idol became clay; Beethoven, his democratic soul outraged by the shattering of his idealization of the man, ripped the dedication from his just completed symphony in passionate tears of disappointment.

In 1806, the year of Roebling's birth, Napoleon, after defeating the Prussian armies at Jena and Auerstadt, entered Berlin in triumph. In 1814, when the French rule ended, Germany faced the difficult problem of reconstruction and of its future reorganization. After the Congress of Vienna in 1815 there remained in Germany thirty-nine autonomous governments—four free cities and thirty-five monarchical states. These joined in a confederation but remained independent.

With the restoration of peace, autocratic reaction set in. The governments of the German states tried to restore absolutism and to obliterate the effects of the French Revolution. The retention of existing institutions, however outworn they might be, became the dominant fetish of rulers. The revolution was dead, and with it the light of hope had died in the hearts of men.

Around the turn of the century a "renaissance" generation had grown up in Mühlhausen and other towns as a fraternal band of ardent young liberals who chafed under the autocratic regimes of the Germany of that era. John Roebling shared this passionate revolt against the reactionary oppression of the rulers of the land. To this group emotion was added his personal rebellion against the bureaucratic regimentation that was strangling his ambition. He not only made one of the rebellious group, he also found himself becoming one of its leading spirits. He discovered that he had a natural capacity for leadership. His warm, magnetic personality drew men to him, and his concentrated enthusiasm balanced by cool, level-headed restraint inspired their confidence. An overwhelming conviction will always beget followers; and Roebling's convictions, once he had arrived at them, were fortified by irresistible logic.

3

There was something new in the air of Thuringia, and all over the world. A new breath of freedom had come across the ocean, a new hope of free opportunity for the human spirit and for individual initiative.

A boyhood friend of Roebling's, an engineer named Etzler, had just returned to Mühlhausen for a visit from America. There, he declared, a man was on his own; his success was limited only by his industry and his talents. There, in the new continent, a man was not tied down by tradition, he was not shackled by century-old conservatism.

"Why don't you come to America?" he asked Roebling. "You would have a much better chance there. Make up a party in Thuringia and take them to America, where they can set up a little colony of their own. If you organized the expedition, your friends would join you."

To John Roebling and his brother Carl the idea was appealing. To be free to work, to build, to achieve! The young men of Mühlhausen gathered to talk about it. A colony in America meant adventure, hope, opportunity, escape.

Yet John could not make up his mind to go. He listened quietly to all the enthusiasm about him. Talk was cheap. An enterprise like that needed planning, careful, long-range planning. Enthusiasm might quickly fade, but plans take time. With his single-track mind he could drive ahead regardless of obstacles, if he were sure he was right; but until then he would wait. For there was change in the air, change that might alter human history, change that might make it possible for a man to shape his own career and attain his highest usefulness even in Saxony and Prussia. Weeks passed. Meanwhile, John returned to his poorly paid, routine work.

Then, in midsummer of 1830, word came of a new revolution in France. The citizens of Paris piled paving stones, trees, barrels, boxes, and carts in the streets for barricades. In three wild days of desperate defiance, the aroused populace drove the king's troops out of Paris and dethroned Charles X. The old Bourbon tyranny was finally overthrown!

The news of the "July Revolution" in France spread into Germany like fire. Liberalism was again on the march. There was new hope for lovers of freedom. John Roebling and his friends eagerly awaited an explosion at home that would shatter official inertia and reactionary oppression. Ardent spirits, inspired by the news from Paris, attempted to stage a similar uprising. But the revolution in Germany was quickly quenched; it was little more than the tremor of a distant earthquake.

Once more the hope of improved conditions in Germany had faded, and the two Roebling brothers and their band of liberals again turned their thoughts to emigration. John wrote to a friend in Hamburg, asking his advice and requesting information about sailing for America. His correspondent replied in mid-August with encouraging reports. Then suddenly there flashed through Europe the news of another revolution. A new nation was born—free Belgium! Word that Belgium had become a free country, released from foreign oppression, gave new hope to Roebling. Perhaps it would yet be unnecessary to leave for America.

Then, just as suddenly, Prussian repression bore down again in full force. The frightened rulers of the German states began tightening their control, attempting to stifle independence of thought and liberty of speech. Surrounded by spies and liable to arrest for the merest word, young men could not breathe freely.

New tyrannical edicts were issued. It was made illegal for a skilled workman or a trained technician to leave the country without official permit. Roebling's friend Etzler was thrown into jail for inciting emigration from Germany. The rest of the men involved in the plan found themselves under strict surveillance, with all their movements closely watched by government spies. It became impossible for them to hold meetings, and their mail was opened by the police. If oppression had been burdensome before, it was now intolerable.

During the fall and winter, Roebling came to his decision. He was going to organize a colonizing party, in conjunction with friends, and leave for America. But he had to move with the greatest caution and secrecy. One false step and he would land in a Prussian prison. He was already marked by the police as the quiet but ardent ringleader of the liberals in Mühlhausen, and only the personal respect and secret sympathy of the mayor saved him from arrest.

4

Roebling proceeded with his usual thoroughness to organize the colonizing party. Giving up his ill-paid job of highway construction in Westphalia, he returned home to get everything ready for the expedition. There Polycarpus, who had long been impatient over the prolongation of John's apprenticeship, was greatly upset because his son, instead of reaping the fruit of his investment by securing a promotion in the government service, had decided instead to go adventuring to the faraway places of the New World. All that money for his education thrown out of the window!

Those were difficult days in the Roebling household. The father rumbled at the son who had betrayed his great hopes. The mother, with tightly pressed lips, attended to her housework. She did not altogether understand what John was doing, but she trusted his judgment, and at least he was getting beyond the walls of Mühlhausen. Polycarpus, on the other hand, could not conceive why anyone should want to make a change. Certainly nothing could lure *him* to America. There was no telling what the wild Indians might do. Moreover, where in all the world could beer like this be found, and what was there in America to compensate for such a loss? Why give up all one's cronies, all the settled comforts and habits and pleasures of life for strange adventuring in a wilderness, among savages and foreigners? Besides, he had heard that new and surprising things were constantly happening in the United States. Everyone knew there was no danger of anything new happening in Mühlhausen.

But he was wrong about his son. There was no lure of glamorous adventure about the enterprise. John's eyes were open to the realities. His homeland also offered him a career, in state service, and he realized that he was surrendering this assured future for an unknown fate. "We are not going with exaggerated hopes," he recorded in his diary; and he declared: "The decision to settle in America must come from a man's own power of will and deed; otherwise he is not suited for America."

In December, 1830, there appeared a printed pamphlet entitled *General View of the United States of North America, together with a Community Plan for Settlement,* issued by "several Germans who are contemplating such settlement and who are seeking other participants." This pamphlet was secretly the joint work of Roebling and Etzler.

John and Carl Roebling had agreed to head the pioneering group which was to go to America, look over the country, select a site, and buy land, so that the others could follow. In writing to the prospective members of the party, John expressed himself with the utmost caution, in case his letters should fall into the hands of the police. An initial list gives the names of fifty men who had decided to join the colonization scheme: farmers, mechanics, clerks, professional men, from Mühlhausen and the adjacent villages. Most of them, however, indicated their intention to follow a year or two later, leaving it to the first pioneering party to locate the settlement and get it started. A larger group of more than two hundred, from Darmstadt, planned secretly to join the Roebling party at Bremen.

John set methodically to work, studying all the books and references he could find on the new country which was to be their future home. At length the plans for the colonizing expedition were completed, and the Roebling brothers and their little band of pioneers left Mühlhausen for Bremen, where they had chartered a sailing vessel, the *Henry Barclay*, to take them and their fellow voyagers across the Atlantic.

5

May 11, 1831, was a memorable day in the history of Mühlhausen, when almost the entire town turned out to bid Godspeed to the departing voyagers. The town officials were friendly to Roebling, and secretly gave him their blessing. At the Lengefelder Watchtower at the edge of the town, John saw many of his neighbors for the last time and caught his last glimpse of the fields and houses, the ancient walls and towers, and the churches and spires of his childhood days. As the stagecoach carried him away on that bright spring morning, there rang in his ears the traditional parting phrase of the good people of Mühlhausen— "*Mach's gut!*"—"Do it well!"—"Make good!" And this heartfelt wish of his home folk he carried with him to his new life in the New World.

Arriving that evening at Heiligenstadt, John left his party and hurried ahead ten miles on foot to Göttingen to catch the morning post for the all-day ride of eighty miles to Hanover. The following day he traveled the remaining seventy miles to Bremen, arriving there with pounding heart to take charge of the assembling expedition—only to learn that he was too late.

The 230 members of the larger party from Darmstadt, instead of waiting as agreed upon, had sailed the day before on the *Henry Barclay*, which they filled to capacity. John Roebling's vision of himself as the leader of a great colonizing expedition was suddenly shattered. Overcoming his first bitter disappointment, he made arrangements for another vessel, the *August Eduard*, which was filled by adding steerage passengers who were waiting for a ship to take them to America.

Four days later, Carl arrived at Bremen with the rest of the party from Mühlhausen, some forty men, women, and children. Etzler had joined this group. Traveling by slower stages, they had been seven days on the road. John's mother had also made the long trip to the seaport to say good-by to her sons—a long good-by, for neither John nor Carl was ever to see their mother or their native land again. Early on the morning of May 22 the party boarded the *August Eduard*. John's mother put in his hands a small sum, the last that she had been able to accumulate through saving and sacrifice.

As the bark sailed out to sea, Friederike Roebling had a heart attack from which she never recovered. She lived barely long enough to receive word of her son's safe arrival in America. Her work was done.

6

The *August Eduard*, a small vessel, carried a full load of ninety-three passengers, including the forty-four men, women, and children of Roebling's party. "The steerage passengers," he commented in his diary, "lie four abreast and partly, where there are children, five abreast, in double berths placed one above the other along the walls of the ship. . . . The steerage has not sufficient height for one to walk there comfortably."

Fortunately for Roebling, he and six others of his group were able to spend seventy-five taler each (about $60) for cabin passage; and in his notes, written for the guidance of future emigrants, he advised all who could afford the higher fare to avoid the steerage.

The voyage was started in high spirits. "Our whole ship's company had been put into a happy mood, which expressed itself in merry choral singing accompanied by a violin and a flute." But soon, to the rolling and pitching of the small ship were added the wretchedness and misery caused by the crowded, ill-ventilated conditions in the steerage. As the leader of his party, Roebling protested to the ship's captain against the practice of keeping all hatches tightly closed; and he succeeded in securing and devising improved arrangements to keep the hold livable for his fellow passengers.

The bark was buffeted by heavy winds which drove her from her course, and at other times there was no breeze to fill the sails, so the voyage across the Atlantic took eleven weeks—a longer time than it took Columbus to sail an uncharted ocean for an unknown land. It was a rough passage, packed with hardships, anxiety, and thrills.

John Roebling kept a detailed record of his observations and experiences on the voyage, setting down everything that might be interesting or instructive to others contemplating a similar journey. He gave it the ponderous title, "Diary of my Journey from Mühlhausen in Thuringia via Bremen to the United States of North America in 1831, written for my Friends." The manuscript was sent back to Germany where it was printed, not at Mühlhausen, where it was forbidden, but at the Roebling Printing House, owned by a cousin, at Eschwege.

Little escaped Roebling's attention. During those long weeks of enforced inaction, his mind was free to roam where it willed. His curiosity was avid. He recorded his thoughts on natural phenomena, meteorology, astronomy, and human nature. He discussed history and sociology, economics and political philosophy. He gave advice on the best routes; on the avoidance and treatment of illness; on provisions, supplies, and money to take along; and on adjustments to life in a new homeland. He recorded his observations on the tackle and rigging of the ship and on the organization of the ship's staff; even on winds and ocean currents, adding his comments on contemporary theories. He praised the home-baked Mühlhausen bread, still tasty after three weeks at sea; and he described life on board, the hardships of the passengers, the personalities of the officers, the strange sights, the adventures, and the intimate human tragedies.

7

During the long voyage Roebling utilized his enforced leisure to study books on the language, history, and customs of the land in which he was to spend the rest of his life. But his preoccupation with his linguistic studies and with his observations on the rigging of the vessel and the details and problems of navigation did not prevent him from thrilling to the natural beauties of sea and sky, from registering moods of the spirit. Never again did John Roebling give the world a glimpse of this aspect of his personality. In this interlude between two life settings, freed from the restraints of the Old World and not yet plunged into the struggles and turmoil of the New, the future bridgebuilder yielded himself freely to the youthful exuberance and carefree excitement of his adventure; and in this mood he revealed the poet and dreamer, an inner self that was later suppressed or hidden under a cloak of reserve.

The artist in him was excited by the magic colors of sunrise and sunset, and his poetic imagination was stirred by the cloud formations vividly illumined by the setting sun: "The glorious colouring of the afterglow upon the clouds became more and more dreamlike, and from their fantastic forms the imagination created all the pictures of a beautiful shore-line in the dreamy distance. Thus I dreamed myself back amongst my native fields in Europe; and the scenes of the last farewell again stood vividly before me. . . . But soon the fascinated imagination dreamed on and already seemed to see, in the magic forms of changing hue, the shores of the new Western World toward which we are speeding."

A vivid picture of the small bark battling a southwester, with high waves pounding the vessel, is recorded in the diary: "It is interesting to stand at the bowsprit and see how the ship, driven by the tensely swelled sails, makes its way by main force through the on-rushing waves. One rolled-up billow follows another; the first rebounds from the ship against the one following. Through their mighty collision they are transformed into foam, which is tinted with lighter and darker blue."

During the third week at sea, more rough weather was encountered, to the distress of most of the passengers. Carl was sick throughout the voyage. John Roebling, however, had resolutely determined not to be seasick, and was keenly enjoying his experiences on board. "The play of the water and the howl of the wind provide entertainment for me. One easily becomes accustomed to the sight of the waves, which, like great wild animals, often seem to leap upon the deck as though to devour the ship, but remain satisfied, however, with giving it such a cuff in the ribs that its wooden frame groans in every joint."

Interspersed among his scientific observations and explanations of trade winds and currents, Roebling set down other colorful descriptions of the strange and beautiful sights that stirred his imagination. "Some nights ago I enjoyed the beautiful sight of the phosphorescence of the sea in all its glory. . . . As the waves rebound from the ship, one perceives in the foam a galaxy of brightly shining stars, which appear as brilliant as the fixed stars in the heavens. Along the entire side of the ship the foam has turned into fiery streaks. At a distance they appear in the dark night as so many flaming masses."

John Roebling would stand on the deck in silent contemplation, awed by the vastness of the sea about him, by the irresistible power of the waves lifting the small bark that carried him, and by the nearness of the blue vault bending over him. There was a feeling of an ever-present personal protection amidst powerful forces that would destroy. And as he traced across the arching sky the steady, unwavering course of the sun, from its birth in the east at dawn to its final glory in the west at journey's end, it seemed to him a flaming symbol marking his life's path from its humble beginning in the Old World to some unknown but inspiring destiny in the New.

And in the silent watches of the night he would be alone on deck, filling his soul with the mystic beauty of the star-studded sky. As the straining ship, cradled in the heaving waves, rolled slowly first to one side and then, as slowly, to the other, the solitary watcher would see the star-filled dome swinging down toward the deck, coming nearer and nearer, until the stars seemed almost within his reach, only to start their upward swing again toward the zenith.

8

The voyage was not without adventure and excitement. On their fourth day out, passing through the English Channel, Roebling saw his first steamship. "It proceeded without sail and labored hard against the billows." Later, schools of porpoises were sighted, and the passengers were thrilled by their first glimpse of a whale. In mid-ocean, a shattered and mournful-looking wreck was encountered, with not a soul on board.

One day a strange-looking armed schooner was sighted. It speedily overtook the *August Eduard* and drew alongside, close enough to permit mutual reconnaissance, so close that shouted orders on the one ship could be heard on the other. The rough-looking crew on the corsair were dark-faced, swarthy men, dressed in bizarre fashion, and shouting in a strange tongue. On the deck was a long cannon, mounted on a turntable so that it could be turned rapidly in all directions. There was a long period of anxious suspense, and then, to the intense relief of the emigrants, the pirate ship sailed away. "The schooner passed by, whereupon it was noticed on our part how more men suddenly appeared on her, who were concealed before. Our captain believes that the pirate ship did not venture to attack our bark, because it discerned many passengers aboard and thus had to expect much resistance and little hope of rich booty."

9

With insight and wisdom rare for his years, the young leader of the expedition anticipated the difficulties of adjustment to a new life. "During the long journey, one has time and leisure to prepare one's self mentally for the New World. All impressions will in reality be new, no matter how familiar we may already have made ourselves with the new continent through reading and hearsay. A different temperament, other human beings, a foreign language and customs, an entirely foreign type of national character, new political institutions—all these will constantly remind us that we are foreigners. The greatest art will indeed consist in shaking off the old European habits of thought and fitting one's self to the New World. And every people demands of a foreigner, who wants to settle permanently amongst it, that he will assimilate himself to his environment as soon as possible."

These are words of unusual comprehension, coming from a youth embarked upon his great adventure. Much heartache and misunderstanding might have been spared if more emigrants, even in later years, had been guided by the same wisdom.

The scheduled time for the *August Eduard* to make the crossing was six weeks, in time to reach America before the Fourth of July. In excited anticipation, Roebling wrote in his diary: "To every freedom-loving man, this date must be significant—a day on which fourteen million citizens of a free land give thanks for their present condition of freedom. We are all anxious to arrive in time to join in the celebration."

Storms, calms, and head winds prolonged the voyage, and hope of reaching port by Independence Day had to be abandoned. After two months at sea, Roebling wrote in his diary: "Still no land! . . . Continuous westerly winds, which occasionally roar like a storm, interrupted by calms. . . . On account of the difference in the reckoning, the captain himself does not know precisely where he is." The dwindling supply of drinking water on board had grown bad and evil-smelling. A child in the steerage, "ever gaily laughing in the days of its health," succumbed to the effects of malnutrition on the long voyage, and was buried at sea. As the ship rolled and pitched violently in a lusty west wind, a passenger carrying a three-year-old girl was thrown across the wet and slippery deck, breaking the child's leg. Waves dashed over the rail, drenching those on board.

One evening a warm westerly wind bore a noticeable fragrance of balsam and pine, a welcome sign of approach to forest-clad shores. A few days later, a flock of small vessels was sighted, and soon a light sailing ship appeared bearing a pilot's flag.

Roebling was moved by his first sight of America. "The shores are becoming more and more interesting," he wrote. "Although low, yet they present a beautiful appearance on account of the handsome groves of trees and stretches of forest with which the bank is still covered—the remnants of the ancient, venerable primeval forests of this new part of the world.

. . . Handsome steamers make their way up and down; countless sloops, schooners, ships and flags of all nations cross one another's path. Here and there in the distance, the dark green of the forests is broken by open places, occupied by fields and farms. . . .

"Today we saw Delaware City on the right bank, a little newly built country town, with clean, substantial houses, amongst which one would seek in vain a dilapidated, dirty hovel, such as one so often sees in Germany in the vicinity of fine homes and affluence."

Finally, on August 6, 1831, the *August Eduard*, with its ninety-two passengers, docked at Philadelphia.

CHAPTER III

AMERICA, 1831

IN 1831 the first train in the United States to be drawn by a steam locomotive made its first run. Modern transportation was just being born. Only the year before, President Andrew Jackson had authorized government projects amounting to the staggering sum of ninety-six million dollars for roads, canals, and railroads. And only six years before, the Erie Canal had been opened, the first link between the Atlantic seaboard and the undeveloped West.

For the young country it was the dawn of a great era of national growth. America was just beginning to be aware of its own vast extent, its resources, its power. The one thing needed was transportation. It was to be the key to the opening of new frontiers. And transportation meant canals and railroads, aqueducts and bridges.

America needed men as she had never needed them before, trained men, technically equipped, to help build her roads and tunnel her mountains and bridge her rivers. The time was right, when John Roebling arrived to build his life in the stimulating and receptive atmosphere of a young and rapidly developing country. He was ready and eager to fit into the picture and to make his contribution.

He promptly grasped the spirit of the new land. "Whence has the multitude of splendid steamboats, mailboats, highways, railways, steam cars, canals, and stages sprung up in so short a time?" he asked. "In part, of course, this is explained by the natural and fortunate position of the country with its manifold resources; but it is principally the result of unrestricted enterprise and the concerted action of an enlightened, self-governing people." In these new ideas—"unrestricted enterprise" and "concerted action" of free citizens—he saw the key to such progress and achievement as the world had never before witnessed.

When John Roebling came to America, at the age of twenty-five, he was no ordinary immigrant. He was equipped not only with a thorough technical and cultural training but, more important still, he had energy, ability, and originality of a high order—qualities which can command a premium in a pioneer country where there is more work to be done than there are men to do it.

He reached the new homeland with less than $400 to his name. This sum, with a like amount belonging to his brother Carl, and with equal shares entrusted to them for investment by their older brother, Hermann Christian, and their friend and fellow townsman, Ferdinand Baehr, made up a total fund of barely $1,600 for the initial purchase of land and farm equipment. A modest capital with which to start a new colony!

2

At the time of Roebling's arrival, the population of Philadelphia had grown to eighty thousand. "The streets," he wrote, "all of which intersect at right angles, are entirely straight, very roomy and in part planted with trees. . . . All houses are three-storeyed and are provided with basements, which, however, are not used for living purposes. The dwelling houses are roomy and are furnished with splendor and elegance, and very neatly arranged. . . .

"Regarding the inner elegance of the residences, and indeed not individual ones but all together, Berlin, for example, offers no comparison and seems to be a city of poor, oppressed people. What is lavished there upon a few public buildings is here lavished by every plain citizen upon his home. . . .

"All the streets are provided on both sides with broad foot pavements, laid in bricks, which afford easy walking. Cleanliness, however, confines itself more to the dwellings than to the streets, where swine are allowed to run around freely, seeking food. This is, indeed, an amazing condition."

What impressed him most, however, was the friendliness of the people. "Philadelphia," he wrote further, "is to be distinguished above other cities on account of its politeness to foreigners. . . . I have spoken to no one yet, even when he seemed to be in a hurry, who has not fully replied to my questions. . . . Nowhere does one see a person in rags; all, even the common workmen, go very cleanly and neatly dressed. Every American, even when he is poor and must serve others, feels his innate rights as a man. What a contrast to the oppressed German population!"

Apparently, even then, the American picture as it reached Europe was distorted. Honesty and square dealing, Roebling was agreeably surprised to find, were more prevalent in America than he had been led to expect. The Americans, he learned, gave him fair treatment when he made purchases and did not attempt to take advantage of the ignorance of a foreigner. Nor, much more surprising, were they a nation of drunkards! He had arrived with the belief that half the population was never sober.

With his background of convention and tradition Roebling was struck by the greater freedom in the manners and customs of the people he met. "The removal of the hat and frequent greetings, which are so burdensome in Germany, do not exist here. Even in the parlor the American keeps his hat on, which astonishes a foreigner and seems like lack of politeness. Everyone wears a felt or straw hat with a broad brim; the German immigrant is immediately recognized by his cap and his pipe. Only cigars are smoked here."

Roebling's small band of followers, who had entrusted their prospective fortunes to his leadership, were depending on his decisions and guidance to help them establish a home. Their resources were limited. Care had to be exercised in choosing the site for their future colony, so that the land might be purchased within their available funds and still offer promise of fruitful returns for industrious cultivation.

The entire party decided to remain in Philadelphia until the place of their future settlement was chosen. Their temporary stay in the "City of Brotherly Love" would give them time to look around and conduct negotiations; at the same time, they would be getting accustomed to the American way of life and improving themselves in the language of the country.

The Roebling brothers found a boardinghouse for their brief stay in Philadelphia. Although the owners and most of the other guests were fellow Germans, the customs and manners were strange to the newcomers. "In Philadelphia we hired rooms in a 'boarding house,' where a person pays three dollars a week for lodging and board inclusive, the latter consisting of 'breakfast, dinner, and supper,' and is very good. . . . At each meal—morning, noon, and evening—the table is set in the English custom, with coffee, tea, butter, corn bread, meat, eggs, roast meat, salad, and the like. The landlady pours coffee and does nothing else, but simply gives the invitation, 'Help yourself.' Nothing further is served. The Americans set to quickly, eat even more quickly, and leave the table without having said a single word."

Thus early, the future citizen learned a typical phrase in the American vernacular: "Help yourself"—a phrase of manifold significance, connoting hospitality and indifference, generosity and greed, acquisitiveness and ambition, self-reliance and independence.

3

The two brothers carefully considered all the possible places for a settlement, and had to exclude most of them as unsuitable. The South had the complication of slavery. The Far West was too inaccessible. In the more developed northern states, land was too expensive. All that remained was the near West—western Pennsylvania, or possibly Ohio.

"Slavery," John wrote, "is the greatest cancerous affliction from which the United States are suffering. The republic is branded by it and the entire folk stands branded with it before the eyes of the civilized world. Grounds enough for us not to go into any slave-holding state, even if Nature had created a Paradise there. It is hoped that slavery will be entirely abolished little by little; there has been talk about it recently in Maryland. . . .

"On the other hand, just as little do we wish to go to the far-distant Western lands, whose boundaries just recently in Missouri and Arkansas have been disturbed by attacks of bands of wild Indians, and where one is cut off from all intercourse and civilization."

After a few weeks, the Roeblings collected their small party of colonists and set out for Pittsburgh. As this was before the days of the railroads, the party had to travel by way of the partially completed Pennsylvania Canal. Leaving Philadelphia, they proceeded by boat up the Schuylkill River to Reading, thence westward by mule-drawn canalboat on the recently constructed canal to Harrisburg, then up the Susquehanna River to the mouth of the Juniata and up this scenic, winding river to Huntingdon. As the canal and its portage railways beyond this point had not then been completed, the remainder of the journey—about 125 miles—was over rough and hilly wagon roads to Pittsburgh. The three-hundred-mile journey, which took four weeks, was fatiguing: changes had to be made in wet weather, and the travelers were constantly exposed to dampness. Carl Roebling became desperately ill with fever and chills, having contracted a bad case of malaria in the damp, mosquito-infested nights on the open canalboats, and he had to be carried to the end of the journey.

4

When the party arrived in Pittsburgh, which was then a new and growing town, Roebling heard of an available tract of land only twenty-five miles away, in Butler County, Pennsylvania. The land was at an elevation of fifteen hundred feet above sea level. Learning that this locality had the reputation of being one of the most healthful spots in the state, John, influenced by the weakened health of his brother, made the decision. With horse and map, he traveled through the wilderness for three days, making a survey of the site. Reaching a satisfactory agreement on price, he promptly concluded the transaction.

The tract, some seven thousand acres, had originally been part of the estate of Robert Morris, financier of the American Revolution, who had purchased the land from the federal government at ten cents an acre. The price agreed upon by Roebling was an average of $1.37 an acre. The initial purchase was only sixteen hundred acres, for which Roebling paid $1,000 in cash, agreeing to pay the balance in two subsequent annual installments. This was intended as a joint investment in four equal parts—for John, Carl, their brother Hermann Christian, and their friend Baehr.

The country was wild and isolated. The tract they had purchased was wholly undeveloped and almost inaccessible; it could be reached only by primitive roads which were practically impassable in winter. The ground was rough and hilly, the soil poor. Only with great labor could it be cultivated so as to yield even a meager crop. It would have discouraged others, but it looked good to these optimistic and inexperienced pioneers.

Of the total company of some three hundred emigrants who had sailed from Bremen in May, 1831, and whom Roebling had hoped to assemble in America, group after group had gone off in different directions. The group who had sailed on the *Henry Barclay* were mainly communists. They could agree neither on their destination nor on their plans, and eventually they split into two parties. The larger party, upon landing in Baltimore, went directly to New Harmony, Indiana, to join the communal settlement of which they had heard so much. But when they arrived there, it was to learn not only that George Rapp's colony was gone, but that Robert Owen's newer socialistic community had also been disrupted.

The other party waited several weeks at Baltimore for Roebling's arrival; becoming impatient, and lacking leadership and common purpose, they scattered to seek their separate fortunes. One group left for Savannah by sailboat to visit a relative who owned a plantation in South Carolina. Months later, Roebling got word that the plantation had been burned down by slaves and that the unfortunate colonists had moved on to Alabama. Their further wanderings are veiled in obscurity.

Etzler and some of his followers, spurning the prosaic possibilities of Pennsylvania and seeking more glamorous frontiers, had gone on to Ohio and Indiana; more of Roebling's party had dropped out at Philadelphia and Pittsburgh. Thus the vision of a large pioneer colony had dwindled away. All who were left to start life on the tract of land purchased for the settlement were the two Roebling brothers and the Grabe family, the latter representing seven hungry mouths to feed. All lived in the one farmhouse purchased with the land, Grabe helping as a farmhand, his wife doing the housework, and their five sickly children running up doctor's bills and driving everybody distracted.

Carefully husbanding the limited remaining funds, Roebling bought a team of horses, two plows, and a cow, together with a stock of provisions for the long winter, when they would be snowed in.

Then followed a year of loneliness and hard work. Determined to turn the seeming initial failure into a success, John Roebling threw all his energy into building a real settlement out of this humble beginning. Other prospective emigrants had promised to come upon receiving his reports of the site selected, and he opened his campaign to let his friends in Thuringia know all about the opportunities in the embryo colony. He enumerated the merits and the attractions of the location, as he saw them: the healthful climate, the good drinking water, the pure air, the "rosy cheeks"; the suitability for diversified farming; the fine pastures for cattle raising, sheep raising, and horse breeding; the low price of land, the high prices paid for produce, and the nearness to market; the rolling country and the fine, distant views; the mineral wealth, the iron ore and coal; and the suitability for manufacturing enterprises, for establishing "a small town where farming and industry could be combined."

He wrote reams of letters to friends in and about Mühlhausen, inviting them to join the colony. To each one he sent exact directions—how to come, what to bring along, what to leave behind. All textile fabrics must be brought. Common tools were to be left behind, because American tools—axes, hatchets, saws, grubbing hoes—were so much better; nobody could cut down a tree with a German ax. Mechanics were needed, and those who knew a trade. Terms of payment for land were made easy. Those who came and were dissatisfied would have their money refunded.

Anxious as he was to build up his little settlement, Roebling did not want to be responsible for any cases of disappointment or disillusionment that might result. He therefore scrupulously emphasized this caution in his letters: "What I here communicate to you and our other friends has not, before God, the purpose of persuading any man to come over to America and eventually to our place. . . . I would not, for any price, persuade any person to come here, even if we should have to be here alone for the rest of our lives."

Hard-pressed for funds after his initial down payment on the land, Roebling wrote to Baehr: "Whether we shall keep the land depends on whether my father will aid us with five hundred dollars. Please talk with my father and my brother about this matter; the money will be safe here and well invested."

Receiving no assistance or encouragement from home, the pioneer persisted in his enterprise, determined somehow to work out the financing of the land venture.

5

Loneliness begets romance. Soon after his arrival in western Pennsylvania, Roebling became a visitor at the farm of a Swiss family named Giesy, near Pittsburgh. There were ten sons—four of whom became preachers—and four daughters, including Helena Giesy, who was then sixteen. Occasional calls grew into frequent visits. Soon the young couple were engaged, and the girl began sewing her wedding gown.

Helena and her brother John had become followers of a Dr. William Keil who was organizing a religious-communal colony to go west, first to Bethel, Missouri, and subsequently to Aurora, Oregon. Roebling refused to go. He had pledged himself to establish a colony, and establish it he would. No one will ever know what happened between the two young people. Helena left with her brother in the covered-wagon train that traveled across two thousand miles of prairies, desert, and mountain wilderness, and finally settled in Oregon. She never married.

In an editorial nearly a century later the *Oregon Statesman* said:

Had such a genius—such a practical genius—as John A. Roebling chosen a different course in his youth in Pennsylvania, and had he become a member of the Aurora colony, and had he, with his vision, made an industrial colony there instead of an agricultural, what a different history might not this part of Oregon have had!

6

John Roebling was a brilliant mathematician, a scholar of considerable reading, a student of transcendental philosophy, an accomplished musician, and at home in three languages—German, French, and English. But what had all this to do with farming? What had a scholar, an artist, an inventive genius, a practical dreamer, a master builder, to do with cultivation and crops, with plowing and planting, with harrowing and harvesting? The mathematician and philosopher turned farmer!

In this strange choice of an occupation the enthusiastic young pioneer was joined by his brother Carl. True, they knew nothing about crops and were unfamiliar with farming, but in the abounding eagerness and confidence of youth they thought they could easily learn.

For the future bridgebuilder, it was at least an introduction to life in America. It gave him an opportunity to get his feet on the ground, to establish a community of neighbors, to learn the topography of the country, to get the feeling of being an American.

After his years spent with books and studies, the hard outdoor life and the exposure to cold, bracing air agreed with John Roebling. He had unaccustomed color in his cheeks, and he felt strengthened and invigorated.

Upon receiving Roebling's first report of the site he had found for the settlement, his friend Baehr replied enthusiastically, describing the plans being completed at Mühlhausen for the early departure of a congenial band of emigrants to join the new colony. They were bringing along complete outfits of tools for their trades: carpentry, cabinetwork, weaving, blacksmithing, and metalwork. They were also bringing various kinds of musical instruments, and music, including church cantatas and a choral mass. Appended to this letter was a postscript, penned in a boyish hand, from Baehr's ten-year-old son, Bernhart:

"Dear Mr. Roebling: I will bring along something lovely for you, and I am so happy that I shall soon see you again. When we get to the other side of the ocean, please come to meet us and show us the way, so that we shall not get lost."

The "lovely" surprise the boy was bringing to Roebling, whom he idolized, was a flute. The gift arrived, but without its eager bearer. On the rough voyage, the boy became seriously ill and died before port was reached. His parents never recovered from the shock. Through life John Roebling kept the letter with the last words the youngster had written to him: "Please come to meet us and show us the way, so that we shall not get lost."

In May Roebling was advised by a banking house in Philadelphia that a credit of some $6,000 in his name had just arrived from Bremen on behalf of a list of thirty-two men—many of whom were already on their way across the ocean, some with their families, to join the pioneer settlement.

The new colonists, headed by Ferdinand Baehr, arrived at "Roebling's Farm" in the summer of 1832. It was a joyous reunion of friends.

7

Just as earlier colonists in America had given names like "New England" or "New Amsterdam" to their colonies, so this small band of emigrants, with pardonable nostalgia, chose at first to name their settlement "Germania." Before many months, however, the name was changed to "Saxonburg."

Following Roebling's surveys and plans, the village was laid out on a hillside, with one broad Main Street on the top of the ridge, running exactly east and west. This was flanked by lots running down a half mile to Water Street, so that each man had a little farm of six to twelve acres to himself. Main Street thus became the center of community life. Water Street, at the foot of the slope, was considered the poorer quarter.

The land was mostly virgin forest, with black oak predominating. Where some of it had previously been cleared, a second growth of small, stunted oaks had sprung up, and here pheasants were found in great numbers. Deer, and even bears, walked down Main Street, and there was an abundance of smaller game.

The first house was built at the highest point of the ridge and was occupied by the two Roebling brothers. It was constructed simply, but substantially, of roughhewed logs. One by one the other homes were built, with all the neighbors putting their strength together.

John Roebling spent his evenings studying books on farming and drawing up plans for local improvements and for the new settlers' homes. Housebuilding in this remote, rocky region was not cheap. A well had to be dug first, and this usually had to be blasted in the solid rock. The earliest homes erected were really log cabins or blockhouses; openwork construction with clay plastering between the timbers came later. Only an elegant house could afford clapboards, as the nearest sawmill was too far away. Timbers were dressed with a broadax and fastened with treenails, for iron nails were scarce. Some of the houses had clay floors.

It was typical pioneer building, using the materials and means available at the site, plus the energy and resourcefulness of the builders. The character of these men went into the houses they built—simple, rough-clad, rugged, foursquare; to be judged not by appearance but by usefulness; without frills or extras, but of honest, substantial construction. Most of those houses built by the early pioneers in Saxonburg are still standing, after more than a hundred years.

When the framework of a new house was ready to be raised, all in the community gathered to help. It was made the occasion for a picnic. The necessities of pioneer life bred a spirit of neighborly co-operation—lending a hand whenever help was needed, and sharing joy or sorrow.

8

In one of the early letters he received from home John Roebling was told of his mother's death. A wave of realization swept over him. She had spent her life and strength to equip him for a career—and he, instead of justifying her sacrifice, had taken up an occupation in which his education and his talents were wasted.

About the same time he learned that his former teacher Hegel had died in a cholera epidemic that was then sweeping over Germany. Two human beings had vitally influenced Roebling's life, and had passed on.

Letters from his sister Amalia—sometimes delayed for months—supplied further news from home. With sisterly affection, she followed the progress of the pioneer colony, which she was hoping to join; and she wrote fondly and proudly of her own "little lambs" who "love their Uncle August"—Fritz, who "wants his uncle to send him a hatchet from America"; Wilhelm, who "is just learning to walk"; and Ottillie—"You should see your godchild, Uncle August; when she comes dancing into the room with her doll, one wants to crush her in a loving embrace."

In a later letter Amalia announced that she had decided not

to come to America. She and her husband—Carl Meissner, a merchant—were now comfortable in Mühlhausen. At last she had female domestic help; and she was unwilling to change to the rigors of pioneer life in a new country where she would have to do the rough work with her own hands.

One of the most distinguished fellow townsmen to respond to Roebling's invitation was Carl Angelrodt, who had served as a member of the provincial assembly of Saxony. When he arrived at Saxonburg in August, 1832, he took one look at the primitive settlement, called Roebling a "swindler," demanded the refund of his land payment, and continued westward—winding up in St. Louis where he later won wealth and prominence in commercial and consular positions. In the meantime he sent adverse reports to prospective colonists in Mühlhausen and vicinity, denouncing Roebling and the site he had selected, and describing the soil, the climate, and the drinking water as not fit for civilized human beings. Further emigration to Saxonburg was almost entirely halted by this flood of discouraging letters.

Soon after, John and Carl received word that their older brother Hermann Christian had abandoned his plan to join them in America. "Not that Angelrodt's scoundrelly reports have dissuaded me," he wrote; but a new situation prevented the promised reunion. Their sixty-two-year-old father, now a widower and feeling fit as a fiddle, was threatening, if left alone, to marry again—and then their patrimony would be gone! Hermann Christian remained in Mühlhausen and finally succeeded to his father's tobacco business.

About this time also, their brother-in-law, Carl Meissner, wrote that he had changed his mind about investing in lots in the new settlement; whereupon Roebling resold the land to another purchaser at a higher price. A family quarrel followed, with name-calling across the ocean, Meissner claiming that the profit on the transaction belonged to him as interest on his investment.

9

Soon after the arrival of the company of settlers Carl became engaged to one of the flaxen-haired daughters of the little colony; and before long they were married.

The following year a smaller party of emigrants from Thuringia arrived at the settlement. Among them was Ernst Herting, who had been a tailor in Mühlhausen. Herting built a blockhouse on Main Street to provide a home for his family, which included a wife and three daughters. Unaccustomed to farming, he had a hard struggle the first few years. He raised hops, which were in demand, and he grew vegetables and flowers in his garden. There was warm hospitality in this simple farmhouse, and much that was quaint and homelike. John Roebling was a frequent visitor.

In 1836, in apple-blossom time, John Roebling and Johanna Herting were married. They settled down to live in the new log house which John had built after Carl's marriage.

Johanna Roebling was an ideal wife for the young pioneer in his years of adjustment and struggle. In a letter to his people at home, he wrote: "The only drawback that my wife might have in Mühlhausen eyes is that she is the daughter of the tailor Herting, which however causes me very little concern, since in this country no one asks about your social rank, inasmuch as here only *one* social rank exists."

On May 26, 1837, in Saxonburg, Butler County, Pennsylvania, a son was born to John and Johanna Roebling. The proud father reverently chose the name that had most inspired him in the history of the young republic, to which he added a modified form of his own middle name; so that this first-born son was baptized Washington Augustus Roebling. Eighty-seven years later, referring to his own birth, Colonel Washington Roebling wrote briefly but characteristically, "I was baptized by the postmaster, Mr. Shilly, there being no preacher as yet —have perceived no ill effects therefrom."

10

Selected for considerations not exclusively agricultural, the location of the little farming colony proved to be far from ideal. The winters were mercilessly cold; the sloping ridge was exposed to the raw, frost-laden winds from the Alleghenies; and the summers brought a heat that burned. The land was poor: the soil was stiff clay, difficult to cultivate, and it required lime before it would yield any crops. The swampy meadow lands needed to be drained. The gnarled and knotty oaks demanded painful clearing and grubbing. The lots were too small for an American farmer, and only the energy and enthusiasm of the new settlers enabled them to accept the hard life and the unpromising yield of the stubborn soil.

But the exertions of the little colony of hard-working, thrifty German settlers soon showed results. In only a few years the primitive hamlet in the backwoods had become a thriving community of industrious and happy people. Many of their friends in Thuringia, Saxony, and Westphalia had been persuaded to come over and join them. There was an abundance of food, and there was work for all.

The dreams of the colonists were coming true. From the hard, stubborn soil they had succeeded in wresting an abundance for their needs; and out of the rough, wooded hillside they had fashioned a settlement of sturdy homes. The colonizing enterprise was a success. John Roebling had demonstrated his ability as an organizer, a planner, a real-estate promoter, and a builder. His faith and determination had created a flourishing settlement in the wilderness, a new town of prospering citizens and happy homes.

But it must be admitted that the future master builder was a failure as a farmer. When he tried to cultivate his own plot of land, he barely managed to make a living. His heart was not in it. His mind turned to inventions that would replace manual labor on the farm, and he gladly relinquished the physical work to others. He began to cast about him for some other enterprise as a side line to supplement his uncertain income. If farming was an odd occupation for a man of his training and talents, the new venture he conceived was even more incongruous. He undertook to breed canary birds for market! The new German colonists in America missed the cheery notes of the yellow-feathered songsters of their native land; and it occurred to Roebling that he could find both pleasure and profit in supplying this demand. However diverting the experience may have been, his short-lived venture in musical ornithology proved most disappointing financially. The elemental biological processes did not yield to mathematical control nor to wishful prognostication, and somehow nearly all of his canaries turned out to be unmusical females.

II

Impatient of things unscientific, Roebling grew restless. His talents were rusting out, and farming was not congenial. Long winter evenings were spent in reading and studying, thinking about bridge construction, and making drawings for new inventions. He found an outlet for his energies in making roads, digging wells, and designing and building houses for his neighbors. However, even these activities could not consume the energies of an active brain.

Here he was—isolated, far from the current of world affairs, at the outer edge of civilization, toiling to stave off physical hunger, but unable to still the growing hunger of his original ambition. Six years of his life had seemingly been wasted.

Such thoughts, gathering force in Roebling's mind, were brought to a head by the shock of a personal loss. His brother Carl was not physically strong enough to withstand the rigors of pioneering and the exposure of farm life; and one day in the summer of 1837, while reaping in a wheat field, he had an attack of sunstroke which proved fatal.

John was deeply affected. His elder brother had been his comrade in hardship and in toil. Carl's death brought all their joint plans and hopes to an end. The lonely survivor no longer had the heart for the task they had undertaken together. How cruel and discouraging fate had been since he started on this strange adventure—mistakes, disappointments, failures, heartbreaks, apparently without end! Surely it was high time to stop drifting and to get back into the course he had originally planned.

The pioneer settlement he had undertaken to establish on American soil was now successfully established; the obligation he had assumed toward his fellow-colonists was fulfilled. With Carl missing, the strongest tie binding John Roebling to the colony was severed. As he helped carry the simple homemade coffin to the little cemetery on the hill, his heart told him that this was a turning point in his life—the sad ending of one chapter and the fateful beginning of another, different and unknown.

CHAPTER IV

THE SURVEYOR

THERE are pivotal dates that stand out in the life of an individual when fate seems to concentrate, in swift successive impact, a critical series of forces and experiences that alter the course and the character of existence. In Roebling's life the year 1837 marked such turning point—for in the space of that single year were packed a succession of significant events and experiences that shook the roots of his being and changed the outlook and the pattern of his life. Impelled by a strange, resistless force, his life had moved steadily and swiftly on its course until it came to the backwater pool of his primitive, rustic existence at Saxonburg. There for six years his progress had been halted—six seemingly unprofitable years that did not visibly advance him toward his goal, but representing an accumulation of unspent driving force. And then there came—all within the span of a few months—the birth of his son, Washington Augustus; the death of his beloved brother, Carl; the securing of American citizenship; and the return to engineering as a career. With new resolve in his heart, he found the outlet he had been hungrily seeking—and then, with redoubled energy, he was once more on his way, following the urge of the mystic compulsion sweeping him ever onward toward his destiny.

The birth of his first son stirred John Roebling to a new and deepened sense of obligation and responsibility. As he felt the infant's soft hands clutch his rough, toilworn fingers, a wave of tenderness swept over him. Henceforth he must provide more than bare subsistence for his little family. He must plan to give his son everything he himself had missed. Above all, he must now take hold of his own life and shape it into a career in which the boy could proudly join him, and which the son could carry forward after the father's course was run.

The death of his brother added new responsibilities also,

for Carl left a widow and two children. This obligation was another reason for giving up the unhappy farming venture and finding work that would better provide for the future of the two families. Roebling set out to fulfill his doubled responsibility.

An important preliminary step, however, was to complete his identification with the land of his adoption. His horizon was no longer limited to the little colony of immigrants he had established at Saxonburg; he was ready to take his part in the larger American scene. On September 30, 1837, John August Roebling became a naturalized citizen of the United States. He appeared on that date before the Court of Common Pleas at Butler and, after renouncing all prior ties and taking the oath of allegiance to his new country, he was duly registered as a citizen. The original document has been preserved.

For America, its traditions and its ideals, Roebling had the highest reverence. To him the attainment of citizenship was a sacrament.

In the life of every immigrant there is an unavoidable element of tragedy—a sense of being a stranger, a continuing problem of adjustment. Roebling's life was not unmarked by this strain. Notwithstanding his deep devotion and his later gifts to the land of his adoption, he had a consciousness at times of being regarded as an alien. With all his soul he wanted to be a part of the new country and to throw himself wholeheartedly into its spirit. But he encountered barriers, strange ways and attitudes that he could not always understand, suspicions and doubts, and sometimes even slights or snubs, fancied or real. In the years that followed he remained deeply lonely, with a sense of not being fully accepted. This perception made him all the more determined to achieve, both in works and in wealth, and thereby to compel the regard of his new fellow citizens. With head held high John Roebling went on his way. He would show them!

2

As the erstwhile farmer commenced a restless groping for new life paths, the inventor in him began to stir. He worked out a new boiler construction for steamships; he conceived a steam tractor to operate agricultural implements; he filed patent papers on these improvements, and then he let his applications lie unclaimed. He built a dye works, but gave it up. He was seeking some way of escape from his unhappy adventure in agriculture, for though he might recover his investment in the farm, he suffered keenly from the failure and futility of the effort. He desperately needed some successful achievement to restore his self-confidence.

During his long evenings in Saxonburg Roebling studied all the new technical books and engineering periodicals he could procure, recording in his notebooks the new ideas and inventions that came to him from these studies. His notes and his patent applications were illustrated with his own beautifully executed drawings. He found himself fascinated by things mechanical: high pressure engines, marine engineering, steam tractors, and agricultural machinery.

With his application for a patent on a "Steam Agricultural Apparatus," comprising a steam tractor operating gang plows, reapers, and other farm equipment, the ambitious inventor built a model of his "Steam-plough" and sent it to the Patent Office.

Other inventions of Roebling's fertile and restless mind followed. Most of them were inspired by the same practical passion for safety which he showed throughout his life—in always seeking to safeguard the construction operations on his bridges and the operating machinery in his manufacturing plants. His early string of patent applications included "a new and improved mode of constructing Steam Boilers and rendering them safe from bursting"; next, a "Self-Acting Gauge" (a safety indicating device to sound an alarm automatically whenever the water level in a steam boiler became dangerously low), on which he was granted a patent; also "a new and useful Safety Valve—discharging steam when the pressure in a boiler becomes too high, and discharging water into the furnace to extinguish the fire when the level of the water becomes too low." Striking out into new fields, with imagination beyond his times, he invented a "Radial Engine," stating prophetically that high-pressure steam was going to become practical and economical for power purposes in the future; and, with a fellow-inventor in Philadelphia, he filed a joint application for "a new mode of applying power for propelling vessels through water which we here denominate Submarine Propellers."

Another application, different but important, concerned a "Spark-Arrester for Locomotives." In those days, frequent fires were started in forests and grain fields by sparks from the primitive wood-burning locomotives, and the invention was designed to eliminate this hazard by arresting and extinguishing the sparks. A patent was granted, and two of Roebling's spark-arresters were later installed on "the noble David R. Porter," a locally famous locomotive on the Allegheny and Portage Railroad.

3

But invention was not enough. Roebling was hungrily seeking an opportunity to work in the engineering profession for which he had been trained. In a young and growing country, there must be a need for constructive talent and technical equipment; indeed, the desired employment was close at hand. Looking about him, he perceived that there in Pennsylvania, practically a frontier civilization, much work needed to be done in transportation—in railroad surveying, in road making, in canal construction, and in river improvement for slack-water navigation.

Until they were eclipsed by the subsequent building of the railroads, the canals of those pioneer days played a dramatic and indispensable role in the opening up of the country; and their story is one of the romantic chapters in the history of life in America. In 1825 the Erie Canal had been completed in New York State. In order to meet this competition, which threatened the commercial prosperity of Philadelphia, the state of Pennsylvania undertook to build a canal to Pittsburgh, connecting by the Ohio River with the rich, undeveloped country beyond; this "Pennsylvania Canal" was constructed in large part between 1826 and 1830. A parallel railroad through central Pennsylvania was also planned, in addition to a system of feeder and branch canals to provide transportation to other parts of the state.

The proximity of the work aroused in Roebling a desire to take part in this pioneer construction program; and soon he found the opportunity for which he longed. The crystallizing suggestion and invitation came to him through a most unexpected channel—in a letter, written in German, from a former schoolmate whom he had not seen for years.

The providential and unexpected letter that was to turn the course of Roebling's life was addressed to him by Edward Thierry, who had been a fellow student at Dr. Unger's Pedagogium in Erfurt. Thierry had sailed from Bremen to New York in 1834, with an ambitious plan to establish in America an "institute of mathematics and architecture"; but soon after landing he found himself friendless and penniless "in a strange land, a strange city; language, customs—all strange," with no one to turn to for help in getting a start. Unable to speak the language of the country, the erstwhile scholar secured work intermittently, successively as a stonecutter in a marbleyard in Philadelphia, man-of-all-work in a wholesale chemical shop, packing clerk in a wine store, laborer in a leather factory—suffering inflamed fingers, swollen hands, aches, pains, bruises, and sickness, from unaccustomed rough work and from exposure to damp and cold, accepting starvation wages of $2.00 a week or even less, and then being cheated out of that hungrily anticipated pay by greedy and unprincipled employers, starving or running up debts at his boarding house between jobs, and having seared into his brain the first English sentence he learned, the sentence he awaited with fear and trembling in each new short-lived job: "I don't want you any more!" Finally, through some engineers he met, to whom he made known his plight, he learned of work on the Sandy and Beaver Canal, then under construction, and got a job there. To him, this was a life-saver. When those in charge told him apologetically that, because of his inadequate English, they could pay him only $1.25 a day at the start, they did not realize that he was trembling with joy at his unexpected good fortune. With this salary, munificent in comparison with anything he had experienced in his two years of desperate struggle for survival, he was soon able to pay off his debts in Philadelphia and even to send money to a destitute cousin who had remained in New York. Now his superior, Edward H. Gill, had asked him if he knew any other engineers who might like to work on the canal. Having just learned that his old friend Roebling was living in the vicinity, Thierry wrote to him from New Castle to ask whether he would be interested in a job on the engineering staff of the canal construction.

THE SURVEYOR

To the restless, ambitious farmer at Saxonburg, this letter came as an answer to prayer. He promptly wrote to Harrisburg, offering his technical services to the officials of the canal project. With his credentials as a trained engineer, attested by his degree of Civil Engineer from the famous Royal Polytechnic Institute of Berlin, he was eagerly accepted; the state authorities were glad to find a man so exceptionally qualified for the position. Following an exchange of correspondence, Roebling was engaged as an engineering aide on the canalization of the Beaver River and on the surveys for the dams and locks on the projected Sandy and Beaver Canal. The care of the farm, during his absence, he turned over to the diligent hands of his wife, who was more competent at that work than he.

4

This initial employment was, unfortunately, of brief duration. The depression of 1837 caused an interruption of the work. After a few months' layoff, Roebling wrote from Saxonburg to E. H. Gill, chief engineer of the Sandy and Beaver Canal, expressing the hope of a re-engagement on any work he might have. Desperate for engineering employment, Roebling put his best art into the composition and phrasing of his letter:

Although in possession of a fine library and my time being principally occupied by interesting studies, I cannot reconcile myself to be altogether destitute of practical occupation, and I shall embrace the first opportunity which offers itself to me to enter service again. But I should decline any offer, if I could entertain the hope of being re-engaged by you, when the times are getting better, to serve under you on the S & B or any other line you may get in charge. It is always with pleasure that I recollect the time of my service on the S & B and the kind treatment and confidence with which you have favored me.

During the time of my present leisure I have improved the ideas of some new plans and constructions regarding engineering, and I should like to communicate these to you after some time and to submit them to your judgment, before I apply for any patent.

The first improvement is a new plan altogether for dams and locks to improve the navigation of large Rivers as the Ohio, Monongahela, Allegheny, & C. by slackwater. These dams and locks will render the shallowest water navigable during the dry season, without being an obstruction to rafting, without accumulating the sediment in the river channel, and without swelling the water much in time of a flood above the former river. I am confident that these dams can be built in a very solid and permanent manner, with an expense but little increased over what common dams would cost.

Another plan of mine has reference to the improvement of the channel of the mouth of the Mississippi R. below New Orleans for which purpose large sums of money have been already expended over a number of years without having obtained any permanent beneficial results from the operations executed.

Another improvement of mine regards a simple contrivance of railroads to make switches and movable rails in turnouts and passings altogether dispensable. This plan will not increase the costs and recommends itself for its simplicity. I have also computed a number of tables being useful in tracing railroad curves in the field. Any engineer, familiar with the use of these tables, will find them preferable to any other tables calculated for that purpose and which I know of.

Thus it appears that Roebling's inventive mind had been busily at work on problems of canal construction, river improvement, and railroad engineering. In whatever branch of the profession he might find employment, he was ready to make original and constructive contributions.

His log house in the wilderness had become a home. Any work on the canal meant long separations from his wife and his infant son. But that meant little to the engineer, driven by his consuming ambition.

Before the end of the year, he secured his second job. This was again on the state waterways, this time on the construction of a feeder canal to lead water to the Pennsylvania Canal from the Allegheny River above Freeport. He was employed on the surveys, determining location, line, and grade, and staking out the work for excavation and masonry construction.

As his ability attracted the notice of his superiors, he was put in charge of a party of engineers in the state service to survey routes for a portage railroad across the mountains, to connect the central section of the Pennsylvania Canal with the western section. Roebling explored the region and laid out three suitable routes. One of these was selected as the line for the Allegheny and Portage Railroad, extending from Johnstown to Hollidaysburg and including difficult and dangerous inclined planes. For two years he was happily occupied on this work, comprising both canals and railways. There was elation in the thought that he was now through with farming and could look forward to devoting the rest of his life to his beloved engineering.

A picture of Roebling's modest scale of personal expenses when he had to live away from home on his canal work along the Allegheny River is preserved in a receipted hotel bill, written in flowing script on a carefully torn half sheet of paper, from the Mansion House at Kittanning, Pennsylvania. Covering board, lodging, and laundry for the period from January 3 to March 27, 1839, the total charges were:

204 meals—9 weeks, 5 days—at $3.00 per week.. $29.10
To washerwoman, for washing and darning,
  82 pieces of wash................... 3.87½
                                      $32.97½

His salary as an engineer during this period was four dollars a day, representing a very comfortable income in those times.

5

Upon finishing the canal work, Roebling was engaged to make location surveys for a possible railroad route over the Allegheny Mountains, to connect Harrisburg and Pittsburgh. It is interesting to record the fact that this pioneer railroad location developed by Roebling for the state was later actually followed in part in laying down the present main line of the Pennsylvania Railroad.

His work on these surveys for the proposed "Harrisburg and Pittsburgh Railroad" evidently gave satisfaction, for he was soon promoted to the rank of Principal Assistant to the Chief Engineer, Charles L. Schlatter, a position which he held for two years. He did not forget his fellow colonists in Saxonburg and he arranged for the employment of a number of them on these projects. His farm he now turned over to neighbors, in order that he might henceforth devote himself unreservedly to engineering. His wife and their son continued to live in the house he had built.

On his railroad surveys, the young engineer was left practically to his own resources, with authority to organize his own corps of instrument men, rodmen, and axmen, and to engage a "waggoner." In August, 1839, Roebling made a request for tents from the Army Arsenal at Pittsburgh and secured three "marquees" to shelter his party on their expedition through the mountains.

Railroad engineering was then in its infancy, and Roebling had no prior instruction and little precedent to guide him. His execution of this assignment affords a glimpse of the railroad-location problems confronting the pioneer engineer in those days when the country was young and undeveloped.

Living almost entirely in tents and traversing some of the wildest parts of the state, Roebling and his party worked through the heat of summer and the cold of winter, starting from Pittsburgh and carrying the survey eastward across the mountain ranges. The young chief frequently went ahead of the party, with his surveyor's compass, to make the advance reconnaissance and to discover feasible passes through the mountains. During the first season, in the fall of 1839, the party reached Laurel Hill, which was generally looked upon as an impassable barrier in the way of a railroad route. It was a mountainous ridge of the roughest topography, where dark recesses and laurel thickets offered a fine abode to deer and bears but where even the most venturesome hunters of the adjoining valleys had rarely penetrated. Roebling proceeded with an instrument to one of the highest points of Chestnut Ridge, affording an extensive view of the western slope of Laurel Hill for a distance of about twenty miles from north to south. He studied the various notches and saddles on the skyline, and then he focused his attention on the breaks formed by the tributaries of the Loyalhanna. "There was our course!" he wrote enthusiastically in his report. His plan for further operations was made accordingly.

Upon completing this first season's work, covering 150 miles of difficult and strenuous survey, Roebling prepared a full engineering report, written in his clear script, and accompanied by a large map of fine draughtsmanship recording the entire route and by larger-scale maps showing the more difficult parts of the survey. The manuscript, now a century old, bears silent testimony to the professional maturity and competence of the young engineer. He did not confine himself to the strictly technical features of his assignment—the bridges and tunnels required,

THE SURVEYOR

the excavations and embankments, the curvature and the grades—he also gave attention to the economic and commercial aspects of the railroad location. He estimated and predicted the future development of the various regions, and he observed and recorded the rich mineral wealth of the country he traversed: the thick strata of coal, the beds of limestone, and the valuable deposits of iron ore. At one point in his report he wrote: "The iron ore on the Laurel Hill is only waiting for means of transportation to be conveyed to the rich coal basins below, where also limestone is to be had in any quantity and, moreover, where an abundance of water power can be furnished by the never-failing waters of the beautiful mountain stream, the Loyalhanna—and certainly capitalists could hardly find a more eligible situation for starting mammoth furnaces on the largest scale, to be able to manufacture iron as cheap as it ever can be done in England."

During 1840 the surveys were extended further eastward toward Harrisburg, continuing through the month of December. Charles L. Schlatter, in his official report published early in 1841, paid tribute to Roebling's care and skill in directing this work, the accuracy with which the levels and the compass lines joined, and the "steadiness and perseverance which overcame every obstacle."

As Roebling's railroad location surveys progressed during the summer of 1841, he kept in touch as best he could with his superior through the uncertain facilities then available for exchange of letters. In June, Schlatter wrote: "You may spend as much time as you please upon the reconnaissance, but get an accurate knowledge of the country. . . . Take receipts for the horses you hire—and I will try and pass the receipts at the auditor's office. . . . I leave everything to your judgment."

It is difficult for us to realize the enforced economy and thrift of those days when funds were scarce. Notebooks were filled with painfully small script, to save precious paper. Roebling's boarding-house bill for the winter of 1840 to 1841 included only one "extra": "Your ⅓ share of fire for room—16 weeks—$5.33"—suggesting freezing temperatures and three men to a room. In one letter, Schlatter advised Roebling not to purchase any axes for the surveys through the woods, but instead to economize by hiring axmen who possessed their own.

In those days, when a single-page letter cost from ten to twenty-five cents, postage was another painful item of expense. In October, 1841, after Roebling had spent months in making a difficult railroad survey through the mountains, and another month at his home in Saxonburg completing the elaborate maps of the survey, his chief, Schlatter, upon the arrival of the charts at Harrisburg, was stunned to learn that the postage on the roll of drawings amounted to $6.00! Hesitating to pay such a large sum—out of his own pocket—for redeeming the shipment from the post office, he promptly wrote to his assistant at Saxonburg: "I do not think it will do to pay so large a sum for the maps unless *you* think they are worth it. If they *are* worth it, I will of course take them out. Write me by return mail." The precious maps were abandoned in the post office on Roebling's advice and he later worked days and nights at Harrisburg to make duplicates in order to save $6.00 for his superior.

The only extravagance Roebling permitted himself in those years was the filing of patent applications for his string of mechanical inventions. With each application, a patent fee of $30 had to be deposited. In order to bridge the gap between his wealth of enthusiasm for new inventions and his lack of funds to pay the fees, he courageously canceled prior patents pending and transferred the credited deposits to the newer and more significant applications.

Thus, from the time of his new start in 1837, Roebling was steadily employed in engineering work—canals, river improvements, portage railways, and surveys for turnpikes and steam railroads—with progressively increasing responsibilities entrusted to him. His few leisure hours, now becoming increasingly rare, were devoted to studies, calculations, and drawings for his inventions. His agricultural venture was a closed book. Henceforth all of his energies were to be devoted to his true career, the lifework for which his talents and training had equipped him and toward which, seemingly, a mysterious providence had been guiding him.

6

With the success of his return to engineering work there came to Roebling a revitalized feeling of confidence, of self-assurance —a sense of justification in his choice of a career, and a conviction that he could handle more difficult engineering problems with equal soundness and resourcefulness, if he only had the opportunity. And with this conviction came a strong reawakening of his original ambition, his youthful determination to become a builder of bridges.

Bridges were needed in America. There were wide rivers to cross, towns on the rivers were growing into cities, and commerce and travel were demanding means of transportation across these barriers. From his years of thought on the problem Roebling felt within himself that he had real contributions to make. At the same time he realized that he could not expect public authorities or private capital to entrust such large and critical enterprises to mere professions of ability, backed by no bridgebuilding experience. He must start more modestly—as an assistant or apprentice to someone who had already achieved opportunity and recognition.

Roebling therefore set about seeking such a connection. Bridge engineering in America was then in its infancy; there were but few engaged in it, and still fewer who had the scientific qualifications or any record of accomplishment.

In those days carpenters were building bridges—covered bridges for highways, and timber trestles and trusses for the pioneer railroads—all without the benefit of mathematics or theory. The spans were proportioned empirically, by native judgment, rule-of-thumb, and trial-and-failure, for a science of bridge analysis and design had not yet been evolved. Longer spans had to depend upon the originality and genius of their builders; and there were very few men who had the temerity to undertake such works, the tenacity and salesmanship to secure the engagements, and the resourcefulness and ability to bring the projects to a safe and successful completion. Without precedents to follow, it was not a routine job; each new span was a challenge involving new problem solving and invention—a task for ingenious, trained minds, for pioneer master builders.

Perhaps the outstanding man of this type in America at the time was Charles Ellet, brilliant, unhesitating, and a master of salesmanship. Through his initiative and personality, and by the boldness and originality of his proposals, he had already achieved a newspaper reputation as a bridge engineer—although he had not yet built a single span!

Ellet was born in 1810 at Penn's Manor, a village near Philadelphia. One of a large family, he was raised to be a farmer, but he early showed an unusual mathematical ability, a phenomenal memory, and an insatiable desire for knowledge. With little more than a grammar-school education he left home at seventeen to start engineering work as a rodman in a survey along the Susquehanna. A year later he was appointed assistant resident engineer on the construction of the Chesapeake and Ohio Canal, his superiors believing that he was much older and had had engineering training and experience. In the meantime the ambitious young man studied steadily, mastering mathematics and the elements of engineering, and learning to speak several languages, for which he had a natural aptitude. Two years of careful saving out of a meager salary enabled him to go abroad, at twenty, to finish his education at the École Polytechnique. A fluent knowledge of French, acquired by hard study in the preceding few years, and his brilliant conversational gifts made him welcome in the finest society of Paris, but he shunned social life in general, cultivating only the friendship of General Lafayette, who took a liking to the gifted young American and enjoyed political discussions with him. Carefully husbanding both time and funds, Ellet saved some of his money for traveling in England, France, and Germany, in order to study public works. He then had to sell some of his books and instruments to raise funds for his passage back to America. After his return home—only twenty-two, but evincing a boldness and originality of thought that marked his entire career—he submitted to Congress a proposal to build a thousand-foot suspension span across the Potomac at Washington! In 1839, he wrote to the mayor of St. Louis, offering to erect a permanent bridge over the Mississippi—a wire suspension bridge with the unprecedented span of 1,200 feet—for $600,000: In ignorance of the peculiar scouring conditions of the Mississippi, he rashly proposed to build his piers on pile foundations. When he declared that a suspension bridge could safely be built with a span of 6,500 feet, the mayor and city council of St. Louis stood aghast at his ideas. In his prophetic grasp of future possibilities in suspension-bridge construction Ellet was a century ahead of his time, but in his immaturity and lack of experience his proposals were half-baked, underestimated, and premature.

In 1840, Ellet advocated the idea of building a wire-cable suspension span across the Schuylkill River at Fairmount in Philadelphia to replace the famous timber bridge, known as the "Colossus," which had been destroyed by fire. This widely heralded proposal, even before it was realized, added to Ellet's prestige and fame.

Word of the projected span reached Roebling, working at the other end of the state. Most exciting was the announcement that the proposed span was to be a wire suspension. The Saxonburg pioneer recalled the "miracle bridge" of his student days, and he had never forgotten his resolve, formed then, to devote his life to developing the suspension idea. Now he saw the possible opening he had so long been craving.

Although Ellet was actually the younger of the two by four years, Roebling naturally assumed that the proposed builder of the Fairmount bridge was the older and more mature engineer. On January 28, 1840, he wrote to Ellet at Philadelphia, eagerly offering his services on the design and construction of the projected cable span:

The study of suspension bridges formed for the last few years of my residence in Europe my favorite occupation; as this matter, however, appeared to be little cared for by engineers in this country, I had no occasion whatever to bestow any further attention on it, while engaged in professional pursuits here.

Some publications of yours, which appeared in the Railroad Journal on the subject of suspension bridges, revived in me the old favorite ideas, and I was very agreeably surprised by the report of your being now actually engaged in making preparations for constructing a wire cable bridge over the Schuylkill at Philadelphia and another over the Mississippi at St. Louis, which latter indeed would form the greatest construction of the kind in existence. . . .

Let but a single bridge of the kind be put up in Philadelphia exhibiting all the beautiful forms of the system to the best advantage, and it needs no prophecy to foretell the effect, which the novel and useful features will produce upon the intelligent minds of the Americans.

You will certainly occupy a very enviable position, in being the first engineer who, aided by nothing but the resources of his own mind and a close investigation, succeeds in introducing a new mode of construction, which here will find more useful application than in any other country.

Should you at some future period be desirous of engaging an Assistant for the construction of suspension bridges, who is competent for the task, and who at the same time would execute with pleasure all the necessary drawings, please bear me in mind.

To this enthusiastic and hopeful letter, Ellet replied in a superior and formal vein:

It has given me much pleasure to learn that you have not neglected the subject of suspension bridges, in pursuing your professional studies abroad; and that you consequently appreciate the merits of that system of construction. It is my intention to endeavor to introduce this improvement in the United States; and shall accordingly pursue those means which appear to me most suitable to convey information to the publick, in the premises, and to extend a knowledge of their powers and the principles of their equilibrium. I have already had one adopted at Philadelphia, and am now engaged in a report for another, and a much more important one, for the Mississippi at St. Louis. You correctly estimate the character of the American people, in supposing that they will not fail to recognize the merits of these structures when furnished with one respectable specimen. Should the system prevail, as I have reason to believe it will, I hope to have the pleasure of engaging the services of one who is familiar with the subject, and whose aid cannot but be valuable.

Clinging to this ray of hope, Roebling forwarded drawings and descriptions of his ideas and continued writing to Ellet, but without receiving any further encouragement. The Saxonburg pioneer began to realize that he had been snubbed. Thus ended Roebling's correspondence with the man who was later to be his rival and sharp competitor in the building of wire-cable suspension bridges. Ellet—audacious, brilliant, cocksure—was to have the earlier start; but his success, like his work, lacked the sound foundation needed to make it stable and enduring.

7

Taking a leaf out of his rival's book, Roebling decided that he must likewise make his ability known to the world. He thereupon prepared an article—"Some Remarks on Suspension Bridges, and on the Comparative Merits of Cable and Chain Bridges"—which was eagerly accepted and promptly published in the *American Railroad Journal*. This monograph, a landmark in the literature of the subject, was Roebling's first publication in the field of his future achievements. Far ahead of his times, he discussed the weaknesses of earlier suspension bridges, their vulnerability to destructive cumulative vibrations caused by winds, and the proper corrective measures to insure the safety of future spans. He explained fatigue of iron from repeated stress, and he pointed out the superior strength, toughness, elasticity, and safety of wire cables over chains. A full century after this pioneer paper was obscurely published, it was discovered by the profession and became a valued item of reference.

Apparently Ellet was premature in announcing that he had been engaged to build the Fairmount bridge, for in April, 1841, the commission was awarded to Andrew Young, a local contractor; and the latter, realizing his own limitations in the new and undeveloped field of suspension-bridge design and construction, promptly accepted Roebling's eager offer to assist him in the work.

John Roebling was in heaven. At last he was to have his heart's desire—an opportunity actually to design a suspension bridge and to participate in its construction. At last he had a glimpse of the goal toward which all of his preparation, all of his ambition, had been directed.

There intervened, however, as often happens on public works, some delays and hesitations on the part of the local authorities. Early in May, Young wrote to the anxious engineer in Saxonburg: "There has been nothing done with regard to starting the bridge since I wrote you last. I think, when the County Board meets after they get home, something will be done—therefore I wish you to inform as many of the members as you will have an opportunity of seeing, that you have entered into an engagement with me in assisting to execute the work—and if the work can be started by the first of June we can complete it this season."

The high hopes that had been raised were soon changed into bitter disillusionment. On June 12 came a heartbreaking birthday message for Roebling: Andrew Young informed him that his contract at Philadelphia had been rescinded and had been given to Ellet. Carried away by pique and disappointment, Young wrote: "I have been most shamefully cheated out of the Contract for the Bridge—by the influence of Johnston the County Commissioner and the course that Ellet pursued—he went underhandedly and offered to take from the men that subscribed, Lots as payment for work done. By this means, a majority in the County Board were influenced to vote for him to get the Contract. It is considered one of the foulest acts that has ever been committed in the City of Philadelphia."

Roebling was stunned. In his unfamiliarity with political manipulations and with the hardened high-pressure methods of the "go-getter" type, he was unprepared for the matter-of-fact transfer of an engineering contract on which he had pinned his hopes. In response to his request for further information, Young wrote him: "I was well aware from the time the bridge was awarded to me that Ellet and his partisans were secretly at work undermining and endeavoring to supplant me, but I had no idea that there was sufficient corruption in the board for them to succeed; the event has proved that I was mistaken. . . . I wrote you two letters desiring you to come on under the firm conviction that the contract with me was about to be confirmed, but at the very next meeting their deep-laid schemes were first made visible and at a subsequent meeting they rescinded the resolutions of the former board and passed a resolution to give the contract to Ellet, 12 to 7. . . . I hope we shall have a chance of letting the publick see that he is not the only man that can build those bridges. . . . There is no doubt, if he makes a good job of the Fairmount bridge, there will be a number built afterwards—therefore if you are agreed to enter into partnership with me in building that kind of bridges when any offer—I will go with you hand & hand."

Roebling felt himself surrounded on all sides by baffling, unyielding barriers. He had thrown himself against them, seeking an outlet to his cherished goal, but the encircling obstacles stood firm, impenetrable, insurmountable.

But there was a way out—and, by the strength of his ambition and the light of his inspiration, he would find it. He had learned his lesson: He must not depend upon others to give him an opening; he must make it for himself! He had the same opportunities as others; and with his more thorough preparation, greater natural talents, and more intense driving energy, there was no reason why he could not do what others undertook to do —and do it better.

John Roebling had learned the lesson of America.

CHAPTER V

THE INVENTOR

WHILE making his surveys over the Alleghenies and in his work on the extensions of the canal system for the state of Pennsylvania, Roebling became familiar with the construction and operation of the portage railways which were used in those days at points where the canal routes were interrupted by mountain ranges. He studied the workings of these primitive railways with their steep inclined planes, for they fascinated him and stimulated his ingenuity.

This system of transportation between waterways involved tracks built straight up the side of the intervening mountain to form an inclined railway of formidable slope. The sectional canalboats were divided and loaded on railway trucks, the passengers were loaded on cars, and the trucks and cars with their loads were then hauled precariously up the tracks on the steep incline by means of huge hemp cables worked by a stationary engine at the top. On the other side of the ridge the trucks and cars were lowered, by a similar installation, to the next canal, and there the boats would be reassembled and the journey continued.

This method of overcoming mountains was a product of American resourcefulness and played a unique and picturesque role in the early history of railroads. A famous portage railway, which Roebling had helped to survey and build, was operated between the two settlements of Hollidaysburg and Johnstown, where the intercepting ridge of the Alleghenies is over two thousand feet high. It was obvious that lifting canalboats over such a barrier by means of hydraulic locks was commercially impracticable, even if possible from an engineering standpoint, and the difficulty was surmounted by the construction of a series of inclined planes. Two engines were installed at the top of each incline; and the cars on the double track were pulled by an endless rope—one car going up and the other going down, to balance their weight. The steam engines at the summits were huge, clumsy machines of only thirty-five horsepower each, a fraction of the power in the compact engine of a modern automobile. When the westbound traveler reached Hollidaysburg, he left his canalboat, took a seat in one of the cars designed to carry him over the mountains, and in it was transported breathlessly up and down the ten inclined planes some thirty-six miles farther along his westward way.

Charles Dickens made his first visit to America about this time, and during his trip he passed over the Pennsylvania Canal route between Philadelphia and Pittsburgh. In his *American Notes*, 1842, the famous author described his journey over the Hollidaysburg-Johnstown portage railway:

On Sunday we arrived at the foot of the mountain which is crossed by a railway. There are ten inclined planes, five ascending and five descending; the carriages are dragged up the former and let slowly down the latter by means of stationary engines; the comparatively level space between being traversed, sometimes by horses and sometimes by engine power as the case demands. Occasionally the rails are laid upon the extreme verge of a giddy precipice, and looking from the carriage window the traveler gazes sheer down without a stone or a scrap of fence between, into the mountain depths below. The journey is very carefully made, however, only two carriages traveling together; and while proper precautions are taken is not to be dreaded for its dangers.

In an early railroad guidebook, published in 1864, we find a description of the famous inclined plane up Mount Pisgah near Mauch Chunk on the Lehigh River. Built in 1827, this was the second railway in the United States, one having previously been laid in Quincy, Massachusetts. During the first year 32,000 tons of coal were carried over this mountain railway, operated by mule power! Until 1845 the empty cars were drawn up the mountain by mules, which were then put on a train of loaded cars going down, in which they all "took a ride," presenting a strange and ludicrous sight to the uninitiated stranger. In 1845 rope haulage was installed, operated by a stationary engine at the top of the incline. The sensations of a passenger are vividly described:

Stepping into a covered car, about one-third the size of an ordinary railroad car, with a safety-car attached behind, and signaling to the engineer at the top of the plane, we commence the ascent, being drawn upward by some magic, but invisible power, into the very clouds. Here, then, we are at the top of Mount Pisgah, having been drawn up the distance of 2332 feet, and being, at the top, nearly 700 feet above the foot of the plane. The novelty of this rare mode of traveling is, however, quite forgotten in the sublimity of the prospect with which the eye, looking from this vast height, is filled. Below us lies Mauch Chunk, which we look upon as a bird or an aeronaut might; towering above it rise the grand spurs of mountains, with the river winding its way among them, and above, range after range, like an aerial flight of stairs, ascend the distant mountain ridges, which, receding farther and farther into the distance as they rise, give us, instead of the contracted view from the foot of the plane, one of the widest reach and magnificence which we have ever seen or shall ever see, though we ascend the White Mountains or the Swiss Alps. The crowning point of the scene is the Lehigh Water Gap, which above the topmost mountain range lifts its perfectly defined walls, through which a passage is afforded into the unseen valleys beyond.

Now we push ourselves away again, descending for nine miles upon a downward grade, our own weight answering for a locomotive. At first, to the timorous, there comes a suspicion of insecurity; but no one has *mortally* been injured since the road was made, and the brake is always sufficient to check the too swift motion of the car. Very soon, though, the novelty and excitement of the ride make us wish that the conductor would put on all the speed in his power. As we glide rapidly around the mountain we have some very beautiful views of the valley beneath us. In about twenty minutes we reach the foot of a second inclined plane, which draws us up to the very summit of Mount Pisgah.

The hawsers used on these portage inclines were cumbersome affairs made of Kentucky hemp. At first they were made about six inches in circumference, but these soon proved inadequate for the work demanded of them, and larger hawsers, nine inches in circumference, were substituted. Some of these cables were over a mile long and cost nearly $3,000 each. As the heavy hawsers became frayed they had to be replaced by new ones, and thus were a continuing cause of great expense, to say nothing of the attendant hazard.

2

As Roebling contemplated these bulky, short-lived, and expensive hemp cables, he realized that they were a clumsy and inefficient means for accomplishing the work to which they were applied. There was a problem for him to solve, a challenge to his scientific mind.

The hawsers on the portage railways had a dangerous habit of breaking without warning. One day Roebling was watching the operation of one of the inclines. A heavily loaded railway truck was being pulled up the steep ascent; suddenly the hemp cable snapped, letting truck and load come crashing down to the foot of the slope. Two men were crushed under the wreckage.

This dramatic incident was the igniting spark of Roebling's genius. All that had gone before was preparation for this crystallizing moment of invention. Up from the subconscious, from years of directed study and from months of intensive contemplation of this specific problem, there suddenly came a realization of the solution. *Wire rope!* And in that moment an invention was born—an invention that was to be the foundation of a great new industry, to revolutionize the art of bridge-building, and to serve the needs of advancing civilization in a thousand indispensable applications.

Since the hawsers made of hemp fibers were bulky, expensive, uncertain, and short-lived, why not substitute for them something smaller, stronger, safer, and more lasting? It occurred to Roebling that if a rope could be made of iron wire so as to be flexible enough to be wound on a windlass, it should cost little more than a hemp cable but would possess much greater tensile strength with about one-fourth the diameter; and, above all, it should outlast a dozen ropes woven from vegetable fiber.

Although it seemed to come in a vivid, illuminating flash, the idea was not entirely new. Somewhere in his subconscious mind was a half-remembered item in a scientific periodical to the effect that some inventor in Saxony had produced a rope by twisting wires together. The details were vague, but Roebling decided to duplicate the feat if possible.

3

The idea of producing a wire rope of great strength combined with flexibility took hold of Roebling's mind, and he proceeded to study the problem and to experiment with it. He presented the idea to the canal officials, but they were skeptical. They had never heard of such a thing as wire rope, and they did not believe it was feasible. But the inventor was not discouraged; he was certain it could be done, and he was determined to prove it.

But to work out the actual process of fashioning a wire rope remained the great problem. No one in America had ever seen one, much less made one. Here Roebling's practical genius asserted itself. With the resources of his trained and fertile mind keyed to the pitch of invention, he worked out the necessary procedure. He dropped everything else and threw himself wholeheartedly into his rope-making experiment. He built a ropewalk on his farm in Saxonburg, purchased a quantity of iron wire, and instructed his neighbors in his newly devised art of twisting the wires into a cable. Willing workers were at hand, adding the zeal of loyalty and friendly interest to their united strength.

The hamlet of Saxonburg thus became the birthplace of a historic accomplishment, although its future magnitude and import were then hardly realized. Roebling was merely following an inner compulsion to prove an idea, to develop it and to make it useful—without dreaming that it was to be the beginning of a great industry and, incidentally, the foundation of a family fortune. He purchased the material from Beaver Falls, where iron wire was drawn, and under his direction a handful of his neighbors made the first wire rope in America.

Colonel Washington Roebling, in his *Early History of Saxonburg*, has left us a description of the procedure his father developed and successfully applied in this initial attempt at wire-rope making: "He had to create the machinery to make the rope. Machinery to make the rope in a factory had not been invented as yet. Recourse was had to using the old rope-walk system. Fortunately for Mr. Roebling his farm possessed a long level meadow, located behind the church, extending to a distance of almost 2500 feet—an ideal location. A small building (still standing) was put up to splice wire and wind it on large reels for running out. First the separate strands were laid up, seven in number, which were then twisted into the larger rope. The twisting machine was all out in the open and operated by hand-power. No horse-power or steam-power was applicable or available in that early day. The strand-making was continuous, whereas a rope itself was laid up only once in a week or ten days."

It was in 1841 that Roebling's first wire rope was made. Although the facilities were crude and the fabricators totally inexperienced, the finished product actually exceeded the inventor's most buoyant expectations.

The pioneer patent application which Roebling promptly filed, covering "Methods of Manufacture of Wire Rope," and on which patent was granted in 1842, might seem primitive today; nevertheless it shows that his dominant thought in the production of wire rope was the maintenance of uniform tension on all the wires of a strand and on all the strands in a rope—a condition that to this day is still recognized as one of the most vital manufacturing requirements for good wire rope.

In September, 1841, Roebling wrote to his friend and superior, Charles L. Schlatter, at Harrisburg, reporting on a railroad location survey he had made that summer and explaining that his completion of the maps had been delayed a few days by his work on his first order for wire rope: "I am just now engaged in the manufacture of a wire cable 600 feet long, 1 inch thick, for Capt. Young in Johnstown, calculated to draw the boats from the basin to the R.R. It will be finished in another week and in the best style. If Young's machinery should be completed in time, then I hope to be able to test its practicability before I return to Harrisburg." In his copybook draft of this letter, Roebling first wrote, "I hope to be able to demonstrate its practicability," but the thought that his assurance might be misunderstood made him substitute the more modest phrasing.

4

In those early days when travelers crossed the Allegheny mountain region by means of canals and portage railways, poli ticians were just as active in their devious ways as they are reputed to be in the more enlightened times of today. Roebling had already had a taste of their proceedings when Young lost his contract, and he his chance to design the Fairmount bridge. Now, when he proposed to substitute his wire cable for the hemp rope on one of the ten inclined planes of the Pennsylvania Canal, he had to overcome the entrenched opposition of both canal-ring politicians and rope makers, and those interests did not yield without a fight. Only after long and persuasive presentation and personal guaranties was he given the opportunity to make a trial demonstration—at his own expense and risk. When at last the wire rope was installed on the plane, a hireling of the opposition surreptitiously cut it at a splice, so that it broke in the test. "Ruined!" was Roebling's despairing thought when he saw the result. Dark disappointment confronted him after all those months of hopeful effort.

But all was not lost. By chance or fate John B. Butler, president of the Pennsylvania Canal Commission, happened to be in Johnstown that day in 1842 when the rope failed and the evidence of sabotage was discovered. James Potts, collector of canal tolls, sympathizing with the unsophisticated victim of foul play, asked President Butler to hear Roebling's story. The future bridgebuilder was a man of few words but, when he finished speaking, his case was won. Butler said, "I appoint you superintendent of Plane No. 3. The State of Pennsylvania is behind you to pay for all you need. Go ahead."

And all that Roebling could say were the words, "God is good."

The inventor's tenacity and courage were now rewarded. After the damaged splice had been repaired, the wire rope on Plane No. 3 proved a complete success; it gave long service under severe, heavy-traffic use, and it became a saying on the road that there never could be such a good rope made again as that first one. In the course of a few years, as rapidly as the driving machinery could be altered, all of the ten planes were similarly equipped, and the success of the new kind of cable was firmly established. Immediately Roebling's wire ropes began to supplant the hemp ones on other canal portages, greatly reducing the costs and the hazards of carrying the boats and their cargoes over the mountains.

With this accomplishment the inventor acquired considerable local fame and rose in professional standing and prestige. Thereafter business and professional letters began to be addressed to him as "John A. Roebling, Civil Engineer," the more distinctive appellation replacing the prior customary title of "Esqr."

5

Saxonburg became the home of this new industry, with Roebling supplying the business leadership, the engineering talent, and the inventive skill. A simple barnlike building served as warehouse and shop, and the long level field behind the church supplied the ropewalk for leading the strands to the crude, hand-operated twisting machine. With all the neighbors helping in their spare time, it was a community industry, providing employment for all who needed it. As word of the new kind of rope and its advantages spread, orders came in quick succession from the various portage systems in Pennsylvania and the adjoining states, bringing prosperity and importance to the young pioneer settlement which Roebling had founded.

Little did he heed the jesting of the skeptics and the prejudiced opposition of the manufacturers of hemp cables, after his first wire ropes, although crudely fashioned by hand, surpassed his own expectations. In the primitive shop, which he had to equip with machinery of his own planning, a period of happy activity now began. Some eight men were continually occupied in making the strands; every eight or ten days a rope was laid, using the outdoor ropewalk and requiring the labor of sixteen to eighteen men for one or two days. From dawn to dark they worked, in order to complete the critical operation. The workers, drawn from the village and the vicinity, could not take time to go home for their meals; instead, they were fed by Mrs. Roebling who, working as unwearyingly as her husband, cooked, on an open fire, the vast quantities of food required for the hungry men.

The making of wire rope was one of the earliest childhood memories of Washington Roebling, who was not quite five years old when the work began. It meant the more frequent presence of his father, who had previously been away from home for long periods on his engineering jobs. The child watched the strand making and rope laying, fascinated; and he found boyish pride and delight in his assignment to run to the houses and farms of the neighbors to call them whenever work was to begin. He wrote of this later in his reminiscences: "The men were recruited from the village and the adjacent farms—quite a task—in which I took my full share. They were always glad to see me, because it meant good pay and free meals for days. Work was from sunrise to sunset—three meals, with a snack of bread and butter in between, including whiskey. Meals were served in the house. My poor overworked mother did the cooking, all done on an open hearth. The American cooking stove was just making its appearance. We did not get one until 1847."

John Roebling was a practical genius. He realized that business initiative and salesmanship must supplement creative talent in order to secure the recognition and adoption of a product. Accordingly he entered into extensive correspondence with the engineers and officials of various transportation companies, proposing to substitute wire rope for hemp cables on the portage railways. He enumerated advantages, supplied technical data, made strength calculations, and submitted estimates. The correspondence reflected the doubts that had to be overcome and the persistence and faith that overcame them.

"Whenever anything new is proposed," remarked Roebling, "the first question is always—'Has it been tried?' or 'Where can I see it in operation?' When such questions can once be met satisfactorily, then the further introduction of any article is simply a question of time and of individual enterprise."

Roebling's persistent sales campaign produced results. Soon his wire ropes were adopted and installed on the famous inclined planes of the Delaware and Hudson Canal Company in the eastern part of Pennsylvania, then on the gravity planes of the Pennsylvania Coal Company in the Wyoming coal region, and last, but not least, on the twenty-two planes of the Morris Canal, in New Jersey, where heavier wire cables, two and a half inches in diameter, were required. On these various installations Roebling's ropes gave long service under the heaviest duty.

From the initial employment of wire cables for the portage railways the field of application soon widened as new uses were developed. The adoption of wire rope in collieries, to serve for hoists on slopes or in vertical shafts, was the next step; and before long the engineer-inventor built his first cableway—a wire rope stretched between two towers and carrying a suspended conveyor for transporting material across a stream. As time went on, there arose a brisk demand for wire ropes for the rigging of vessels, for ferries, towlines, and dredges. John Roebling was launched on his career.

6

Through years of close co-operation the friendship between Charles Schlatter and his principal assistant grew ever more cordial, each sensing the loyalty and dependability of the other. Roebling was put in charge of the contracts for section-boat planes and trucks, with authority to change the designs to assure safety; and he was directed to prepare the plans and estimates for the boat slips at Hollidaysburg and at Johnstown.

In April, 1843, Schlatter wrote from Columbia, Pennsylvania, to his principal assistant, who was then working on the new machinery installation at Hollidaysburg: "The wire rope arrived here this morning, and is indeed a perfect specimen of neat and compact workmanship. . . . The legislature has driven me out of the State by their penurious laws. My salary is reduced to $1000—and I am obliged to travel all over the state at my own expense.—I was obliged to look out for work elsewhere, and received the appointment of United States Superintendent of the Harbour at Chicago on Lake Michigan. The post is a permanent one, and all travelling expenses paid. I start with my family in four or five days by the way of New York. My regret at leaving my native State I cannot attempt to describe—but the Legislature of Penna. will eventually drive every man out of the state, who acts honestly, by actually starving them. . . . I rejoice indeed to know that your energy and perseverance have at last conquered the prejudices of ignorance and envy and that your rope has been pronounced successful by all. . . . I mentioned the subject of a wire rope at the Ferry to Mr. Butler, and will speak to McAllister relative to it the first time I see him."

Throughout his life Roebling had a faculty for making

friends, winning them and holding them by his loyal, sympathetic, and magnetic personality. They felt the sincerity of his regard, and they knew that he would gladly do anything he could to help them; and the genuineness and faith of his friendship evoked a like response from them. His former chief was one of these loyal friends. He showed almost as keen an interest as the inventor in securing the widespread adoption of wire rope. Through his connections with the authorities, at Washington and at his own post in Chicago, Schlatter was instrumental in securing favorable consideration for the new product. For his own harbor improvements he gave substantial orders for wire rope for pile drivers and dredging machines, and through his acquaintance with shipowners he was also helpful in encouraging them to adopt the new product. In November, 1843, he wrote enthusiastically to Roebling from Chicago: "I do assure you that all of us here at the Fort have the wire rope fever, and we are planning many ways to introduce tillers and rigging upon the Lakes."

Roebling, alive both to professional obligations and business opportunities, wrote an article on wire rope for the *American Railroad Journal*. It was printed as the leading article in the issue of November, 1843; and the editor, D. K. Minor, sent an acknowledgment expressing his satisfaction and inviting further contributions. In this letter he stated that he had interviewed Peter Cooper and others in New York as to the possibility of finding some capitalist who might be interested in the manufacture of wire rope: "I communicated your wishes to a gentleman in this city engaged in the manufacture of wire on a limited scale, Peter Cooper, Esqr., a very candid man—& asked him if he could point me to any one who would be likely to engage in the business. He said he did not know of any one—though he thought the time *would* come when it would become an important branch of manufacture. He assured me he would give me any information in his power & would gladly receive any communication you might feel disposed to make."

This introduction, through correspondence, to the great industrialist, Peter Cooper, was significant, since his personal interest and advice were later helpful.

Roebling wrote to I. & C. Washburn at Worcester, Massachusetts—then the leading makers of wire in America—inquiring as to their possible interest in manufacturing wire rope for him on some profit-sharing basis. They replied: "We hardly know what to say, knowing nothing of the merits of wire rope and never having seen any." Thus, through hesitation, the firm missed the greatest opportunity in its history.

Following the enthusiasm of receiving and filling an order for wire rope, the occasional difficulty of collecting payment was a source of disappointment and heartache. In one instance Roebling addressed a letter to a "Captain W. Howard of the U. S. Marines" at the Astor House in New York:

"Sir: You have received Wire rope for Bowsprit-rigging, made and delivered to order and you have caused me to write duplicate receipts twice, but the money has never been paid. This affair has caused some little suspicion on my part and I have come to the conclusion that either gross negligence or roguery is at the bottom of it. I have looked to you for payment but find it useless to wait any longer. I must therefore apply to headquarters and *will* find out who is to pay. I would inform you of this to give you an opportunity to settle provided you feel inclined to trouble yourself about it."

Whether the young wire-rope manufacturer, unaccustomed to the delays and complications of governmental red tape, was needlessly upset in this instance, or whether he was actually duped by an impostor, is not known.

7

Following Schlatter's resignation, a politician by the name of John Snodgrass secured the appointment in charge of the Allegheny and Portage Railroad. Snodgrass, after having succeeded in making a place for himself by squeezing the former chief engineer out of office, now directed his newly acquired authority against all who had been friendly to his predecessor. All further adoption of wire rope on the portage inclines was discouraged, forceful hints were dropped to subordinates to make trouble for the wire rope already installed, and payments for past orders and for services were held up. Bitterly disheartened, Roebling resigned from his superintendency at Hollidaysburg and returned to Saxonburg.

Expressing his unrestrained opinion of Snodgrass, Schlatter

wrote to Roebling: "Not the least cause of my indignation was his double-dealing and mean conduct to yourself." In February, 1844, to cheer the young inventor when he was discouraged by the failure of the state to pay him for his rope installation on one of the inclined planes, his friend wrote him prophetically from Chicago: "Let one thing console you and keep you cheerfully in the path which you have chosen, and which your indomitable courage and perseverance will render sure, that is, that you will be before long at the head of the list of those benefactors to mankind who employ science to useful purpose—while the name of John Snodgrass will only be remembered as having done all in his power to crush your inventions and your fortunes."

The following month James Bowstead, an employee on Plane No. 10, wrote that he had lost his job after personally making a perfect installation of Roebling's wire rope and preventing a certain John Black from tampering with it: "Everything worked admirably. I had just come to the conclusion that the rope would last seven years, should I stay that long to take care of it.... The couplings work perfectly—I put them all on myself. John Black wanted to put them on but I would not let him.... Snodgrass, Campbell and McGuire came down on Wednesday last. I had then hauled up a great many cars. Snodgrass walked down the plane and he said to the men that the couplings were too small and that he was afraid to ride down. He never spoke to me. I then came to the conclusion that the wire rope was doing too well for them—the cars were coming up the plane too fast, and they did not like to see it. I also came to the conclusion that I would have to go from here, but I determined first to do you all the justice that lay in my power. So the next day my discharge was sent up.... I would advise you not to put a rope on No. 2 till they alter the machinery; for they have, for one thing, put the double groove sheave on the wrong side...."

Two months later, John O'Neill, employed on Plane No. 3, wrote to Roebling: "The Wire Ropes on this road are fully equal to your most sanguine expectations.... Snodgrass stopped at the foot of Plane No. 3 a few days ago and after looking *wisely* at the rope he remarked to John Black, who was with him, that it would scarcely last this season and of course

Mr. Black agreed with him. I happened to be present and I took the liberty of differing with them. . . . Snodgrass has a contract for carrying the mail from Chambersburg to Pittsburgh. This, it is generally hoped, will take him away from here together with all his understrappers; they are universally disliked here."

Fleeing the pettiness and oppression of Prussian bureaucracy, John Roebling came to America with visions of a land of highminded men joining in co-operative effort to advance humanity's ideals. We can surmise his disillusionment and heartache when he learned that, even in the land of democracy's hope, men were not always actuated by the highest principles and motives but were subject to all the imperfections of the human race. These things must have hurt, but he kept his thoughts to himself. We learn of his disheartening experiences only from the letters of others. While the inventor's friends vigorously recorded their indignation, Roebling silently continued to manufacture wire rope for other purchasers.

And then, to add to the heartache of disillusionment, came news of an outbreak of intolerance—the "Know Nothing" movement, composed of men secretly banded together for the one purpose of arousing the passions of prejudice against all foreign-born citizens. Forgetting their own foreign origin and the newness of their own roots in the land that had given sanctuary to their fathers and grandfathers, the "Native Americans" proceeded to raise a barrier of hatred and discrimination against all subsequent arrivals. Perhaps the earlier settlers thought the continent was becoming too crowded. With the slogan "America for Americans" one generation of refugees from foreign oppression was declaring war against another, inflaming prejudice and preaching discrimination. Feelings ran high when this movement began to manifest itself a century ago. In May, 1844, John O'Neill wrote to Roebling:

"I suppose you have heard the news of the riots in Philadelphia. This is a melancholy affair. The principles of this Native American party are wicked and, in my opinion, unconstitutional in themselves. I am told they attempted to hold meetings in Harrisburg and Pittsburgh, but the city officials would not allow them. After the riot, a collection of $20,000 was subscribed for the sufferers in Philadelphia. The mob was composed of all the ruffians and cutthroats in the city. It is said that General Cadwalder is to be impeached. The Governor soon put a stop to it; he swore there should be no *blank cartridges* used next time."

Early in 1845, with a change in the administration at Harrisburg, the source of scheming persecution in the affairs of the canal was eliminated. Wire ropes were finally ordered for Planes No. 1 and No. 6, and Roebling received pay for his back claim of two years' standing in the amount of some $1,500.

8

Even before he completed his first rope of twisted wires, the engineer-inventor conceived an entirely different method of utilizing wires to make cables, and in March, 1841, he applied for a patent:

"The nature of my invention consists in the combination of any number of wires, laid parallel to each other, so that they form a round cylinder, and occupy the same position respectively for their entire length; all the wires to be uniformly strained in thus collecting them, and the whole to be wrapped with wire, either at intervals or throughout the whole length. . . . The wrapping may be made of annealed wire, so as to form a close cover.

"There are two different methods by which the uniform strain of all the wires composing the rope can be attained. . . . When all the wires are suspended so that they deflect equally, they will be equally strained by their own weight. A change of temperature cannot affect the process injuriously, as all the wires will rise or sink in proportion as they are contracting or expanding. . . . All the wires being placed parallel to each other, uniformly strained throughout and not twisted, the greatest strength will be obtained by the least quantity of wire."

In the beautiful simplicity, precision, and clarity of this presentation the clear thinking of a logical, creative mind is revealed. All of Roebling's mental faculties were now brilliantly focused on the problem of fashioning wires into ropes and cables. True, this particular patent application lacked direct applicability; parallel-wire construction was not destined to displace his earlier invention of ropes made of wires spirally twisted. The significance of his new conception, however, was that it embodied essential elements of the best method of building cables for suspension bridges! Possibly the young inventor did not realize it at the time, but in working out the problems of parallel-wire construction at this early date he was anticipating his major contribution to bridge engineering—the method of cable construction developed by him and used ever since for the building of the world's greatest spans.

9

Roebling's success in introducing his wire rope soon transformed Saxonburg from a small farming community into a thriving industrial town. As more and more orders were received, farmers were metamorphosed into mechanics, and an unanticipated era of prosperity dawned. He had supplied not only the idea, the inspiration, the invention, but also the practical vision to realize its potentialities and the practical ability to sell the new idea to the world. The one complemented the other, and both were essential for full accomplishment.

Now, after a century of increasing applications, we can better appraise the great significance of this invention of John A. Roebling's fruitful brain, for wire rope has become an essential element in our common industrial life and is used today in a thousand ways hardly anticipated at the time of its conception. It has proved invaluable in almost every field of activity and progress, from logging in the forests and quarrying in the mountains to building huge skyscrapers in cities and big spans across our rivers.

In this, the first of his achievements, Roebling set the pattern for those that were to follow. He had conceived something that was needed and bold and new, and he had wrought that vision into reality; he had met the hesitation and the doubting of authority and, by his own confidence and courage, he had answered the questions and the doubts; he had inspired the faith and loyalty of his friends and associates and, in his success, he had justified and rewarded that faith; he had created something that was needed in the march of progress and, by his own energy and persistence, he had won, for the product of his genius, the recognition and acclaim of his fellow men.

To Roebling all this brought a feeling of inner exultation. It was the forward step his spirit craved, the test for which his ambition hungered, the transition from futility to success. Any whispering doubt as to his own ability, his place in the scheme of things, was now set at rest; and that mysterious inner awareness of purpose, of talent, of power, was now justified and strengthened.

Roebling had found himself. The years of backwoods life, of frustration, and of groping adjustment were over, and he was again in the current of strenuous daring and doing.

He had achieved. He had triumphed.

But there was no song in his heart. In the strains he had undergone on his way toward achievement, a change had taken place in the man. The years of baffled struggle, of tragic heartaches, of bitter disillusionment, had killed something in him. Any lingering trace of youthful romanticism was now obliterated, or deeply suppressed. The rebellious liberal, the poetic dreamer, the optimistic pioneer, was now a personality of the forgotten past. Henceforth John Roebling was grim, hard, determined—the Man of Iron.

CHAPTER VI

THE BRIDGEBUILDER

ALTHOUGH his inventive genius, coupled with his practical initiative, had just produced a significant and revolutionary accomplishment—sufficient in itself through subsequent development to bring him fame and fortune—John Roebling was not content to rest on his laurels, as many another might have been. Instead, without loss of time, his boundless energy was seeking new worlds to conquer. He wanted more from life than mere wealth or renown as a successful manufacturer of wire rope. He wanted to be a builder of bridges; and, until he could satisfy this passionate, compelling inner urge, no other success would bring a sense of fulfillment.

With his development of wire rope, Roebling had become the inspired apostle of the power of wire. His faith was as the faith of the saints proclaiming the gospel. He believed—and he was ready to prove—that in wire the solution of the major problems in bridgebuilding would be found, permitting greater spans and heavier loading than had ever before been attempted or even conceived.

The thought of using wire for bridge construction had occurred to others. Roebling knew of those earlier attempts. The first wire suspension span in America, and perhaps in the world, was a light footbridge over the falls of the Schuylkill at Fairmount in Philadelphia, built in 1816 by White and Hazard, local wire manufacturers. This pioneer cable span, hailed as one of the great curiosities of the time, was 408 feet long and 18 inches wide; it carried a light plank floor and was calculated to carry not more than eight persons with safety. The cables were fastened at one end to two trees, at the other to the top windows of the wire mill. Each of the two cables consisted of three brass wires! The bridge was built in fifteen days, at a total cost of $125; a toll of one cent was charged for each person crossing the span. Within the year, as might have been expected, this flimsy span collapsed under the winter load of ice and snow. Its predecessor, a chain suspension bridge, had suffered the same fate.

Obviously that was not the way to build enduring bridges. During the crowded years from 1840 to 1844, in such spare time as he could find between his engineering employment and his rope-making enterprise, Roebling concentrated on the problems of future suspension spans. And while he was fortifying himself to tackle the problem, fate was once more providing, close at hand, the very opportunity for which his ambition hungered. The canal had become his *deus ex machina*.

Roebling's wire ropes had replaced the hemp cables on the portage inclines and were now helping to carry canalboats over the mountains. But just as a canal might at some points encounter the obstacle of mountain ridges across its route, so, at others, it must hurdle rivers. In such a case, the canal had to take the form of a huge wooden aqueduct, holding water enough to float a loaded flatboat. Such an aqueduct was a big and expensive undertaking; it was hazardous during construction and vulnerable when completed. Frequently the ice in the river would crush and demolish the piers in midstream and the supports on the banks, with the result that the entire structure would sometimes be carried away with the ice jam and the canal itself would join the ice and float down to the sea.

Pondering on this problem, Roebling found the solution. It occurred to him that if an aqueduct were *suspended* across a river, it would need neither props in the stream nor posts along the bank, thus allowing the river to go its way without obstruction and eliminating the attendant hazard; even for crossing a wider river, the suspension construction would minimize the number of piers and leave higher underclearance; and if he were to use *wire cables*—so much stronger and more efficient than iron chains—for the suspension, then maximum economy and safety would be secured. The glowing mental picture of the "miracle bridge" of his youth suggested the form of construction, and the inventive achievement of his mature years suggested the material to be used.

Soon the opportunity presented itself. In the winter of 1843-44, the ponderous structure built to carry the Pennsylvania

Canal across the Allegheny River at Pittsburgh was badly damaged by ice, leaving the aqueduct so unsafe as to render its removal and the erection of a new structure necessary. Baffled by their recurring problem, the canal officials decided to invite plans and proposals in open competition. Word of this opening came to Roebling in May, 1844, through R. Townsend & Co., Pittsburgh, who made the iron wire for his ropes. In the last paragraph of a business letter, they wrote: "We send you by mail this day a paper with a proposal from the Aqueduct Committee offering a premium of $100 for the best plan for a wooden or suspension aqueduct. Now is the time for you to strike; the aqueduct is at present impassable."

2

The future bridgebuilder had learned his lesson. This time he was not going to depend upon others—engineers or contractors—in the hope that they would secure the job and engage him as an assistant; instead, he would create his own job by going after it for himself. Not only had he learned this lesson from his unhappy experience four years earlier; in the intervening years he had also gained in self-assurance and in professional stature. He worked days and nights in his home at Saxonburg making the studies and the calculations, the drawings, the tabulation of materials required, and the estimates of cost. On his pioneer farm he built a model, a small suspension span using wire for cables, in order to demonstrate his theories and prove his calculations.

The three things Roebling had and was ready to give to his work were originality, courage, and determination, and these three things went into his next achievement. The problem of *suspending an aqueduct* with its enormous load had never been tackled. The problem of suspending ordinary, light bridges from wire cables was partly solved; a few cable spans had been built in France and Switzerland; and Ellet, schooled in Paris and following the work of the French engineers, had built one span of that type in America, the Fairmount bridge, in 1842. But Roebling was ready to outdo the contemporary European engineers; he had conceived *a new method of building wire cables*—the method that he later followed in all of his own masterworks and that has remained standard practice for the world's greatest spans to this day. In the first place, instead of a number of separate small cables, such as were used in the French bridges, Roebling, for greater effectiveness and protection, proposed to consolidate his small strands into large, compact, cylindrical cables. Even more important, however, than this innovation was his departure from the French method of erecting cables, in which the separate cables were laid up in straight lines on shore and then brought to the bridge and hung over the towers; for he realized that this procedure left the individual cables unadjusted for curvature and therefore made it impossible to secure uniform stresses in the wires. Recognizing these defects in the method developed by the French bridgebuilders, Roebling devised his own original system of building suspension-bridge cables—the "air-spinning method." In this procedure each wire of a cable is drawn across and hung in its proper curve, *in the air*, between its points of suspension—the successive wires being regulated in the same position, length, and sag, thereby automatically assuring uniform tension—and later all of the wires of a predetermined number are bound together to make up a strand. A strand is like a skein of thread wound upon the two hands: gather the two divisions of a skein together along the length between the hands and bind them with thread—and you have the cable strand. (Instead of hands, iron "strand shoes" are used.) Finally all of these strands or bundles of wire are squeezed together and consolidated into a single compact cable. Another innovation was the *wrapping of the finished cable*. In order to protect the cable wires from the weather, he would cover the completed cable with a tight wrapping of soft annealed wire, which, with a subsequent paint coating on the outside, would produce a hermetic seal to exclude moisture. The clear, logical reasoning of a trained scientific mind had led to the conception of these new ideas; and, by the same logical reasoning, the pioneer knew that he was right. Given a chance, he would show the world.

3

Upon completing his design studies and model tests at his home in Saxonburg, Roebling went to Pittsburgh to lay his plans and his calculations before the engineers and officials in charge of the aqueduct, and, as he stood before these men who held his future in their hands, he prayed for eloquence to convince them that they should give his ideas a trial. He knew his subject, he marshaled his facts, and his earnestness and faith carried conviction. Frankly and objectively he outlined the difficulties; he admitted that what he proposed to do was without precedent; but, appealing to the American spirit of courage and enterprise, he maintained that the advantages to be gained justified some boldness and possibly some risk in undertaking the venture.

Many were the questions asked him. Had anyone ever built a suspended aqueduct before? *Then how did he know it could be done?* How did he know it would be safe? How did he propose to build the cables? How did he know that all of the wires would share uniformly in the load? Had anyone ever built wire cables in that manner before? *Then how did he know it could be done?* How did he calculate the size of the cables? How did he know the strength of the wires? How was he going to protect the wires from rusting? Had that ever been done before? *Then how did he know it could be done?* How did he calculate the load on the spans? Why did he not include the weight of the canalboats? If the spans did not carry the weight of the boats, what did carry them? What would be the effect of temperature changes upon the cables and spans? How would he prevent vibrations and swaying of the spans? How long would the structure last? How much would the work cost? How did he know it could be done for that figure? Would he undertake the contract himself? What prior experience had he had as a bridgebuilder? *Then how did he know he could do it?* Where would he get his materials? His workmen? How long would it take him to complete the contract? Could he carry on his operations during the winter months? Would he guarantee completion of the structure within the time limit? Would he agree to build it for his own estimate of cost?

To all of these questions John Roebling gave clear, straightforward answers, for through years of concentrated thought he had prepared himself for this test. When he saw that his inquisitors looked puzzled or incredulous, he did his best to clarify the unfamiliar points and to resolve their doubts. He told them of his studies and calculations, of the span model he had built on his farm, of his experiments and tests on wires and wire cables. When the officials showed their indecision, hesitating to risk the large investment on a new and untried method of construction, Roebling courageously offered to take the contract to build the aqueduct safely and successfully, according to his own plans, at his own closely figured estimate of cost. And then, with all the earnestness of his being, he told these men of his life ambition and asked them for this one opportunity to get started toward his goal.

Who could resist the unquenchable urge of a human soul toward the goal of its aspiration? Some of the officials were impressed by the passionate conviction of the man, others were won by his lucid and persuasive presentation, and still others were influenced by the practical consideration that Roebling's plan was the most economical of all the proposals that had been received. And so the pioneer, guided by his vision and fortified by his years of preparation, courageously won his battle and secured entrance into the lifework for which he was born. With his thirty-eight years, as he drew upon the forces of his past and the strength of his future, he stood then at the height of his powers; and in him the spirit of progress of the times was embodied.

Impressed by his earnestness and faith, the commissioners agreed to give his proposal serious consideration. Roebling's hopes ran high. Although forty-three other proposals had been received from interested contractors, he learned that his bid was the lowest. In the meantime his plan was meeting with opposition from his competitors and was assailed as impracticable by other engineers. Those were days of critical suspense for Roebling as he realized that his entire future career was at stake.

Finally, one day in August, the glad news came that his plan for the aqueduct was officially approved and that the contract for its construction had been awarded to him. There was jubilation in Saxonburg. The whole town rejoiced in the opening that had at last come to their leader. His own excitement was tempered by a realization of the grave responsibility he was assuming, both as an engineer and as a contractor. As he proceeded to organize the work, placing orders and hiring men, he was fully aware that the result would either ruin him completely or put him in the forefront of contemporary American engineers.

4

The contract stipulated that the work must be completed in nine months, including the winter season. A rapid river current and a severe winter of frost and ice increased the physical difficulties of the undertaking. For the first time in the history of bridge engineering large cables for suspending a span were *strung in place*, wire by wire, in their final position—the pioneer application of the manner in which future suspension-bridge cables were to be constructed. The novelty of the method of construction created problems to be solved, and the unavoidable imperfections of untried machinery added to the difficulties.

Six straining winter months followed, the most strenuous six months of John Roebling's life. In cold and sleet, in wind and snow, the pioneer builder was up on the towers, out on the frail catwalk, swinging aloft in a boatswain's-chair hung from the cable strands—directing and taking part in every step of the operation. New methods had to be invented as he went along and unexpected snags had to be overcome. The new ideas had to be made clear to the men, and they had to be broken in to the novel and unfamiliar tasks. A loyal crew of workers was recruited, men who quickly caught the enthusiasm and courage of their leader. Many of them were his Saxonburg neighbors. They were accustomed to working under his inspiration, and they repaid his personal interest with their loyalty and devotion.

The engineer-contractor was operating on a narrow or vanishing margin of profit; in his anxiety to prove his ideas he had cut his bid to a minimum. In the course of the work he found that a substantial item of cost was the toll charged him for stone and other materials transported through the very canal for which the aqueduct was being constructed, and he therefore made application for a refund of these tolls; but he promptly found himself blocked by legal red tape. The answer to his petition was a formal resolution passed by the canal commission at Harrisburg: "*Resolved:* That the application of John A. Roebling for having refunded to him by the State $111.50 toll paid upon stone transported and to be used in the erection of piers for the Suspension Aqueduct over the Allegheny River at Pittsburgh on the Western Division cannot be allowed—there being no power vested in the Board to refund Tolls."

5

Finally, after the most anxious period of his entire life, the master builder came out conqueror over the bitter cold and the raging river. In May, within the specified contract time of nine months, the difficult and novel undertaking was completed. John A. Roebling had built the world's first suspension aqueduct. In spite of the advance condemnation of the idea by the leading engineers of the day, the great work was a pronounced success. The structure comprised seven spans of 162 feet each, punctuated by the intermediate stone piers and towers. Strung in festoons from tower top to tower top, but in unbroken length from the masonry anchorage on one shore to the masonry anchorage on the other, were the two powerful cables, each 7 inches in diameter. From these two cables, each composed of 1,900 parallel wrought-iron wires a little over ⅛ inch thick, the aqueduct was suspended—a sturdy timber flume, 16½ feet wide at the top, 14 feet at the bottom, and 8½ feet deep—wide enough and deep enough, when filled with water, to carry the boats of the Pennsylvania Canal. Bracketed out from each side of the huge timber flume was a planked towpath, 8 feet wide, for the indispensable mules that provided the motive power for the canalboats.

The total weight of the water supported on the spans was over two thousand tons! The engineers and contractors who had scoffed at the project of a suspended aqueduct had predicted its downfall as soon as the water should be let into it. But when, on May 22, 1845, the vast load of water was admitted into the suspended flume, the cables held firm, the spans remained unyielding, and the water did not burst through the planking into the rushing stream below. On top of the spans floated the canalboats, high and safe above the current of the Allegheny River. The success of the work, dumbfounding the skeptics, brought new laurels to the pioneer engineer. His reputation as a designe of suspension spans was now established, and his career as a bridgebuilder had begun.

The total contract price for the reconstruction of the aqueduct, including the removal of the old ponderous structure and the repairs of the damaged piers and abutments, was the relatively small sum of $62,000. By economical and efficient management of the work, Roebling had succeeded in making a modest profit. Out of the contract price, he netted approximately $3,500—little enough to pay him for a year of intense effort covering both the engineering and the construction, the invention of new methods, the design of the bridge and of the erection equipment, and the management, the responsibility, and the gamble of the contract.

6

With the encouragement of the editor the bridgebuilder prepared an illustrated account of his suspension aqueduct for the *American Railroad Journal*. An anticipatory editorial heralded the accomplishment as "the *first* work of the kind ever constructed"; and, in the next issue, Roebling's article appeared. It is unfortunate that this pioneer contribution was missed or lost by the profession, for it contained solutions of many problems that perplexed succeeding generations of suspension-bridge designers. In modest fashion, but with technical thoroughness, the proud builder described his first accomplished work. The most significant feature was the construction of the cables, following the new ideas he had developed. The cables, built by his method, were "solid and compact"; and the important improvement of a "continuous wrapping" of wire around the cables had its first application.

The method of construction used by Ellet on the Fairmount bridge had been criticized by Roebling, and a controversy had arisen between the two engineers. Roebling maintained that a suspension bridge could be built possessing inherent elements of stiffness and durability by a proper combination of its own elements, and that by laying the wires in compact cables of larger diameter, some of the integral resistance of a solid bar could be obtained. Ellet held it to be impracticable to combine the wires into a cable of large diameter so that each wire would bear its proper share of the burden, and that therefore the wires must be laid in cables of small diameter, adding to their number as additional strength was required. So the contest rested, until Roebling's completion of the Allegheny aqueduct afforded an effective demonstration of the principles for which he had contended.

For this work Roebling had also reasoned out a new form of anchorage, which he patented and which he used for all of his subsequent bridges. In fact this anchorage design became a standard form and was used in many structures long after his time. The new construction provided permanently sealed and protected anchor chains, requiring neither inspection nor maintenance, instead of the earlier French system with the anchor bars in open galleries to permit periodic examination, care, and repairs. Here is the master builder's description of his improved design:

> The extremities of the cables do not extend *below* ground, but connect with anchor chains, which in a curved line pass through large masses of masonry, the last links occupying a vertical position. The bars composing these chains . . . are forged in one piece without a weld. The extreme links are anchored to heavy cast iron plates which are held down by the weight of masonry. The stability of this part of the structure is fully insured, as the resistance of the anchorage is twice as great as the greatest strain to which the chains can ever be subjected.
>
> The plan of anchorage adopted on the aqueduct varies materially from those methods usually applied to suspension bridges, where an open channel is formed under ground for the passage of the chains. On the aqueduct, the chains below ground are imbedded and completely surrounded by cement. The bars are painted with red lead; and their preservation is rendered certain by the known quality of calcareous cements to prevent oxidation. If moisture should find its way to the chains, it will be saturated with lime and add another calcareous coating to the iron. This portion of the work has been executed with scrupulous care, so as to render it unnecessary on the part of those who exercise a surveillance over the structure to examine it. The repainting of the cables every two or three years will insure their duration for a long period.

Another original feature was Roebling's new method of securing the cables against any tendency to slip forward in the saddles on which they rested on the tops of the towers. This he accomplished by incorporating short wires, thus forming slight swells in the cables to fit into corresponding recesses in the saddle castings. That this ingenious idea was somehow lost by the profession is witnessed by the fact that, some forty years later, another eminent bridgebuilder, Gustav Lindenthal, reinvented the idea and secured a patent on it.

Still another interesting feature of this first of Roebling's suspension structures was the fact that he made the flume or "trunk" of such rigidly braced and substantial construction as to be self-sustaining—carrying its own weight as a huge beam from pier to pier. Consequently the only load that the cables really had to carry was the weight of water within the flume. The relatively high cost of cable wire and the large beam depth naturally supplied by the flume made it economical to let the timber structure carry as much of the load as possible. When the canal was full, the loaded weight of one span was 420 tons; of this total load, the trunk carried its own weight, 120 tons; and the cables carried the weight of the water, 300 tons. The resulting tension in each cable was less than one-fifth of its tested strength. Following his terse but lucid summary of the concepts underlying his design and resulting in a structure of assured strength and rigidity, the builder wrote: "The plan of this work therefore is a combination which presents very superior advantages, viz. *great strength, stiffness, safety, durability* and *economy*." Into his first accomplishment as a span-builder had gone much pioneering thought and supreme creative effort.

Roebling concluded his article on the suspension aqueduct with a most significant and prophetic statement: "This system, for the first time successfully carried out on the Pittsburgh aqueduct, may hereafter be applied with the happiest results *to railroad bridges*, which have to resist the powerful weight and great vibrations which result from the passage of heavy locomotives and trains of cars." Even at that early date, the pioneer span-builder was already contemplating the feasibility of a suspension bridge capable of bearing the weight of railroad trains!

The local population was proud of its Allegheny aqueduct. The Pittsburgh *Gazette* enthusiastically recorded the observed strength and resistance of the structure: "The strength of the

Suspension Aqueduct was fully tested on Monday afternoon. We noticed no less than six line boats, heavily laden, upon it at one time—one boat on each span. The quantity of water displaced by them was very large, and the motion communicated to it was of that kind to test the strength of the cables as severely as it is possible to do." The general public did not understand that the weight of the canalboats did not add to the load on the cables; but the motions of the large quantities of water displaced did constitute a maximum test of the stability and rigidity of the structure.

Whereas its predecessor had lasted a total of only seven years, the suspended aqueduct built by Roebling successfully and safely carried its load and its traffic through the ensuing years, without a break, until the entire canal was abandoned in 1861.

7

Stimulated by his first success, John Roebling's ambition drove him to further conquest. The suspension aqueduct was a proof of his strength; and now, to come closer to his goal, he longed for a chance to build a suspension *bridge*. Soon the opportunity came.

Located at the junction of the Allegheny and the Monongahela rivers where they meet to form the mighty Ohio, and wedged between the two confluent tributaries, the growing city of Pittsburgh depended upon bridges for its physical connection with the country on either side. The most important of these crossings was a toll bridge over the Monongahela River at the foot of Smithfield Street, a covered timber structure of eight spans, which had been erected in 1818, after designs made by Louis Wernwag.

Work on Roebling's Allegheny aqueduct was drawing to a close when, one day in April, a fire swept over the city, leaving forty acres of ruins where in the morning had stood the greater number of Pittsburgh's business houses. The blow was stunning; for a time it seemed that it would be fatal to the city's prosperity. But soon the native energy of the pioneer population reasserted itself, and the work of restoration began.

Among the valuable landmarks destroyed in the fire was the old covered bridge at Smithfield Street. A new structure was urgently needed to re-establish communication across the Monongahela, but the toll-bridge company was in no condition to incur heavy expense. Everyone seemed to be ruined, and it was questionable whether the necessary funds could be raised.

Roebling, with his reputation as a span-builder now established, seized the occasion to make and offer a plan and estimate for a wire suspension bridge, in which the two abutments and the seven dilapidated piers of the burnt bridge could be utilized. Having all the mechanical appliances required for the execution of the work, together with the skilled and unskilled workmen, still with him at the aqueduct job, he was determined not to miss this long-sought opportunity. He therefore made his estimates at a figure just sufficient to cover actual cost, leaving his proper remuneration to future works. His estimate was so low that the owners of the toll bridge accepted, and John A. Roebling was commissioned to design and build the new spans of the Monongahela bridge.

On May 1, twenty days after the destruction of the old bridge, work of preparing for the new one was commenced by the bridgebuilder and his crew of workers. The abutments and piers of the old structure had been greatly damaged by the fire. The injured portions had to be torn down; and the old stonework had to be strengthened before the new masonry was laid.

Before completing the piers and abutments, Roebling made an earnest effort to be allowed to raise the level of the bridge ten feet—the extra cost of which he estimated at $10,000—but without avail. Quite a bitter controversy arose on the subject. The up-river interests called for the increased clearance. The local interests feared that giving a greater headway over the channel might enable boats to ascend the Monongahela direct to Brownsville, and that Pittsburgh would lose the business that came to it as a terminal. In those days Brownsville was the point where the "National Road," one of the main arteries of westward travel, struck the western waters. From that up-river terminal, handsome packets brought the traveler to Pittsburgh. This travel formed an important item in the city's prosperity, as the passengers generally laid over at least one night in Pittsburgh before taking steamers for their western homes, this delay giving the local merchants and manufacturers an opportunity to secure many good customers. Naturally the local interests were averse to doing anything that might impair this advantage by moving the head of navigation up to Brownsville.

In June, 1845, when the construction was in its initial stage, the Brownsville *Herald* charged that the Honorable William Wilkins, president of the toll-bridge company, was opposing the rebuilding of the bridge at an increased height because it would let boats pass up the river, to the injury of Pittsburgh's flourishing business as a terminal. Neville B. Craig, the able editor of the Pittsburgh *Gazette*, promptly published a reply in which he ridiculed the insinuation.

This is sheer folly [he wrote]. Pittsburgh from her size and wealth, her geographical position, her situation at the terminus of the Pennsylvania Canal, and as the converging point of roads and trade, and means of intercourse with a wide extent of country, is eminently a point for commencing and closing voyages. In this respect no other places on the Western waters equal her, except St. Louis and New Orleans. How preposterous to suppose that the raising or lowering of a bridge is going to affect our trade! We would be glad to see the bridge raised to give our Brownsville neighbors the fullest opportunity of rivaling Pittsburgh.

Later in the same month Craig printed an editorial in which he expressed the wish that the bridge might be raised to the level of Smithfield Street, as Roebling had recommended, so as to improve the approach. But the controlling reasons given by the owners for not raising the height of the spans were the low condition of the company's finances, the gloomy state of affairs in general, and the fear of getting into trouble by saddling themselves with an additional debt of $10,000—at that time, and all circumstances considered, a fearful amount. The construction of the bridge was therefore continued on the engineer's original plan.

On each pier and on each abutment Roebling erected two cast-iron towers standing on stone pedestals and firmly anchored at the base by bolts going down into the masonry. Each tower was composed of four strong iron columns inclined inward and connected all around by artistic lattice panels secured by bolts. The lattice panels were shaped to leave an open portal which permitted the continuation of the sidewalks from one span to another.

On top of each tower rested a massive casting from which a short "pendulum" was hung, made of parallel iron bars. To the lower end of each pendulum the cables of the two adjoining spans were attached. The point of junction of the consecutive cables was thus hung from the top of the tower by means of the pendulum, free to swing to and fro on its upper pin. This was one of the most interesting innovations in the design.

The old abutments, one on each shore, had to be converted into anchorages for the cables. The anchor pits were dug within the abutments, filled with cement mortar, and then the anchor plates with the first links of the anchor chain were let down into the cement. On top of the anchor plates the masonry was commenced and carried up. All the links of the anchor chains were built solidly into the mass of stone until both reached the surface. Emerging from the masonry, the anchor chains extended diagonally forward and upward to their connections with the pendulums of the end spans.

While adopting the unique system of pendulums for this bridge, Roebling did not recommend their general use. They were an ingenious expedient, necessitated by the conditions. The purpose of the pendulums was to provide an automatic means of equalizing the pull of the cables on the two sides of a tower. By thus eliminating the strains in the towers from unbalanced cable pull, Roebling made it possible to use the slender towers to which he was limited in building on the narrow piers of the old bridge.

8

After a century of progress in the industrial arts it is difficult for us to picture the difficulties which the bridge engineer encountered through the limited resources at his command in those early times. The hand-pulled whipsaw was used to cut the floor beams of Roebling's first spans. Screws were still cut in the lathe by hand. Iron-fabricating works were little more than a one-man forge. Steel was almost among the precious metals.

The order for the ironwork for the Monongahela bridge was given to W. H. Everson, a pioneer in this future industry of the Pittsburgh district. The anchor and pendulum bars were made by this master craftsman himself, in his small forge consisting of a not very heavy helve hammer and a heating furnace. There he could be seen daily, with his leather apron on, handling the tongs, at the very spot on which, some forty years later, stood one of the great iron-fabricating shops of Pittsburgh, with Mr. Everson still at the head. The modest "Pennsylvania Forge" grew into the "Pennsylvania Iron Works."

The finished bars were made of the best iron to be had in those days, and Everson's knowledge of its quality was the real test to which the iron was subjected. In addition, all of the material used in the structure was closely examined and tested by Roebling by all the means then at hand—for he well knew that, in so light a structure, quality became of the greatest importance; and, so far as he could, he allowed nothing but quality materials to enter into the construction.

While the work was progressing on the abutments and piers, the wire for the cables was being made. The manufacture of this material suffered from the serious limitations of the art in those early days. From the light rods then available wires could not be drawn longer than one or two hundred feet, whereas nowadays bridge wires are drawn in continuous lengths of three thousand feet. Each wire was tested by Roebling to twelve hundred pounds tensile strength. It was also held in a vise or in the pliers and bent to a right angle, then bent over to the same angle in the opposite direction, then straightened up. If it stood this test without sign of fracture, it was accepted; if not, it was rejected.

In this case, in view of the limited time and the short lengths and small size of the cables, Roebling decided to spin them on shore, in separate lengths for each span, to be subsequently hoisted from flatboats to their place between successive tower tops. Although the cables for this bridge were premade on land, they were not laid straight as in the French system. Instead the master builder used his air-spinning method.

The ground on the South Side above the bridge site was, in those days, open fields. There the workshops were erected and the cables made. Two high platforms were built on opposite sides of a meadow, and on each platform a horizontal iron pin was fixed. A guide wire was stretched from pin to pin, giving the exact length and sag of the cable. Each cable was formed of a continuous wire, 140,000 feet long, the thousand short wires composing it being laboriously spliced together. The long continuous wire was then drawn from the reel and carried, by a suspended wheel, back and forth from one pin to the other. By means of nippers to catch the wire and a half-windlass to apply the pull, each wire, as it was strung, was drawn up to match the sag of the guide wire.

When 750 wires had thus been strung and adjusted, they were clamped together and then wrapped from end to end with smaller annealed wire, to form a compact, round, protected cable, 4½ inches in diameter. The wrapping was done by an ingenious hand-operated machine, devised by Roebling. Each cable, although constructed on land, was thus built in a suspended curve identical with the curve it was to occupy in the bridge.

When the sixteen cables were thus completed, planks were laid on the ground and, by means of rollers and hauling pulleys, the cables were moved to the riverside and put on flatboats which had been coupled together end to end. The boats were then towed out and anchored in position between the piers.

Between the tops of the towers, from pier to pier, Roebling hung the cables, each span being supported by two separate cables. The towers had to be well guyed to avoid pulling them over while the cables were being hoisted to their places. From these cables the graceful structure of eight spans was suspended.

In addition to providing for wagon traffic, the roadway also carried two lines of car tracks for horse-drawn "rapid transit." The sidewalks were outside of the cables. The structure was so built that any part of it was easily accessible for repair.

A novel feature was the "cradling" of the cables—hanging them with their planes inclined toward each other, five feet closer together at their lowest points than at their points of suspension at the tower tops. Roebling did this to increase the resistance of the suspended system to lateral disturbances or swinging vibrations.

The roadway of each span was supported by iron suspender rods hung from the two cables and, in addition, by a system of *inclined stays* made of iron rods, attached at the top to the same points of suspension as the cables and connected below to the timber beams of the roadway for a considerable distance on each side of the piers. These diagonal stays constituted a most significant innovation in suspension-bridge construction, contributing an important new factor of strength and rigidity to the spans. Roebling employed the same idea in all of his subsequent bridges, including his masterworks at Niagara, Cincinnati, and Brooklyn, thereby securing for them a new high measure of safety and stability against the forces that wrecked the weaker suspension spans built by his contemporaries.

In addition to the inclined stays for supplying stiffness to the design, the builder provided a deep timber railing or "parapet" on each side of the bridge. These were fashioned in the form of lattice trusses with heavy timber chords on top and bottom. The object of adopting and providing this form of deep railing—which was really a "stiffening truss" in embryonic form—was to increase the resistance of the spans to deflections and undulations. The necessity of providing adequate stiffening constructions in suspension-bridge designs was one of the most important lessons that Roebling taught the profession. Earlier spans built by others failed because this essential principle was not realized. And a century later, when bridge engineers—overconfident—forgot the lesson and began to slight this prime requirement that Roebling had established and emphasized, disaster resulted.

Roebling's recorded calculations for his Monongahela bridge show that he figured the stresses producible by a wide range of contingencies, including the weight of a "hundred head of cattle" on any span; the weight of four six-horse teams, "each loaded with 104 bushels of coal," on any span; also the weight of horsecars, crowds of pedestrians, wind pressures of hurricane violence, and all other conceivable strains to which the spans might be subjected.

9

The erection of the cables and the construction of the spans over the river had to be done in the freezing temperatures of an unusually severe winter. The work was accomplished without any accident "with the exception of one of the workmen who was seized with fits and killed by falling off a pier." The bridge was completed and opened to travel in February, 1846, eight months after it was begun. But it had been used once before, to meet a critical situation.

One Saturday morning in January, as the bridge was approaching completion, the Monongahela burst its winter fetters and the turbid stream was soon covered from bank to bank with a grinding, growling flow of ice cracked into pieces a foot thick and running "from the size of a hat to the sweep of a town lot." Immediately there was an active scurrying and elbowing of the river craft for safe berths. Flatboats, barges, and even saucy steam packets joined in the general scramble for safety. Presently the ferryboats, steam or horse driven, were snugged up under the lee of a rock or a point, on the other side of the river. Anxious and worried crowds lined the bank; for it was market day, and thousands of country folks caught on the Pittsburgh side could only look on helplessly, shivering in the damp cold and thinking how nice it would be were they at home, with a fresh backlog on the kitchen fire, toasting their feet and enjoying a warm meal. But there was little prospect of getting across. No boat could navigate in the tumbling and twisting mass of ice. The crowd thickened, until the upper line of the wharves was packed with wagons and horses from Liberty Street to Smithfield.

But someone had a lucky thought. Perhaps the new bridge, though not yet completed, might be used in the emergency. A deputation waited on Mr. Roebling. After reflecting a moment and passing a hurried order to cover a gap or two in the planking, he lifted the barrier, and the word was given, "Pass on." Whoop! Hurrah! The crowd broke up and began to move. On they marched, wagon after wagon, as fast as Captain Hart's nimble fingers could take the change, and soon the bridge was one unbroken line of wagons, horses, men, women, and children. And beneath all this mass of vehicles, horseflesh, and humanity, the unfinished bridge stood up without a quiver. With great trepidation the worthy treasurer of the company, John Thaw, walked to and fro until the whole stream of market wagons and other vehicles had passed safely over. It was a joyous relief for the trapped and belated market folk hungering for their homes and their firesides, and it was a great triumph for the builder of the bridge.

The following month the bridgebuilder wrote proudly and

enthusiastically to his friend, the editor D. K. Minor in New York, reporting the completion and opening of the new structure: "The Monongahela bridge is now open for all kinds of travel, and in full blast. The bridge surpasses the expectations even of its friends, and has silenced the opposition forever. The appearance of the whole structure is very pleasing now, and will be more so when painted and finished. The greater part of the coal consumed in Pittsburgh passes over it, in four and six-horse teams, besides a great deal of iron, and other heavy freight; from this you may judge that it is fairly tested. . . . The Fairmount bridge is well known here, and is in disrepute, on account of the vibrations to which it is liable. The stiffness promised for the Monongahela bridge has been fully attained, and pleases the public much—as many prophesied differently. The principal feature in which this structure differs from other suspension bridges is the process of making the cables, one cable in place of a number, a continuous solid wrapping laid on perfect by machinery, in place of mere bands put on by hand, at intervals; also the difference in the wooden structure, and the general arrangement of cables, suspenders and stays, which in this bridge are all so arranged as to produce the greatest stiffness with the least weight of material."

The length of the work between the abutments was 1,500 feet. The whole cost of the construction, executed by Roebling as engineer and contractor without profit, was $55,000. It is probable that no other bridge in the world, of the same length and capacity—carrying a double carriage-way, two streetcar tracks, and two sidewalks—has ever been constructed so economically.

The accomplishment was notable as the first suspension bridge scientifically designed and constructed to provide against all the forces with which a bridge must battle. Said Colonel S. M. Wickersham, a prominent engineer of the time: "This surely entitles Mr. John A. Roebling, its architect and constructor, to a position in the front ranks of Civil Engineers."

In the issue of the *American Railroad Journal* presenting the featured article by Roebling on "The Wire Suspension Bridge over the Monongahela at Pittsburgh," illustrated with a large lithographed plate of the designer's beautiful drawings of the structure, the editor paid an enthusiastic tribute: "It is with more than ordinary pleasure that we lay before our readers the drawing and description of this work of our esteemed friend and correspondent, *John A. Roebling*, C.E. . . . The aqueduct at Pittsburgh by Mr. Roebling has at last encouraged us to hope that wire bridges are soon to replace the costly, ugly and unsafe structures so common throughout the country. The aqueduct is hardly finished before we hear of a new work of Mr. Roebling, and next that it is finished and in use. This work, the Monongahela bridge, is a credit to Pittsburgh and to the ingenious constructor. As the traffic is a very heavy one, the test to which this beautiful structure has been subjected will, we think, convince those who have heretofore doubted, that the suspension type may be adopted, not only with entire safety, but also with great economy, on *railroads*."

As the fame of Roebling's work spread, requests for full particulars came from engineering authorities in various parts of the country. One was from D. H. Mahan, professor of military and civil engineering at the United States Military Academy at West Point—and, incidentally, father of a boy then five years old who later became a famous admiral and the world's foremost authority on the influence of naval power. Professor Mahan wrote to Lieutenant Rodman of the United States Army at Pittsburgh: "I some time back addressed a letter to Dr. McDowell of Pittsburgh requesting that he would put it into the hands of the engineer of the suspension aqueduct there for the purpose of procuring for me a drawing and description of it. A few days ago I received, through one of the cadets, two drawings, one of the aqueduct and another of the Monongahela suspension bridge, but without any other reference to these structures from which I could make out an accurate description. As I am desirous of mentioning some of our American structures in the text book in civil engineering that I am now preparing for the cadets, may I request you, as one of my old pupils, to procure for me answers to the following queries and transmit them to me at your earliest convenience, as the work is now in press." And he went on to enumerate a comprehensive list of engineering data desired on the two structures. This information Roebling cheerfully supplied when the request reached him. The two technical accounts promptly appeared in the 1846 edition of Mahan's classic textbook, *Civil Engineering*, one featuring the "Monongahela Wire Bridge" and the other reproducing Roebling's description and drawing of the "Wire Suspension Canal-Aqueduct-Bridge over the Allegheny River at Pittsburgh."

10

In the years following 1846, steamboats and barges could be seen proudly sailing up and down the bend of the Monongahela River, passing under the spans of the graceful suspension structure. The Pittsburgh dock line was filled with steamboats in their berths. The opposite shore was still wooded; and in the distance could be seen the first clusters of tall stacks, emitting rolling billows of smoke from furnaces and iron mills. The suspension bridge was the sole link between the two shores, and its aspect was that of a string of weblike spans hung from the row of sturdy pylons.

Several times during the existence of the bridge the rivermen had occasion to be thankful that Roebling had installed his system of pendulum suspension; for in consequence of this provision they were able to pass under with their boats, whereas without it they would have had to await the fall of the river. In cases where they lacked a few inches of headroom under the channel span, they would have all wagons stopped in the two contiguous spans, thereby depressing them and causing their cable pull to lift the channel span enough to let them through. On one occasion William Robinson, a local river pilot, raised the channel span fourteen inches by means of heavily loaded coal teams and passed his boat under.

A curious thing developed in connection with the lattice trusses which stiffened the bridge while serving as outside railings. The truss was finished with a broad top rail or cap piece running lengthwise of the bridge, covering the upper ends of the planks forming the lattice work. Owing to the rapidly varying deflections of the bridge under moving loads, this top rail would work somewhat loose and slide, and several times the friction thus caused actually set the wooden rail on fire! The movement was of slight extent, almost imperceptible, but so rapid as to raise the temperature to the kindling point.

The structure, during its years of service, was often sorely tried—sometimes when it was crowded with people viewing a steamboat race and sudden rushes would be made from one side of the bridge to the other. Such conditions afforded a real test of the designer's foresight in providing the various features that assured enduring stability. The bridge continued in service for thirty-five years, carrying the heaviest kind of street traffic, tram cars, steam rollers, and eight-horse teams pulling heavy trucks loaded with iron and machinery. The multiple-span arrangement, although quite satisfactory for an aqueduct with its loading constant or uniform in all spans, is under a disadvantage in a suspension bridge carrying variable loading. Despite the system of inclined stays which Roebling had installed, a loaded span sometimes deflected as much as two feet, with a corresponding smaller rise of the adjoining spans. Not only the designer but the profession generally profited by this experience.

Notwithstanding its one technical defect, necessitated by the financial limitations and excusable in the state of the art at that time, the bridge was an object of admiration for its graceful lines and pleasing proportions and for its strength under progressively increasing loads. The bridge outlived its builder. Finally, in 1883, to meet new traffic requirements, the historic structure was taken down and replaced by a new bridge.

11

While he was building the Monongahela bridge, Roebling received a letter from his old friend Edward H. Gill, who had been his superior on his first engineering job in America and who was now located in the young and growing city of Louisville in Kentucky. "I have just time to inform you," he wrote, "that the city authorities are about commencing waterworks for the purpose of supplying this place with water, and will be in want of a competent engineer to take charge of the work. . . . There are no engineers about here that know anything about it. You might come down and we could make plans, estimates & c. for the work, and if you wished you could get charge of it. Please write me by return mail. . . . This is a fine city. . . ."

Although Roebling was glad to know that his former chief thought of him and had so much confidence in him, he was strenuously occupied with plans and activities in his own

INCLINED PLANES ON THE MORRIS CANAL

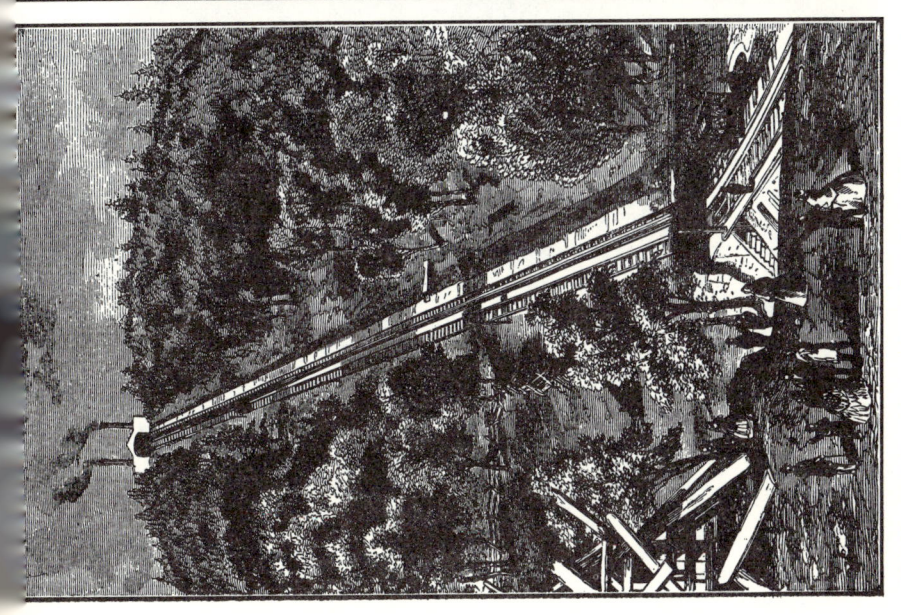

INCLINED PLANE UP MOUNT PISGAH

THE DELAWARE RIVER AQUEDUCT AT LACKAWAXEN
Built in 1848, it is still in use as a highway bridge.

chosen field and would not allow himself to be diverted by a glimpse of different pastures. His life plan afforded no time for waterworks; all of his strength and energy must go to his ruling passion, suspension bridges.

Following his successful completion of the first suspension aqueduct, the fame of this new type of construction spread, and soon the designer's services were in demand for other canal crossings where similar problems were involved.

A transportation project with a fascinating history embracing a century of epochal commercial development of the state of New York was the Delaware and Hudson Canal. The initiative for this enterprise was largely supplied by the powers of observation, the foresight, energy, and persistence of William Wurts, a businessman of Philadelphia. Without being a professional geologist, or even a hunter or trapper, he was fond of taking long tramps and excursions through the valleys and along the streams of northeastern Pennsylvania, sleeping in the forest wherever night overtook him, and subsisting on provisions from the knapsack of a chance hunter or on the trout which he lured from the cool streams of the Lackawaxen and the Lackawanna. As early as 1812, while on one of these excursions, he was attracted by the shining black stones which he noticed cropping out of the ground here and there. His interest aroused, and believing they had value, week by week and month after month he followed and examined the outcroppings, noting and mapping their location, until he had traced the outlines of the great northern and eastern anthracite coal fields of Pennsylvania. On his return he carried specimens to Philadelphia; submitting them to his brothers, Charles and Maurice, he finally induced them to go back with him and look over the ground. By repeated trials they became convinced of the value of the strange new rock as fuel.

Then began the heartbreaking struggle to convince others and to establish the black mineral as a marketable commodity. The three brothers labored under desperate obstacles—often defeated, but never giving up. The frail rafts on which they attempted to embark a few tons of coal were driven by the rapids against projecting rocks and were sunk; and when they finally succeeded in bringing a ton or two to Philadelphia, their hard-won cargo was spurned as being of little or no value. But land was then also of little value in the almost inaccessible wilderness where the deposits of coal were found. With faith and courage, the brothers made extensive purchases of tracts at prices ranging from fifty cents to three dollars per acre—coal lands whose value later rose a thousandfold. The incalculable value of this discovery is indicated by the fact that the entire anthracite area of Pennsylvania is limited to about five hundred square miles; and from this area, commencing with the first shipment of 365 tons of Lehigh coal in 1820, more than as many million tons have since been mined and shipped. After more than ten years of adverse fortune and staggering difficulties the day of success began to dawn upon the efforts of the Wurts brothers. In 1822 the General Assembly of Pennsylvania passed an act "to improve the Navigation of the Lackawaxen River," empowering Maurice Wurts to clear and improve the channel, to establish dams, locks, or canals, or to complete a system of slack-water navigation, and to establish tolls; and the Legislature of New York State then passed an act incorporating the projectors as the "Delaware and Hudson Canal Company." The preamble of this charter recited the desirability of introducing "stone" coal to New York State, the existence of extensive beds of this coal owned by Maurice Wurts, and the petition of certain citizens of New York for water communication between the Hudson and Delaware rivers. In the spring of 1823 Maurice and William Wurts secured the services of Judge Benjamin Wright, chief engineer of the Erie Canal, and instructed him "to have a proper survey or running level carried over the country from tidewater of the Hudson River, at the mouth of the Walkill, up the valley of the Rondout, and thence over the Delaware River, and thence up the same to the confluence of the Lackawaxen, and thence up the Lackawaxen to a point as near to the coal mines as possible." On the basis of the ensuing surveys and estimates, a total of $1,500,000 had to be raised by the investors, to cover the construction of the canals with their locks, inclined planes, and aqueducts. Thus the project of "Wurts' crazy ditch" was initiated. In 1825 construction was commenced, and four years later the Lackawaxen section was opened for navigation, with 7,000 tons of coal passing through it the first year. Despite financial difficulties, successive extensions and improvements were made during the years that followed, increasing the volume of commerce until the project became a most successful venture, yielding, after 1850, profits that were described as "enormous," ranging from 10 to 24 per cent of the capital that had been courageously invested. It is interesting to note that for many years, from 1833 on, all locks on the canal were closed on the Sabbath.

In the period around 1846 the Delaware and Hudson Canal was transporting hundreds of thousands of tons of anthracite. To handle the growing volume of this business, the company authorized the enlargement of their canals and locks for boats of 120 tons, those previously in use measuring only 30 or 50 tons. With these improvements, the canal—which was the only line of transportation between the Lackawanna coal fields and the Hudson River—was increased to a capacity adequate to handle the future business of two millions of tons a year.

There was a bottleneck, however, at points where canal and river crossed and their traffic mixed. At the mouth of the Lackawaxen, the Delaware River was crowded with floating craft. At the same time that the anthracite-laden canalboats were attempting to move across the river, the lumbermen of the upper Delaware were running hundreds of pine and hemlock rafts down the stream past Lackawaxen. The canal people had a dam at the spot to form a pool, so that their boats could be floated from the Pennsylvania to the New York side in still water. But, with the rafts running, there was turmoil, damage, loss, and litigation on account of the recurring mixups.

Roebling knew the conditions on the river and on the canal, and he was acquainted with the baffling problem at Lackawaxen. "*Build the canal above the water*," he said. The Delaware and Hudson engineers looked over his drawings for a four-span bridge, with three river piers. Mr. Lord, the company's engineer, went to Pittsburgh to examine Roebling's suspension aqueduct over the Allegheny, and he then advised his company to adopt the same plan. The location of the canal was modified to suit the proposed new method of crossing the rivers, and the span-builder was told to go ahead. Two aqueducts were authorized, one over the Delaware River at Lackawaxen, the other over the Lackawaxen River, two miles above its mouth.

Roebling went ahead and completed his plans. In the fall and winter of 1848 he was building the Delaware aqueduct, stringing two 8½-inch cables over the saddles on top of the masonry piers. From these cables—using *wire rope* hangers for the first time—the canal "trunk" was suspended, with a planked towpath on each side.

Near the end of the year the suspended aqueduct was finished and was ready for the opening of navigation in the spring. Distinguished engineers from many parts of the country were on hand the day the water was let in and the first boat sent through. Crowds lined the banks below and cheered at the unusual sight—a canalboat moving smoothly across from bank to bank, high above the river, high above their heads! The cables and the piers sustained their heavy burden safely, and the Delaware and Hudson officials were happy with the solution of their problem.

There was one complaint at first. Although the aqueduct piers were about 134 feet apart, they made the rafting channel in the river a hazard. The steersmen were "put to it" to get their unwieldy "doubles" between the piers without damaging the sides of the rafts. But soon the rivermen began to take pride in their newly acquired skill—when they got through without "scratching," they bragged of their accomplishment.

12

In the following spring Roebling completed the suspension aqueduct over the Lackawaxen River, a similar but shorter structure, with only two spans. In each of these new works he followed the same general design and methods of erection he had already developed, except for such progressive improvements as experience taught him.

Pleased with the progress on their first two aqueducts, and even before these structures were completed, the canal officials commissioned Roebling to design and build two more suspension crossings—this time in New York State. The bridgebuilder wrote home: "I have contracted with the Company for two more aqueducts, one over the Rondout River; the other over the Neversink River, of 170-foot span, requiring cables of 9½-inch diameter, *large enough for the support of a suspension bridge over the Niagara River, at the site in contemplation below the Falls, of 750 feet span.*"

In April, 1849, Roebling commenced the construction of the High Falls aqueduct over the Rondout, near its junction with the Hudson. And a few months later, he started work concurrently on the Neversink aqueduct, three miles above Port Jervis. This had a single span suspended from two 9½-inch cables, of 3,024 wires each—the largest cables that anyone had thus far constructed. These two aqueducts were completed in the fall and winter of 1850.

The canal aqueducts built by Roebling were essentially permanent works, as only the woodwork of the flume would require occasional renewal.

For seventy years the Delaware and Hudson Canal served its farsighted purpose. But, like all other intermediate steps of human advance, the canal and its works finally yielded to the inevitable march of progress. On November 5, 1898, canalboat "Number 1107" was loaded and cleared, and this was the last boat carrying the company's anthracite to pass over the canal. Soon thereafter the entire bed of the canal, the locks, and other works were sold for a nominal sum to private parties, to be occupied by railway lines and for other uses.

Following the discontinuance of the canal, the suspension aqueduct over the Delaware River at Lackawaxen, although fifty years old at the time, was promptly converted by its new owners into a highway toll bridge, the bed of the aqueduct serving as the roadway for the viaduct. The ancient structure, built a century ago for other duty, is still in use, carrying modern automobile and motor-truck traffic.

The other structures of the old Delaware and Hudson Canal were taken down for salvage. Here and there an undemolished lock wall, aqueduct abutment, old anchorage and cable, or a section of the abandoned canal prism are the only physical vestiges of that famous waterway of a bygone age.

In 1921 some samples of the wire were taken out of the old cables left in the abandoned ruins of the High Falls aqueduct. Laboratory tests showed a tensile strength even higher than the original strength reported by Roebling. The wires were of wrought iron, and although they were ungalvanized, they showed no material corrosion after their seventy-two years of service and exposure.

13

In the years following 1850, when canal transportation occupied a prominent and important place in the nation's life, these suspension aqueducts attracted universal admiration and gave their pioneer builder a high standing among the engineers of his day.

And thus Roebling realized his ambition—he had become a builder of suspension bridges. Six of them he had built in six years, five carrying canal aqueducts and one carrying highway traffic. In these projects he had found the opportunity for which he had been hungering, a chance to try out the new ideas that had been crowding in his mind during the preceding impatient years. And all of those ideas had proved successful—as had also the structures to which they were applied. Six suspension structures successfully completed! No other bridgebuilder of the time could point to such a record. The erstwhile immigrant and farmer had become the world's foremost exponent of a new art of bridge construction.

But the pioneer was still impatiently seeking new worlds to conquer. The thought that the cable-suspension system might appropriately be used for railroad bridges had become a fixed idea in his mind. It was obvious that a suspension aqueduct was nothing but a suspension bridge carrying an enormous load. If his wire cables could carry thousands of tons of water in a canal flume, they could equally well carry the weight of trains and locomotives. This conception was the seed of John Roebling's next great achievement.

CHAPTER VII

A LONELY BOYHOOD

WHILE the father was strenuously and ambitiously engaged in launching his professional career—working on canal construction and railroad surveys, inventing and manufacturing wire rope and securing its adoption, and building his first suspension spans for aqueducts and highway traffic—the son was quietly growing from infancy to boyhood in the secluded and primitive environment of the pioneer village of Saxonburg. And during these formative and impressionable years the boy saw little of his father, for the latter's engineering engagements and undertakings necessitated extended absences from home.

In this simple community Washington Roebling spent the first twelve years of his life. Through these early years the youngster, with the seriousness of an oldest child, was keenly registering the scene about him, with its dominant theme of struggle and toil, of poverty and deprivation. Conveniences were few, and comforts unknown. Lonely years, rough roads, and freezing cold were outstanding memories of a sensitive childhood. For the boy it was Spartan training that left its lasting influence on his personality and his character.

In the air of Saxonburg there seemed to be some magic potency, conducing to longevity. The two infant children left by Carl Roebling when he died in the summer of 1837 lived to be nonagenarians; and the first-born son of John Roebling, despite years of suffering in middle life from the crippling affliction of caisson disease, survived to attain the ripe old age of eighty-nine.

What an autobiography he could have written! For he had the material, vividly stored in his memory, covering nine decades of dramatic history and personal heroism; and he had a gift for honest, colorful narrative, for suggesting a picture of life in a few words, while revealing the stamp of his own strongly marked personality in his telling of the story. Despite his advancing years, his memory of persons and events was of unusual clarity. His style was pointed and pithy. Through his physical ordeal he had learned to be sparing of words in his speech, and he wrote as he talked. To the same heroic endurance of bodily suffering may be attributed some of the personal characteristics reflected in his writings: a certain caustic humor, a philosophical acceptance of life, an intolerance of sham and pretense, an unrestrained candor of statement, a shrewdness of judgment and insight, and an appreciation of true human values.

Unfortunately, however, Colonel Roebling never wrote an autobiography, but the few writings he left are illuminating. One of these is his paper on the *Early History of Saxonburg*, written in 1924, when he had reached the age of eighty-seven. This narrative of his boyhood reminiscences was prepared at the request of the Butler County Historical Society, and was read at a public meeting held in the Memorial Church at Saxonburg. As the Colonel, because of his physical disability, could not come to Saxonburg, his monograph was read by proxy. The audience, which filled the church auditorium, included many of the descendants of the original settlers.

2

After relating the story of the migration and the selection of Butler County as the site for the colony, the Colonel described the layout of the pioneer settlement: "Next in order came the laying out of a village . . . so that each man had a little farm to himself as is the custom in Germany. . . . As late as 1845 a black bear walked down Main Street—he got away. . . . The two Roebling brothers kept the two largest reservations for themselves. The first house was built at the highest point and was occupied by the two brothers until they married. The house still stands. . . . For the time being my father had given up engineering and passed as a farmer—that did not last long."

The Colonel then described his father's campaign to build up the infant settlement for which he had secured an option on seven thousand acres of land: "The main object of the Roebling immigration had to be tackled now—namely the sale of lots to future settlers, who had to come from Germany, because these lots were too small for an American farmer. At any rate this location was not very desirable—the winters were very cold. ... The direct road to Pittsburgh was atrocious, especially in winter time. Many a time have I made that trip—one winter my toes were frozen. . . . My father would have made a good advertising agent. He wrote at least a hundred letters to friends in and about Mühlhausen, extolling the virtues of the place— its fine climate—the freedom from restraint—the certainty of employment, etc. Many accepted and came."

Referring to the death of his uncle Carl, the Colonel wrote: "This somewhat upset my father's arrangements, but he had already abandoned farming, finding that his temper was too impatient for the slow plodding work of a farm. My mother was much better at it. Many side issues were tried to help along. A dye house was built but given up. Some from the Harz in Germany raised canaries and sold them. A good mechanic could always find work in Pittsburgh and come home to Saxonburg."

When John Roebling's venture in the breeding of canaries had proved musically unproductive and financially unprofitable, he wisely, though regretfully, transferred the canaries to his father-in-law, Herting, a delightful old German whose little farm at Saxonburg was a colorful wilderness of flowers and fruits and vegetables, and whose dogs and cats and birds were the most licensed members of his household. Herting made canary breeding pay, though, truth to tell, he cared little whether it paid or not so long as the birds sang to him and filled his home with their music. And for him they warbled and trilled from morning till night.

In the atmosphere of quaint charm and warm affection of his grandparents' home Washington Roebling spent much of his early life. That it was just a crude blockhouse and a tiny farm too small to yield a subsistence made little difference in its homelike cheer; the fragrant garden, the singing birds, the fascinating keepsakes that filled the house, and the inner radiance that expressed itself in joyous laughter and heart-warming love left an impression on the growing boy and supplied his childhood years with the warmth and sunshine they needed. In this household two proud grandparents and two doting aunts welcomed the wistful child with tender smiles and loving embraces, and his lonesome mother rejoined the happy family circle during her husband's long absences on engineering work. Not driving ambition for great achievement but peaceful gladness and kindly affection filled this simple home. In his reminiscences, written some eighty years later, Washington Roebling referred to his father at times as "Mr. Roebling," but to the poor, struggling erstwhile tailor, Herting, he always applied the more intimately reverent name of "Grandfather."

Another home in which the boy found eager welcome and affection was that of their friend and neighbor, Ferdinand Baehr. These good people, heartbroken and heart-hungry after the loss of their boy, sought to fill the void by opening their arms to the son of their closest friend. All of this left its impression on the youngster.

"Our immediate neighbor," he wrote, "was Ferdinand Baehr, from Mühlhausen, expert mechanic and wool carder, and well-to-do, who had sacrificed his home to strive for unknown benefits abroad. He built a larger house than ours and painted it pink. On the way over to this country his only son died of seasickness. The mother never got over the shock. As a makeshift, to hide his grief, Baehr took great interest in me as a boy. I was his daily visitor. He had brought over fine books, beautifully illustrated, and many knickknacks interesting to a child. He established some carding machines adjoining the house, driven by a white horse; bought wool in quantities and was well-to-do. (In those days there were no stocks or bonds to invest in.) His wife's brother Eisenhardt was his manager. Eisenhardt had been in the battle of Waterloo, member of a Nassau regiment which held Hugomont against the attacks of the French. The French bullets rattled like hail against the big oaken doors which were never battered down. Day after day this story was related to me while sitting on the wool sacks. When I called for the twentieth rehearsal, he lay dead on a sack.

"Baehr died about 1844, leaving the house to a friend on condition that he would take care of the wife after his death. She lived for twenty-five years after that. His greatest lament was that he should die without having seen a steam engine."

3

Washington Roebling did not know his father as a farmer, for the unhappy agricultural venture was given up for engineering when the boy was born. This early work in the state service, taking John Roebling away from his home, is summarized in a few lines in his son's monograph, concluding with its memorable climax: "While making these surveys he became acquainted with the workings of the Portage Railroad with its many inclined planes, transporting section boats over the mountain. In this connection it occurred to him that he could replace the heavy hempen haulage cables by a much lighter and stronger wire rope. This marked the beginning of wire rope!"

This is the only place in his entire narrative where the Colonel used an exclamation point. The making of the first wire ropes was the most exciting and dramatic experience of his boyhood memories. It not only brought his father home, sometimes for weeks at a stretch, but it also enabled the child to see his father in action; and the boy, as his years permitted, was given a chance to feel that he was participating in an achievement. At eighty-seven he still recalled the thrill of the four-year-old lad when his father was authorized to make the first wire rope for the portage railway.

John Roebling recognized that the financial returns from engineering work proper were precarious and limited; he therefore recommended, as an ideal combination for the technically trained man, an active interest in some other enterprise for profit, permitting pursuit of his chosen profession for inner satisfaction and a sense of fulfillment. This practical philosophy —its truth and its wisdom—was remembered through life by the son: "My father always held it as a necessity that a Civil Engineer (one of the poorest professions in regard to pay) should always, when possible, interest himself in a manufacturing proposition. The rope business being established, his ambition prompted him to greater efforts."

Except for the few succinct and impersonal summaries of his father's engineering accomplishments during the Saxonburg years, the Colonel's memoir was devoted to delightful intimate glimpses of the life and the personalities of the little community as seen through the eyes of the boy and as vividly recalled by the unfaded memory of a remarkable old age. "Being the 'Roebling boy' [he wrote], I had the entree to all houses, to wonder over the many heirlooms the people had brought over —curious old clocks, old Bibles and books, quaint pictures, novel utensils of copper, brass or china, long German pipes. My grandfather Herting had a wooden traveling box with covered top, inside of which a picture of the battle of Navarino was glued, showing the burning of the Turkish fleet—that was a treat. A similar picture depicted Marshal Blücher driving the French over the Katzbach."

Another home that fascinated the child was that of Bernigau, the cabinetmaker: "Bernigau made all the coffins—also wonderful furniture and fancy articles for which the Pittsburgh demand was great. He had no small children. His very particular wife always made me take off my shoes before I could enter her house."

4

When Washington Roebling was six years old, there walked into his life—and into the life of Saxonburg—the most fascinating personality of his boyhood memories, a stranger named Julius Riedel.

"One day in 1843, a new character appeared on our front steps—a short stout man, with a rifle and shotgun slung over his back—a carpet bag in one hand and leading a hunting dog —Rolla by name—by the other. A huge pair of spectacles adorned his face. He had walked up from Freeport, was exhausted and done up. My father took him in and practically took care of him for the rest of his long life. Julius Riedel was his name. Having no profession or trade, it was hard to place him, so he was made a sort of tutor for me, teaching the three R's in German. Later on he established a small German school and married my mother's sister, raising ultimately a family of six who all drifted to Trenton in course of time. Riedel was the son of a Lithuanian Baron with a large family. This Baron was a cousin of Field Marshal Blücher. At the time when Napoleon escaped from Elba and entered on his fatal Waterloo campaign, Julius was sixteen. The father asked the Marshal to let Julius ride along with him, to which assent was given—so Julius came

A LONELY BOYHOOD

to be near Waterloo, within sound of guns if not in the thick of the fight, and also entered Paris with the Allies, where he remained a while before returning home and commencing life over again, which was hard, as he had a splendid German education but no profession to make a living. Riedel's early experiences made a great impression on my youthful mind.

"In 1844 there was a vacancy in the church—no preacher. It occurred to my father that Riedel could fill it, so he was installed at once. He was gifted with a good delivery, a wonderful sympathetic voice that moved the hearers to tears. The church was crowded to capacity—young couples yearned for his services—even the deathbed anticipated his consoling funeral orations. His sermons were taken from Heinrich Zschokkes' 'Hours of Devotion' in seven volumes, which were better than he could write. No one, however, knew the catastrophe that was impending. One day a delegation arrived from the Lutheran Synod in Pittsburgh, and demanded his removal on the ground that he was no preacher, not the graduate of a Theological Seminary—had no license to preach; that the marriages that he had performed were illegal, and must be performed over again; likewise the baptisms; but the dead could remain in their graves. The congregation was in dismay—finally gave in and bade a silent farewell to poor Riedel. They never had another preacher like him. After some time a place was found for him as teacher in Columbus, Ohio. When he departed with his family, I was dumped into the same wagon and dropped at Pittsburgh to live with a German teacher named Henne."

Humor and pathos are blended in this account of Riedel's election to fill the village pulpit and his subsequent disqualification by the preachers' union. Affectionate admiration for the itinerant teacher and incredulous indignation at his dismissal retained their vividness after a lapse of eight decades.

The story of Uncle Riedel was briefly concluded: "The stay in Columbus did not last long. After some small terms of employment with my father, Riedel moved his family into our old house when we all departed for Trenton. He died in Saxonburg in 1891 at the advanced age of 91. He had filled a large horizon in my early youth. At all times he was a strict adherent to German etiquette. No one could enter his room without taking off his hat and making a profound bow."

In a single terse sentence the Colonel told the story and paid his tribute: "He had filled a large horizon in my early youth."

Thus Washington Roebling's education began at the age of six, when he was put under the tutelage of Julius Riedel, the newly arrived immigrant who so strangely walked into their lives. Successively tutor, schoolteacher, and preacher in the small community that was spellbound by his charm and his talents, this impoverished son of a Lithuanian baron and cousin of Field Marshal Blücher married Leonora, a younger daughter of "tailor" Herting, and thus became brother-in-law to John Roebling. Thereafter their families and their interests were strongly linked. The Riedels had a son, Edmund, who later became superintendent of the main rope shop in the Roebling works at Trenton, where he lived to the age of eighty. The two cousins, Edmund Riedel and Washington Roebling, looked a good deal alike, showing that the Colonel favored the maternal rather than the paternal side of the house.

When his beloved tutor lost his church engagement and had to leave Saxonburg to find a livelihood, the seven-year-old boy started his schooling at Pittsburgh, the nearest large town. This was a tender age to start boarding-school life away from home; and the difficulties and hardships of the occasional all-day wagon journey between Saxonburg and Pittsburgh—twenty-six long miles, rough hilly roads, rain and mud and snow and ice, hunger and cold, frostbitten face and frozen toes—were remembered throughout his life. In pleasant weather it was an adventure; but in the raw winds of winter it was an ordeal, mitigated only by the anticipation of the home fireside and Christmas tree awaiting the midnight arrival:

"My trips from Saxonburg to Pittsburgh, where I had been sent to school at too early an age, by way of Freeport and the Canal, or in a wagon belonging to the cabinet maker Bernigau —who always kept a store and drove to town at intervals, exchanging country products for city wares—were a great pleasure in summer time. Starting before sunrise, the first stop was at a coffee house, six miles out, price 6¢ for all the cups one could drink. When half-way out, near Deer Creek, we took dinner at Mrs. Staly's—arriving late in the afternoon. The time was passed in making German verses and poetry. The Pittsburgh rendezvous for Saxonburg teamsters was always at Madam Dubail's on 8th Street, an Alsatian who kept an excellent Inn with ample yard room for teams and horses. This good woman often reinforced my starving pouch with delicious tidbits.

"The return trip, especially in the winter time, was something dreadful—uphill all the way, snow or mud axle deep. A short cut by way of Lawrenceville was often frustrated because the ferryman refused to come over for us. Then we had to turn back, adding six miles to the length of the trip. The road through Sharpsburg was a bed of mud, churned up by the many iron wagons. The hill at Deer Creek was so steep that the load had to be transferred. Finally Saxonburg was reached late at night—all half frozen and hungry. The waiting Christmas tree was the recompense for all this distress."

5

The Colonel's memoir gives a cross section of the population of Saxonburg and an impressionistic picture of the life of the community. All of the settlers were Lutherans, except "one poor Irishman—Mularky." How this solitary son of Erin came to be in the settlement, or what he was doing there, remains a mystery. Other glimpses of the Saxonburg scene are given:

"The people who arrived in style were the Helmbolds—had driven in a coach and four all the way from the East. Their farm, just east of ours, was the largest and best. The coach was not used very long. . . .

"As a rule the Saxonburgers had large families—a great help to our early settlers. It was an orderly town—no murders or arrests. . . .

"The people were strict Lutherans, but not strict Sabbatarians, so that Sunday afternoons were devoted to reasonable enjoyment. . . .

"The early emigrants were usually well educated; knew Latin, belles lettres, history, and smatterings of law; and their wives knew how to cook and practice economy."

The primitive community had its social life. The Colonel, from his lingering recollections, sketched the collective recreations and the limited pleasures of those early days. Social gatherings in the homes, picnics, and dances were the principal forms of entertainment. "Lager beer had not yet arrived; ale was not to the German taste; wine was too dear; hence Monongahela rye whiskey became the staff of life." No party was complete without music. "Bernigau played the violin; Wickenhagen the violoncello; Neher the cornet; Roebling the flute and clavier." Amateur theatricals were staged in Aderhold's tavern. And the most memorable outdoor festival of the year was the three-day rifle-shooting contest. "At an early day, Aderhold, who kept the hotel and baked bread, instituted an annual festival for sharp-shooters, lasting three days, attended by great crowds. I believe they are still continued, although Aderhold has been dead these fifty years. He died while pushing some dough into his oven. My uncle Riedel made a poem on the sad event, which is preserved."

Guided by the Colonel's jotted reminiscences, we can follow the lonely youngster in some of his ramblings to the places that fascinated him in the pioneer village.

Up on Thorn Creek was the old swimming hole, where Saxonburg's miller—Bauer, from Bielefeld in Westphalia—had built a gristmill with a huge wheel driven by water from a dam in the creek. Everyone carried his grain there to be ground. The dam was washed away several times, and patiently rebuilt. The cool millpond, shaded by maples and willows, was a delightful spot, especially alluring on hot summer days. The youngsters of the village gathered here to fish and swim.

Right on Main Street was the smithy—always fascinating to a small boy. It occupied the larger part of Stuebgen's house. Through the open doorway one could see the flames leaping in the forge as the bellows were pumped, and then watch the sparks flying from the glowing metal under the ringing blows of the hammer.

On the lower end of Main Street was Aderhold's establishment—the town's hotel, bakery, ballroom and theater. The place was a Mecca for Pittsburghers seeking rustic amusement. From the house there was a fine view of Thorn Creek Valley. Aderhold, who came from Bleicherode in the Harz, was a leader in the social life of the community. Washington Roebling recalled him sympathetically. "A beloved man. . . . He lost a child on an Ohio steamboat. . . . He died baking bread."

A mile out on the Pittsburgh road was the farm belonging to Horn, one of the most prominent men in the settlement. The ground was high and was known as "Horn's Heights." Here young Roebling would come to enjoy the panoramic view of the village and to watch the man and his sons at their work, for Horn was kept busy with orders for the products of his skill. He was a coppersmith of reputation, and made all the kitchen utensils for the settlement. Tinware was but little used in those days. He was also a sheet-iron worker and could not begin to fill his orders. "It was a pleasure to see him work. His sons helped. Such men were the backbone of the community."

Across the road from Horn's was Knoch's blacksmith shop, where the father and son could be seen in their leather aprons, working at anvil and forge on odd jobs of horseshoeing and other ironwork for the community. This was also one of the homes with an enchanting list of old things to please a boy. Young Knoch subsequently married Washington's cousin, Amelia, daughter of Carl Roebling; at a very ripe old age they were still living "across the road from Horn's."

Down on Water Street was the cabin of Mayerstedt, a weaver, who was in poor health. This friendly weaver had "two remarkable pictures: of Ithuriel and his spear, and another Archangel, which were my delight to see."

6

People, as well as their homes and their lives, fascinated the observant youngster and were sympathetically recalled by him after the passing of fourscore years. In his paper on the early years of Saxonburg, the Colonel included a list of over sixty families making up the settlement in his boyhood, with thumbnail sketches of their lives and personalities.

There was Lamb, the butcher. "A jovial, genial citizen—our best butcher, who came to the homes Christmas time to butcher the pigs, stuff the sausages and have a little party afterwards. He was also a First Aid man and vaccinated me. A staff to lean on in misfortune."

And there was Neher, the carpenter, who had been Bernigau's

assistant in coffin building and cabinetmaking, and who then married the widow and succeeded to the business. "Expert on the cornet, especially on moonlight nights."

Then there was Gosewisch, who was an artist and painter, the only one in Saxonburg. "Made a specialty of designs on curtains, silks and hangings—peacocks and birds of paradise—much admired. Moved after a while to where people could afford to pay for artistic work." A drawing made by Gosewisch in 1835—a view of Saxonburg from Horn's Heights, the only existing picture of the pioneer settlement when it consisted of but twelve cabins and blockhouses—was discovered in Germany in 1921 and was purchased by Colonel Roebling.

Young Washington found a special attraction at the home of the Gerstners. He was fascinated by the beauty of the eldest daughter, four years older than himself. Gerstner farmed a little, and worked for the Roebling family much of the time. His wife was a sister of Lamb, the jovial butcher. The Ladies' Library Club of Saxonburg met at this home, with Julius Riedel presiding. Miss Gerstner was sixteen when the twelve-year-old Washington left Saxonburg for Trenton. The Colonel wrote in his reminiscences: "A most delightful family, where I spent much time. . . . The oldest daughter married Kleber of the Pittsburgh piano firm. . . . Mrs. Kleber was a handsome woman, retaining her beauty until she died—in 1913—at the age of 80." The sturdy old vine-covered house in Saxonburg which John Roebling built more than a century ago for his wife and children is today owned by the Klebers—descendants of the Gerstner daughter whose charm Colonel Roebling admired through a lifetime.

After Riedel's debacle the village had a Lutheran pastor, Clement, who came from Gotha, not far from Mühlhausen. The Colonel remembered him: "A most amiable man, from whom I took a few lessons, receiving a lump of sugar for good behavior each time."

Bollert was the village shoemaker, who made semiannual rounds to the various homes "to repair shoes beyond repair," as the Colonel put it. "He always carried a glass globe filled with water to temper the candle light falling on his work."

One of the early settlers was Tolly, who came from Leipzig in Saxony. He was a first-class coppersmith, making a specialty of copper steam pipes for steamboats. He was employed in Pittsburgh, but came home to Saxonburg once a year to be with his family. The small blockhouse he built, being located at the foot of the hill, was swamped in heavy rains. He also owned some acres out on the Pittsburgh road, which no one wanted to farm. "I knew the family," Colonel Roebling wrote. "The boys were just boys, but the last boy and girl rejoiced in the names of Theokar and Sheradina."

Besides Bauer the miller, there was another Bauer, not related. "He was a cable-maker employed by John A. Roebling to lay cable wire for aqueducts," wrote the bridgebuilder's son. "While working on the High Falls Aqueduct in New York, he was drowned bathing in the Rondout."

Schwietering, who "had horses and wagons," was one of the teamsters who hauled wire for Roebling from New Brighton and Pittsburgh to Saxonburg and then hauled the finished wire rope to the canal at Freeport. In Washington's memory this teamster was associated with a glamorous distinction: "He had a beautiful daughter who became the leading lady of a circus troupe."

One of the most poignant lines in Colonel Roebling's *Early History* is his reference to one of the poorer farmers, named Ohl: "Became old and tired of living, like some others in Saxonburg."

Several of the farmers in the colony were quite prosperous. As a rule they were the ones who had come to Saxonburg with ample means to make a good start.

The Helmbolds, "a fine family," had a big farm adjoining John Roebling's on the east. They farmed on a large scale, with excellent herds of cattle and sheep. "Were among the first families—had brought over a piano on which my father often played," Washington added in his reminiscences.

The Colonel, recalling the boyhood glamour and thrill of the early days of the pioneer settlement, wistfully expressed regret for the passing scene: "As the years went by, cultivation improved—woods were cleared, so that the whole expanse of Thorn Creek Valley is now visible from the cemetery high ground—but the old charm of woodland is also passing. . . . Strangers who visited the place as a matter of curiosity were disappointed—they found it a quiet little country village. The glamour of former days had vanished. . . . I have been back twice, for a day or two, once in 1858 and later in 1868. Such subsequent visits only interfere with the glamour of earlier impressions."

During their pioneering years in Saxonburg, six children were born to John and Johanna Roebling. The second, a daughter, died in infancy. Washington Roebling had to help look after a brother, Ferdinand, five years younger than himself, and three sisters—three, seven, and ten years younger. This was enough of a family to fill the little cabin and to tax the strength of the "poor, overworked mother." After the family moved to Trenton, three more sons were born.

In his reference to Dr. Koch, "Saxonburg's only homeopathic doctor," the Colonel vividly preserved an unpleasant recollection: "Pulled out my best *sound* back tooth. I miss it yet." And, in the same association of ideas, he added: "Homeopathy and my father's hydropathy were the vogue in Saxonburg." To John Roebling cold water was the lustral liquid par excellence, the grand catholicon for all ailments. To all his friends he preached the virtues of his cold-water cures, and to those of his own household he copiously and freely administered the treatment whenever they complained of not feeling well; though, in all fairness, we must record the fact that when guilty himself of being ill, he took his own medicine. Every book ever published on hydropathy, he bought and studied; and he applied its teachings according to his own notions, not with the tentative, cautious doses of a physician, but in the large, generous, voluminous proportions of an engineer. As was quite natural—for after all, what was hydropathy but a branch of hydraulics?

7

In the Colonel's reminiscences of his boyhood, something that we should be glad to find is missing. On the illuminating subject of life in the Roebling home, on the significant question of the relations between father and son, the record is strangely silent. Is there a trace of disappointment, a tinge of bitterness, in the memories of old age looking back through the years to the period of childhood? The boy grown to man's estate understandingly forgives the early physical hardships and deprivations chargeable to poverty and to circumstances; but when the more precious things that belong to the natural heritage of childhood are missed, life somehow can never make amends.

Concerning a father's companionship—sharing confidences, recounting adventures, joining in play and in laughter, wandering hand in hand through the woods, going together on camping, fishing, or hunting trips—concerning all those joyous, priceless memories that a boy normally cherishes through life, Washington Roebling's reminiscences are silent. In view of the otherwise lucid and precise memory of details exhibited in his 1924 chronicle it is strange to find a lapse of accuracy in a most unexpected place. The Colonel gives the wrong date for his father's birthday. Forgetfulness is said to be a revealing clue to suppressed emotions or subconscious inhibitions. We remember the birthdays of those we love.

John Roebling had known little of visible love and affection when he was a child, and the austerity of his boyhood scene supplied the stern pattern of his life as a parent. His own ascent had been rough and back-breaking; hence, untaught in the gentler and wiser precepts of Froebel, he adhered to the Hegelian dictum that children must be trained in the hard way. They must not be coddled, they must not be pampered; they must be driven, they must be "broken in." Poverty and ambition had been the twin spurs goading him to an effort that had won him the race. Proudly conscious of his own accomplishment under strain and sacrifice he insisted upon the same hard curriculum for his sons. As tenderness and warmth had been denied him or had been crowded out of his life, so there was little demonstration of affection in the home he ruled.

Following the German tradition, the father was lord of his household. To this was added the natural egotism of the self-made man—the man who had overcome obstacles by his own struggles and powers until he had won success and the homage of his fellows. His knowledge, like his professional achievements, had been hard won. In the fields of science and technology and other concrete branches of learning he knew what he knew. But when, yielding to a common human failing, he carried the same pride of knowledge, the same conviction of infallibility, to other domains of thought, he needed his parental authority to bolster his oracular pronouncements and to silence any timid questions of his listeners. He loved to argue, but an argument with a member of his family rapidly became a one-sided harangue. Any ventured differences of opinion would be crushed with loud and angry parental authority.

A sermon or an address once heard, Roebling could recall for years afterward almost verbatim; and this he would amplify into an interminable discourse, with his wife and children as a constrained but respectful audience. The fact that they did not understand what he was talking about mattered not at all. He would talk at them by the hour, while the poor victims would struggle to keep their eyes open, blinking in the illumination of his monologue like young owls in the unaccustomed sunshine.

The laws of the Roebling household were Draconian, and any infraction was followed by prompt and unsparing retribution. The switch, kept for terror rather than for use, or the mother's slipper, tearfully and apologetically applied on graceless youth, may be recalled in later years as a memory, half whimsical and wholly forgiving. But in the Roebling home the birch rod meant business! Too often it was applied by the father in anger—and such punishments are hard to accept and equally hard to forgive.

CHAPTER VIII

SO LITTLE TIME

JOHN ROEBLING worked feverishly, as though realizing that he had so much to accomplish and so little time in which to accomplish it. The master builder treasured every hour as though he had a premonition of destiny. His culminating work was a race against death.

How and where he found time to do all that he did, to attend professional meetings and scientific conventions, write voluminously for technical journals, practice the flute and piano, study metaphysics and put forth his conclusions in lengthy manuscripts, invent machinery and equipment and make his own drawings and models for the Patent Office, survey canals and build portage railways, create and develop a new industry, design aqueducts and bridges and himself superintend their construction—how he achieved all this in his limited lifetime bewilders the imagination. Every hour of his day was scheduled, every minute of his time was jealously guarded. Each night, before retiring, his notebooks and his daily journal had to be written up in the minutest detail, if it took till morning.

His rule was to cancel a conference if the man with whom he had an appointment was five minutes late in arriving. On one occasion, during the Civil War, General Frémont sent for him and kept him waiting in the anteroom. Finally, growing impatient, Roebling sent in a card on which he had hurriedly penciled these words: "Sir, I am happy to do any work you want. But waiting in idleness is a luxury I never permit myself." It was this driving sense of fleeting time, of the irreplaceable pricelessness of every minute, that enabled Roebling to do the work of ten men—to compress the accomplishments of many lifetimes into one. With tireless energy, deep into the night, he calculated, he made plans, he wrote.

With uncomprehending amazement Roebling's neighbors saw the restless man energetically active from early morning to midnight hours as, with a self-imposed sense of duty, he seemed to begrudge himself each speeding second. They asked: If we are mortal, then why the strain? And if we are immortal, then why the haste? Perhaps, unconsciously, Roebling had the answer: Because man is mortal, he must strive to the uttermost if, through his work, he would gain immortality.

2

During the period of his crowded and strenuous activity between 1844 and 1850, the bridgebuilder found time to prepare papers on engineering and economic subjects and to deliver some of these before learned societies and civic organizations. Railroads, transatlantic cables, power transmission, and other developments of future progress received his prophetic analysis.

A century ago the further expanding growth of the nation was awaiting the development of means of transportation. There were a few farseeing men who sensed all this, who realized the impelling need and the unlimited potentialities of the great railroad systems that were to be. And one of these men—directing the gaze of his fellowmen to the opportunities of the future—was John Roebling.

Although he had been working for most of the preceding ten years on the construction and improvement of canals and their mountain and river crossings, he had the perception and the vision to realize that the real future of American transportation was in the development of its rail system. At that time the era of railroads was just dawning and only small lines here and there were in actual existence.

In 1847 Roebling read a remarkable paper before the Pittsburgh Board of Trade on the courageous though timely subject of "The Great Central Railroad from Philadelphia to St. Louis." The dissertation, running to twelve thousand words, must have cost Roebling many days and nights to write. To his audience the subject was one of vital interest. The complete paper was promptly published in a special edition of the *American Railroad Journal*, together with a companion article, by the same author, entitled "Location of the Central Railroad through Pennsylvania." The masthead of the periodical featured the sensational contribution in large type.

The principal theme was Roebling's plea for the building of a railroad through Pennsylvania. He showed why such a transportation line was necessary for the future development of the state, and especially for the city of Philadelphia if it was to preserve its share of commerce in competition with New York, Boston, and Baltimore; and he outlined the engineering problems involved, estimated the cost of the project, and showed how the entire undertaking would be economically justified, yielding a profit on the investment. "We may now project works with unerring certitude in advance of population and traffic," he declared. "In place of railroads growing out of commerce and wealth, the former are now creating the latter. Like a magic wand, railroads open the slumbering resources of the nation."

The engineer proceeded to develop his theme in detail. He pointed out that canals could not supply the growing needs of transportation, especially as they could not be operated during the winter months. After citing what the other eastern centers were doing to advance their trade with the West, he foresaw that the city of Philadelphia and the state of Pennsylvania would be left far behind in capturing western commerce unless a forward policy were adopted. This meant constructing a system of railroads to tap the fertile valley of the Ohio River and, passing through the states of Ohio, Indiana, and Illinois, to gain access to the Mississippi at St. Louis.

He then laid down the key principles which should guide the new railroad development—principles that are as fundamentally sound and valid today as when Roebling propounded them:

No two roads shall be made when one can accomplish the business.

The whole country shall be divided into railway systems, with main trunks forming direct communications between the most important commercial towns, and lateral branches extending through the adjacent country, also connecting with the main trunk lines.

The main lines shall be so located as to interfere as little as possible with each other.

The main part of the travel should be accommodated by passing through the principal centers of population.

The freight business should be attracted from the country itself by branch lines and intersecting improvements which will discharge directly and save transshipment and commission.

The character of the roads as to lines, grades and superstructure should be adapted to the magnitude of the trade; this trade should be estimated with a due respect to the future advance of population, opening of new resources and greater expansion of business generally. In consequence of the increased facilities of transportation offered, the expense of construction may be increased in proportion as the annual charges of transportation, etc., are thereby diminished, with due allowance for the increase of business which may reasonably be expected in consequence of the greater perfection and capacity of the road.

In this analysis Roebling anticipated, by half a century, the modern science of railroad economics. In conclusion he wrote: "If we proceed in the location of the Great Central according to these principles and secure a connection with the West, before it can be done by our rivals, we shall be certain of success."

This pioneer paper was stimulating and prophetic. Seven years after it was published, the Pennsylvania Railroad Company completed and opened for service its line from Harrisburg to Pittsburgh, following in part the route Roebling had surveyed fifteen years before; and three years later the railroad gained control of the entire route across the state. Subsequently, as Roebling forecast, the line was extended through Ohio, Indiana, and Illinois to reach the Mississippi at St. Louis. Today the Pennsylvania Railroad System, representing an investment of over $3,000,000,000, has eleven thousand miles of line and carries annually thirty billion ton-miles of freight.

In his companion paper, dealing with the choice of routes to be selected and the engineering problems to be solved in the construction of the proposed line, Roebling concluded with a prophecy that revealed his grasp of the situation and his remarkable vision in predicting the successful career of the Pennsylvania Railroad even before that railroad was born: "In conclusion, I may yet remark that the completion of the railroad will, like a great national event, form one of the most remarkable epochs in the history of Pennsylvania. . . . The 'Great Central' railroad is destined to become the future highway of an immense traffic, not ephemeral in its nature, but most stable and permanent. It will vastly contribute to the wealth of Pennsylvania, and will place the future prosperity of Pittsburgh and Philadelphia upon a basis which cannot be shaken."

Sixteen years before, John Roebling was a poor immigrant, entering an unfamiliar mode of life in an alien land. Now, in thought, in language, in vision, and in accomplishment, the boy from the sleepy old town of Mühlhausen, the son of a placid, unambitious tobacconist in faraway Thuringia, had become a voice "as one speaking with authority"—a spokesman for the spirit of America.

3

Throughout his life, the bridgebuilder showed far-seeing perception of future developments. He was a pathfinder and prophet of progress.

One of the developments which he ardently preached was transoceanic communication. In an article published in the New York *Journal of Commerce* in 1850, he set forth his conviction that a transatlantic cable was a highly necessary and perfectly feasible project. He presented ample reasons for his opinion and furnished a detailed scheme of the method to be used in constructing the multiple insulated cable and in transporting and laying the submarine line.

No plan [he wrote] which proposes to suspend a line above the bottom of the ocean, which contemplates stations in the ocean, and which relies on floats and buoys, can, on account of the very frailty of these provisions, seriously claim our attention. . . . My plan proposes to lay down a *Submarine Telegraphic Wire Rope*, commencing either at New York or Boston, or vicinities, say the eastern end of Long Island, and terminating at Land's End, the most western point of England, or at Bristol or vicinity. . . . The machinery on board the steamer, which plies off the rope, is to be so constructed that the speed of the latter will be under perfect control. It is also important to provide means for recovering the rope in any depth, if by any unforeseen accident it should break while descending. . . . In conclusion, I would suggest to memorialize Congress, praying for a survey of one or more lines between New York and Great Britain, to take soundings at regular intervals, so as to obtain a correct profile of the bottom of the Atlantic.

In this article Roebling was a pioneer in forecasting transoceanic communication. It was not until four years later that Cyrus W. Field became actively interested in the idea, beginning a struggle which lasted twelve years before success was finally attained in establishing a cable connection between the two continents.

4

Roebling's reasoning in matters within his technical range was always clear, simple, and explicit, and sustained by facts. He took nothing for granted. His arguments were drawn from his store of scientific knowledge, with a mathematical accuracy and fitness that carried with them a conviction of truth. He was impatient at ignorant or biased obstruction of his projects, but always courted a discussion by those who brought sound theoretical or practical grounds for opposing his views.

The bridgebuilder's only recreation, aside from his music, was his indulgence in philosophical studies, a manifestation of the lasting influence of his teacher Hegel upon his intellectual life. His son, many years later, recalled a manuscript volume of "a thousand written pages," bearing the imposing Hegelian title "Roebling's Theory of the Universe." Anything of this description must have been evolved during the long and tedious winters of the pioneer farming years, when enforced idleness and unhappy restlessness drove the scholar to find an outlet for his pent-up mental energies.

A search among the family records and papers has failed to locate this metaphysical tome. Instead, a much briefer paper has been discovered—compact, systematic, and in beautiful script—the only extant example of John Roebling's excursions into philosophy. Bearing the more modest title, "A Metaphysical Essay on the Nature of Matter and of Spirit," and neatly filling some ten pages—not "a thousand pages"—this essay is far from being formidable and repelling.

In this manuscript, which dates from Roebling's Saxonburg years, we see a mathematical mind seeking to apply the methods of scientific reasoning to the spiritual intangibles of life and, unconsciously, reaching above cold, logical processes to arrive at a satisfying creed. Step by step he formulated his questions and developed an answer to each, starting rigorously with the elementary and the known and ending with the infinite and the unknowable:

"What is matter or body? . . . What is Soul or Spirit? . . . The properties of Spirit are then: Infinity and omnipresence, penetrating all, independent of space and dimensions, undividable. . . . A Spirit cannot be born, cannot grow, cannot decay, cannot die. . . . There is but One—all powerful, all present, all wise and intelligent, pervading the whole universe. This One is independent of space and time, therefore has existed and will exist forever. . . . There is no chance. All is law and premeditated design, all is system and order. . . . There is one great book of Revelation—the book of Nature. . . . The properties of our mind, its power of reasoning, comparing, imagining, willing—aye, its free will—are effects produced by the influence of a spiritual agency. . . . This agency is everywhere, pervading the universe, ready to act upon a mental organism wherever it presents itself, as gravitation acts upon all ponderable matter. . . . Our reasoning faculties are inspired by the Great Spirit of the Universe. . . . As far as our bodily and mental organization is concerned, we form individuals; but we are all influenced by the same Spirit. . . . Our individuality will therefore cease with our existence. . . . The arrangement of the whole universe is evidently calculated for progressive although slow improvement. The eventual destiny of rational creatures on this globe is the enjoyment of unlimited collective happiness. Collective happiness however is not always possible without partial suffering. . . . As far as our mental and moral faculties are in our own power, we are free agents and responsible for our thoughts and actions. Every sentiment of our mind forms a note in the music of creation and will, either as a dissonance or consonance, depress or swell the harmony of the whole. Our thoughts act like pulsations upon the mental ocean; once produced, their effects will be perpetuated. . . . No effect can be destroyed or lost in the Universe."

This youthful essay illustrates the man's lifelong pattern of seeking to reduce every problem to logical reasoning. Through this mental excursion outside the realm of the technical, Roebling effected a reconciliation of his scientific thinking and his religious searching so as to arrive at a simple all-embracing philosophy, a concept of individual life in harmony with an eternal, creative, universal Spirit. Nothing startling, nothing mystifyingly profound. But in the resulting creed the pioneer found satisfaction, for it answered the questions of his youth by harmonizing his analytical reasoning and his spiritual perception.

5

In his later years, after he had found his lifework and in it had found absorbing satisfaction, John Roebling no longer needed to seek an outlet for his intellectual restlessness in metaphysical speculation. Essays on "a Theory of the Universe" or on "the Nature of Matter and of Spirit" became things of the past; they belonged to a time and a world that he had left far behind him. In that forsaken past there had once been an old man named Hegel who had talked of abstractions and unrealities. In the new life and its strenuous activity all of that speculative philosophy of a bygone age seemed remote and meaningless. No longer did Roebling have time for mystic and ponderous tomes. He wanted to find his mental refreshment in clearer and more succinct form—something closer to real life. And he found it in Emerson. This wise and luminous thinker, whose philosophy had an understandable and practical side, was a man after Roebling's heart.

The truths that the master builder had evolved out of his own experience and wisdom he found again in the writings of this favorite philosopher. "Whoso would be a man, must be a nonconformist." This urge to be true to his own spirit and to his own code, despite the risk of being misunderstood, John Roebling manifested even before military demigods. The preciousness of time and the sacredness of unremitting labor had become his dominant rules of life. For these also he found confirmation in Emerson. And with what delight did he read Emerson's essay on *Nature!* The engineer-philosopher found that his own thoughts stood nearer to those of the contemporary sage of Concord than to those of the German Hegel. The wisdom of Emerson rang clearer, his philosophy rang truer.

6

The character of the man is revealed in his portrait. There was something striking in John Roebling's appearance, which the uninitiated might at first interpret as unfriendliness, but which in reality was only the outward aspect of a constant and intense inner concentration. Under the high forehead and deep under the shaggy brows lay his overshadowed gray eyes, eyes that rested on people and on things, firm and clear, earnest and penetrating. The strongly carved features, the firmly set lips, the rather haggard face with the cheeks somewhat sunken, the erect carriage of the tall man—all these betokened a strong personality and an inflexible power of determination.

Always engrossed in his plans and his work, he was sparing of speech. When he did relax in conversation, he reversed the initial outward impression of taciturnity, for he spoke frankly and cheerfully, with a fresh and wholesome humor, giving himself naturally and without reserve. Those were the moments when he took a vacation from his duty-driven, task-burdened self. Normally, however, he was serious and tense—a spiritual replica of his mother, who had given her strength for her son but who stood dry-eyed as she bravely sent him off to his destiny.

JOHN A. ROEBLING

ROEBLING'S DRAWING OF THE MONONGAHELA BRIDGE

PITTSBURGH IN 1854, SHOWING THE MONONGAHELA BRIDG

PART TWO
THE FOUNDATION

CHAPTER IX

THE INDUSTRIALIST

THE YOUNG and thriving industry which John Roebling had started in his little backwoods community in western Pennsylvania had outgrown the limitations of its birthplace and needed to find a new location for future expansion. As orders for wire rope kept coming in from portage railways, canal companies, and shipowners, the volume of business soon strained the capacity of the little hand-worked shop and ropewalk in Saxonburg, and orders were lost because early deliveries were impossible. Difficulties of transportation imposed an additional handicap. For five months each winter the canals were closed and no shipments could be made.

For greater accessibility to the growing eastern market, and to reduce haulage and shipping costs, a new site had to be found for the infant industry. Roebling realized this. Saxonburg, remote and secluded, was not the place in which to expand his growing enterprise or even to preserve the lead which he had already acquired. He had long contemplated a removal.

The suitability of Trenton for the proposed wire-rope factory had been suggested to Roebling by Peter Cooper. In 1830 Cooper had built the "Tom Thumb," the first locomotive in America, and fifteen years later he decided to go into the business of rolling railroad rails. He had selected Trenton for his own manufacturing ventures; and in 1847, together with his nephew, Abram S. Hewitt—afterward mayor of New York— he organized the Trenton Iron Company.

Peter Cooper's encouragement was not entirely disinterested, as he had a shrewd eye open to the possibility of securing profitable relations with the new venture. Knowing that Roebling had been accustomed to purchase his supplies of iron wire from wire mills, the Trenton Iron Company naturally assumed that he would continue to do so, and therefore that, as local manufacturers, they would enjoy a large share of his

135

orders. But their hopes were destined to be unfulfilled, for no sooner was Roebling's little mill established in Trenton than he began immediate preparations *to draw his own wire.* Cooper and his associates had underrated the Saxonburg pioneer's business sagacity and foresight; their credulity would have been staggered if anyone had suggested to them in those days that Mr. Roebling's insignificant venture was destined in the ensuing years to grow to such mammoth proportions as ultimately to eclipse their own plant, not only in the manufacture of wire but also in total volume of business.

Trenton was then a small town that showed little promise of future growth; but its potentialities and advantages as a manufacturing site were soon apparent to Roebling. Eastern location, transportation facilities by both rail and water, availability of skilled labor, and proximity of related industries—all these pointed to Trenton as an ideal site for the new enterprise.

In August, 1848, upon reaching his decision, Roebling purchased twenty-five acres of land in what was then open country, about one mile from the center of the city of Trenton. The price was $3,000. The ground was undeveloped farmland with no buildings on it. Skirting one side of the property were the Delaware and Raritan Canal and the Camden and Amboy Railroad, affording convenient shipping facilities.

2

Preparations were promptly made for the erection of a mill, a ropewalk, and a house for the family. Roebling was unable to remain in Trenton to supervise these building operations, as he had to attend to his engineering engagements on the construction of the suspension aqueducts of the Delaware and Hudson Canal, besides needing to wind up his affairs, including the sale of his house and land, in Saxonburg. He therefore put one of his most trusted assistants, Charles Swan, in charge of the construction work in Trenton.

Although leaving the supervision to Swan, Roebling himself remained architect, engineer, inventor, designer, and draftsman of the work. He personally planned and designed every building of his new wire mills and shops, and he invented and designed nearly every piece of machinery that went into them. And how well the founder planned and wrought is attested by the fact that machines he designed and built are still in use in the Roebling plant today—after nearly a century of service! Some of the machines he designed and constructed, in the early days of his rope making, are still turning out rope, and good rope.

During the winter and spring of 1848–49, the work of preparation of the ground and the building of the Trenton plant proceeded rapidly under Swan's capable supervision. The bridgebuilder kept in constant touch with his factotum by correspondence. Writing from Saxonburg in February, Roebling gave instructions and advice about a well which was being dug at the new site and with which Swan was having trouble, owing to the appearance of quicksand and consequent caving in. Writing again in March, he commended Swan's action in halting a proposed encroachment upon his land by the projected opening of a new highway. In April he wrote from Two River, Pike County, Pennsylvania, where he was then building the Lackawaxen aqueduct, to express his regret that some trees, which Swan had planted, had not been watered and consequently might not grow, and incidentally remarking that in the course of another week he would be going to High Falls to commence a new suspension aqueduct. He warned Swan against employing a certain Jonathan Rhule, if he should appear in Trenton, as he was an "Ellet's man"—referring to the rival bridgebuilder. "I do not want," Roebling wrote, "any news carried between myself and Mr. Ellet.—Ellet has promised Rhule work for no other reason but because I discharged him."

From High Falls he wrote in June, warning his superintendent that in laying brick in mortar he must not forget to have the bricks well wetted or saturated with water. In July he urged Swan to push the building of the house and also the other work, "so that we can commence making ropes on the first of September." He said that he expected his family to leave Saxonburg for Trenton in September, and he authorized Swan, if there should be a public auction in Trenton or near it, to buy some *used* furniture "if good yet, and cheap, for one or two rooms."

In August he wrote that he had found several orders for wire rope awaiting him, "all of them very urgent"—and "to hurry up our works *as fast as it can be done.* . . . We must commence making rope *before* the first of October. I have promised to ship a rope 3000 feet long for Charleston, S. C., on the first of October."

Early in September he asked Swan, on his next trip to Philadelphia, to look at some good stoves for cooking as well as common use. "I must," Roebling wrote, "get a cooking stove, and one for the lower room, before my family arrives." At last the Roebling family was to enjoy the new luxury of an American stove.

The letters followed one another almost daily. Writing from High Falls, Roebling introduced a new German machinist, and incidentally mentioned the initial work on the Neversink aqueduct. "The bearer is Ernest Ermelling, the machinist, who is to work with you. You will find him an intelligent, willing & skillful hand. You may give him instructions to work the Engine, so that he can attend to it while you are engaged at other work. My principal object is to put him at the Wire drawing. . . . To morrow I have to go up to the Neversink to lay the second Anchorage. I expect to start for Trenton this day week." Later in the same month he transmitted two letters for Ermelling, methodically charging him for the postage paid. "Enclosed I send you two letters for Ernest Ermelling, one from Germany. Charge him 34 cs. of Postage."

Writing next from the Astor House, on lower Broadway in New York, Roebling stressed the need of securing some good men experienced in wire drawing, and he counseled Swan, if any such skilled workers should appear, to "do as if you know all about wire." He stated that he himself had made arrangements in New York to engage "German wire-drawers, in case any should arrive here, also one or two German machinists, who frequently arrive and get no work. Should any such men come to Trenton we must keep them, put them to the rope business and teach them wire drawing. Such hands will learn easily and are satisfied with common wages, say, 85 cents a day."

A few days later, he sent a letter from High Falls introducing a mechanic, one Carl Lange, who he judged would be a good hand at rope making. "Lange has been getting 85 cent wages here, which I will continue to the fifteenth of October, then reduce to 75 cents for the winter."

What a heart-constricting revelation of the penury of the times! To Roebling and to his contemporaries, this was ordinary business shrewdness or necessity. Fortunately the times have changed and, with them, our concepts and standards of industrial relations.

John Roebling was investing every dollar he had, and probably all that he could borrow, to build his plant and to establish an industry; during the initial period of struggle and trial he had to make ends meet, or else all his plans would go down in a financial crash, carrying with them the hopes of those who depended upon him. He was also reacting to the scarring and hardening influences of his earlier experiences—his childhood scene, when all of life was dominated by a fevered passion for thrift, when penny pinching was identified with a drive toward a grimly resolved goal; and his pioneer farming years, when desperate effort and the utmost economy were the necessary twin conditions for survival. The man's life had been warped, and he had not yet found the larger pattern.

Continuing his letter, Roebling warned Swan not to permit Lange to smoke during work. "He is an inveterate smoker, and I believe he was the cause of a fire which broke out in my shanty yesterday but was fortunately discovered in time. I have now strictly prohibited smoking upstairs amongst bedding, where there is danger, and I want you to insist on it and carry out the same rule at Trenton, without exception. I would much rather discharge a good man than lose a building by his carelessness. Tell Fritz that he must either quit smoking upstairs or quit work. Do never depart from this rule. Last winter we had a most dreadful accident in this neighborhood, when a large shanty burned down and three men lost their lives, the whole caused by smoking amongst bedding."

The new wire and rope mill at Trenton was of wood construction, with sleeping quarters for the workmen in the upper story or attic. Roebling, the son of a tobacconist and raised in a home always filled with the fragrance of tobacco, was not himself a smoker; he would not divert any of his precious moments of time or periods of concentration to the relaxing pleasures in which more normal human beings indulge; and his lifelong passion for safety far outweighed any sympathetic understanding—if he had any—of the craving for the nerve-soothing solace of a pipe or even a Pittsburgh stogie. On his memorable voyage on the *August Eduard* to America in 1831, a set of rules he drafted to govern the conduct and the treatment of the steerage passengers included this drastic provision: "Whoever smokes, or strikes fire, or has a light in the steerage, will have his pipes, tobacco, or lighting apparatus thrown overboard immediately, and besides, such a passenger will be condemned to bread and water."

3

Mrs. Roebling with her five children, ranging in age from twelve to two, and only some ten weeks before the birth of another child, bravely made the eastward trip across Pennsylvania and New Jersey in six days, traveling by canal and *railroad*. They had never seen a train before. The father, tied down to his engineering projects in the East, could not spare the time to escort his family on this difficult journey to their new home. Grandfather Herting and the mechanics Kunze and son accompanied the family on their pilgrimage to Trenton.

A few years after the pioneer wire-rope manufacturer sold his home and all of his lands in Saxonburg, oil and gas were discovered under the property, yielding unexpected wealth to the new owners. In a reminiscent letter written by Colonel Roebling a few months before he died, he recounted the story of this strange twist of fate—the fortune which the family unknowingly had in its grasp and blindly let go. "So it goes in this world," was his resigned comment.

In his *Early History of Saxonburg*, the Colonel also told of the blow suffered by the little community in the loss of the wire-rope industry, and of the subsequent compensation of fate: "To say that the removal of the Roebling interests from Saxonburg proved a death-blow to the village is very near the truth. Fortunately the discovery of oil in the neighborhood, a few years later, tempered the force. But by that time Mr. Roebling had sold all his lands. . . . On the same meadow where rope was made, a gas well was struck. . . . Just south of Thorn Creek notable wells were struck and enriched their owners. . . . In our house we employed a man of all work named Conrad Seibert. He had acquired a small holding there—when oil was struck he made a fortune and had the good sense to keep it. . . . Under our house is a vein of coal, untouched, because there is so much coal in the neighborhood easier to get at."

When the pioneer founder of the settlement, in his enthusiastic promotion letters written in 1831 enumerating the advantages of the site, included the potential mineral wealth of the locality, he was writing more prophetically than he knew. But fate decreed that this buried wealth should remain undiscovered until he himself had abandoned the site to his fellow colonists, to seek his own fortune elsewhere.

With the construction work speeded under Roebling's driving energy, his wire-rope works at Trenton were ready to start partial operation early in October, 1849. Not until then was it generally known that the new plant would do its own wire drawing. A month later the Trenton *State Gazette* reported further progress:

> Mr. Roebling will soon commence the manufacture of a rope about 3600 feet long, which, when finished, will weigh six tons. It is ordered by a South Carolina railroad, and is to be used on an inclined plane near Augusta, Georgia. . . . Mr. Roebling's railway, on which the machinery is about 4000 feet long, is now completed. This machinery is so perfect that every strand of wire rope will bear its due proportions of the whole strain. A steam engine of twelve horsepower drives all the machinery. We could not give our readers an intelligent description of the machinery if we should try.

4

In the new home in Trenton, one evening in December, Roebling's seventh child and third son, Charles Gustavus, was born. Thus destiny brought into the world, in the same place and at the same time, an industry of potential greatness and the future genius of its development and fruition. As the first-born son was to give his health and his strength to carry his father's dream span to completion, so this younger son was to give his brains and his energy to carry the industry forward after the father's death, to expand and perfect the wire and rope-making project beyond the dreams of its founder. Charles G. Roebling lived in Trenton all his life, and there he died. After an engineering education, he devoted himself to the engineering and manufacturing end of the business; became President of the company; built the near-by industrial town of Roebling to accommodate expanding production needs; built a new wire mill, a new rolling mill, and a new wire-rope factory; erected a wire-cloth factory and a large rod mill; started the manufacture of copper wire and insulated wire; played an important part in the design of the equipment and the procedure for moving "Cleopatra's Needle" from Egypt to its site in Central Park, New York, in 1880; built the Oil City suspension bridge and the cables for the Williamsburg Bridge; served as a member of the New Jersey legislature; and, after losing his son —Washington Augustus Roebling, 3rd—in the *Titanic* disaster in 1912, died six years later, the wealthiest of the three brothers. Transcending even his ability, vision, and industry were his impulsive generosity and shy but warm human sympathy.

Of the birth and death of this beloved younger brother, Colonel Washington Roebling wrote to a sympathetic friend and neighbor, Mrs. Stockton, in 1918:

"One winter evening, on December 9, 1849, I was sitting next a room where a woman was in labor; presently I heard the faint cry of a new-born child. Last Saturday I heard the last sigh of that same child. When I think back on the great work, the vast accomplishments and undertakings achieved between those two sighs, it surpasses my understanding. May he rest in that peace of which he saw so little during his lifetime."

5

In the installation and adjustment of the machinery the founder personally took a leading and active part. Around Christmas time he suffered a painful accident, with more serious consequences narrowly averted. While testing some of the new machinery, his hand was accidentally caught in one of the ascending pulleys. He was suddenly drawn upward some six or eight feet, when one of the workmen fortunately saw him and stopped the machinery. Roebling fell to the ground, having sustained some mangling of his hand and bruising of his arm. This unfortunate accident left a permanent injury; the pioneer's left hand was maimed, ending his ability to play the flute and the piano.

During the years that followed, as the bridgebuilder was compelled to be away from home for months at a time, he depended upon Swan to take care of all his interests in his absence, not only as superintendent of the mill but also as trusted confidential representative, with full authority to attend to all matters—manufacturing, commercial, and personal—subject only to Roebling's direct orders. The latter, during his extended absences from Trenton, wrote to his aide almost daily, giving him the most minute instructions relative to the running of the mill, the handling of the business, and the interests of the family; and the faithful agent was required in turn to keep his employer informed as to all transactions and occurrences. Moreover, while the father was away, Swan stood *in loco parentis* to the children; they recognized his authority, as they sensed his loyalty and devotion; and even after they were grown, they corresponded with him when they themselves were away, making him their confidant.

And thus, for the next twenty years, Charles Swan played an important part in the Roebling affairs. He was one of those loyal, trustworthy, patient souls, wholeheartedly devoted to the interests of his employer and willing and anxious to subordinate himself to the wishes of his chief. The position was a difficult one, and only a person of the rarest qualities could have filled it; for Roebling was no easy man to get along with, and always demanded an unqualified and unquestioning obedience to his orders—any slip, variation, or infraction was visited with his hot and instant displeasure.

The letters, numbering several hundred, written by John Roebling to Swan during the years from 1849 to 1869 were devoted almost exclusively to the work at the wire mill and told little or nothing as to his own doings. Rarely were there any allusions to personal and family matters; business was the absorbing theme—what Swan was to do, and what he was to refrain from doing. In these letters, the master builder drew sketches for new machinery he wanted to have installed, and he gave exact measurements and specifications and the places where the equipment or parts could be obtained. He described new processes he wanted to have tried out, and he gave detailed instructions and precise formulas for them. He cautioned Swan about extending credits to certain customers, and directed him as to lawsuits. From those files of letters, the whole business of the wire mill, for nearly twenty years, can be reconstructed, except for the short periods, marked by gaps in the correspondence, when the bridgebuilder was at home.

Keeping a close control of his accounts, Roebling carried his ledger and cashbook with him on his travels and entered debits and credits, instructing Swan which bills to pay and which to hold up. He enclosed checks to be deposited to his account. His mail was forwarded to him with daily letters from Swan, and thus he kept in close touch with all of his business affairs and with everything of importance at home. All this he did, in minute detail, while engaged in the active and exacting work of building aqueducts and bridges. His energy knew no bounds, his industry never flagged. He read the current technical and scientific journals and made notes of new ideas, inventions, and processes; these he preserved for future reference. Every night, methodically and systematically, he wrote up his diary—of conferences, activities, events, instructions given, and personal expenses.

With a peculiar passion for meticulous equity, he kept a careful record and account, in his private ledger, of the sums expended for his children—every dollar spent for their support, for their doctor bills, for their schooling, and every nickel advanced to them for their personal expenditures; and then he wrote in his will that the full and exact amount charged against each child during his lifetime was to be deducted from that child's equal share of the final inheritance.

Roebling did not correspond with his wife; apparently she never learned to read English, and he was too busy for nonessential correspondence. When he wanted her to know something of special interest in one of his business letters, he would add a line, "Please read this to Mrs. R."

In the hundreds of letters he wrote to Swan, there were rarely any passages disclosing the personal and human side of the bridgebuilder. Never in a single instance was the lighter element present—no humor, no description, no anecdotes—all was business, nothing but business. The man was so engrossed in his immediate tasks that he had neither the time nor the inclination for anything else. The questing youth who wrote the sensitive diary of his ocean voyage in 1831—with eyes and mind and heart open to all things of human, esthetic, and imaginative appeal—had since then allowed all those refreshing branches of his personality to become atrophied in order that every ounce of his vital energy might be concentrated in the main trunk of a relentless and unsparing ambition. With his receptive and observant mind, as he traveled from place to place in his professional career, he must have had many unique experiences and he must have met many interesting people. But no inkling of those things cropped out in his letters. Like a runner in a heart-straining championship race, he had no time to look to left or right, but directed every nerve, thought, and sinew toward the goal. He was as a man possessed. He had become self-contained, self-centered, self-driving—with no interests apart from his work.

6

Roebling was at home most of the time during the year 1850. His growing family was now established in the house built for them by Swan, and the little factory was in full working order with ample demands for its products. The planner and builder had just completed the third and fourth of his aqueducts for the Delaware and Hudson Canal, and was actively filling in his time while waiting for his next bridge-engineering engagement. Roebling was not resting. He was thinking and planning and figuring for the next span he hoped to build, a structure larger and more important than any he had previously constructed: the proposed Niagara Railway Suspension Bridge. In his spare time he took part in other intellectual activities.

An item in the local newspaper reported his participation in a public discussion on geology:

At a meeting of the Trenton Philosophical Society held on Thursday evening, the question was discussed, "Is there sufficient reason for believing that the earth was formerly in a fused state?" Mr. Roebling took part. He alluded to various phenomena of geological formation, which seem to prove the fusion theory, such as the crystalline structure of the primary rocks everywhere, which led to some further remarks.

It was at this time also that Roebling developed and published his proposal for building a transoceanic cable. Two days after its publication an announcement of the article appeared in the Trenton *State Gazette:*

John A. Roebling, Esq., proprietor of the Wire Rope Factory, has furnished the *Journal of Commerce* with a long and ingenious article on a transatlantic telegraph. Mr. Roebling has had much experience in the construction of wire-cable suspension bridges and aqueducts and in the manufacture of wire ropes. He considers the construction of a line across the Atlantic entirely practicable, on which he thinks very large dividends may be expected. Many things which seem preposterous have proved not only practicable, but eminently important and valuable.

During this period also, Roebling found time to develop some useful new inventions, supplementing earlier inventions on which he had secured three patents in 1842 and 1843. Between 1846 and 1867, he patented his design of anchorages for suspension bridges, his apparatus for spinning cables for suspension bridges, a "closing head" for a stranding machine for making the strands for wire rope, the idea of using wire rope for transmitting power, a new marine type of boiler, structural girders made up of steel and wood in combination, a passenger car constructed entirely of iron, and two improvements in "rail chairs" for supporting and protecting rail joints in railroad track.

The inventor's vision of new and useful services to which wire rope could be effectively applied was always far in advance of prevailing conceptions and practice. When he proposed to transmit power over a distance of 3,500 feet by means of wire rope running over large sheaves, many people shook their heads and said that his enthusiasm had outstripped his practicality. He proceeded to demonstrate the feasibility of his idea by installing such a transmission at his Trenton plant, using a wire rope 7,000 feet in length to operate a ropewalk; the success of this installation, economical and efficient, had a far-reaching influence. Soon after his patent was granted, wire rope transmissions became widely adopted for the delivery of power over long distances.

Until superseded by the development of electrical power transmission a generation later, the wire-rope system of transmitting power offered unique advantages. Combining his capacities of invention, engineering, and promotion—functions which after his death were divided among his three sons—Roebling prepared an illustrated article on this interesting application of wire rope, featuring its possibilities and supplying the technical information required. In his introductory summary, he wrote: "The use of a round endless rope running at a great velocity in a grooved sheave, in place of a flat belt running on a flat-faced pulley, constitutes the transmission of power by wire ropes. The distance to which this can be applied ranges from about 60 feet up to about three miles. . . . In point of economy it costs one-fifteenth of an equivalent amount of belting and one twenty-fifth of shafting."

The Seventh United States Census, that of 1850, was the first to include a survey of the industries of the country, and we thus have available the government record of the Roebling business during the first year of its operation. The report gives the names and ages of the members of the family, together with ten Roebling employees who were also included as members of the one household, making up a total of eighteen, all of whom, with the exception of the six Roebling children, appear to have been born in Germany. The total investment in the wire-rope plant, including land, buildings, and machinery, was recorded as $20,000. The motive power was steam and the equipment included two cupola furnaces. The employees were listed as twenty males, no females; and the monthly cost of labor was $500. The raw materials used in the year, including fuel, were 300 tons of iron, valued at $35,000, and 400 tons of coal, valued at $1,200. The annual product comprised 250 tons of wire rope, valued at $40,000, and other finished articles valued at $1,000. These figures indicate that, during the first year of adjustment and initial operation, the new plant showed no profit—no return on the investment and no compensation for the creative and directive energy of the founder. The raw material was disproportionately expensive, the breaking-in of the men and of the machinery consumed time and money, and the volume of work was small for the size of the plant. But great things grow from small beginnings.

That cash was scarce and sorely needed is revealed by Roebling's correspondence from Trenton in 1850. In May he wrote to William H. Campbell, Superintendent of the Allegheny and Portage Railroad, at Summit, Pennsylvania: "Since the Appropriation Bill has been passed, I expect you will soon be enabled to pay off the old scores which Mr. Power left unsettled. Two sets of Ropes are not paid yet, viz: 6100 ft. delivered in May

1849 on Plane No. 6, 3494 ft. delivered in Aug. 1849 on No. 1, ... $5068.91." And after several more letters covering additional shipments of wire rope, he wrote again in October: "I am greatly in want of money and should be glad to hear from you that you can settle one of my bills." A week later he wrote: "It is certainly wrong that I should be kept out of money, which is justly due, for so long a time." And the following week he wrote again: "I am anxious to receive your draft, being much in want of funds."

The winter of 1851-52 was unusually severe, bringing suffering to the poor of Trenton. When an appeal for charitable relief was published, the first to respond was John A. Roebling. The collector of the relief fund immediately published a grateful acknowledgment in the *Star Gazette:* "Mr. Editor:—Permit me to acknowledge through your columns the receipt of ten dollars from John A. Roebling, Esq.—the first fruits of the appeal made last evening for the poor of Trenton. May many more such gifts be sent in. 'Before the ashes on the hearth-stone of poverty have grown cold, or the tears on the cheek of suffering have had time to dry.'"

A number of his Saxonburg neighbors followed Roebling to Trenton to find employment in his plant. Toward his fellow countrymen—in fact toward all his employees—he always maintained a loyal helpfulness and a warm personal interest. Even in his early lean years, he provided shelter for the homeless and work for the jobless. And as he prospered, his faithful workers prospered with him.

7

From its modest beginning in 1849, when the first wire-rope mill at Trenton succeeded the primitive ropewalk at Saxonburg, the Roebling industry has grown to giant proportions. A mere enumeration of the present mills, shops, furnaces, laboratories, and buildings would fill pages, and a list of the different items manufactured—in different forms, uses, and applications of wire and wire rope—would fill a good-sized catalogue. The products of the Roebling plant have found applications beyond the imagination of men at the time the industry was founded. From a hairpin to a harbor-defense net, from a delicate hairspring in a watch to the powerful cables and suspenders of the Golden Gate Bridge, the range of products and their uses is staggering.

In 1850, the first year of operation, twenty men were employed and the annual output was $40,000. Twenty years later, at the death of the founder, one hundred men were employed and the product of the plant approximated $250,000 annually. In the next forty years, when the three sons carried forward the business their father had started, the industry continued to expand until it had eight thousand employees and the output ran far into the millions of dollars. The humble little factory built in 1849 had to spread its buildings not only over the surrounding acres, but across what were then neighboring farm lands, until, forced to expand by the pyramiding demand for its product but constrained by the soaring land values that had grown up around it—and of which it had in no small measure been the creator—the enterprise had to go pioneering again. In 1905 it moved ten miles down the Delaware to establish a newer and larger nucleus in a new town which it created to its own needs and ideals.

John Roebling, upon his arrival in America, planned and founded the little village of Saxonburg in Pennsylvania. Seventy years later, his sons, carrying on the tradition, planned and built the model town of Roebling in New Jersey. A sandy tract, unattractive and unproductive, was transformed into a beautiful residential community. Where tangled weeds and bushes grew, today a charming park overlooks the Delaware River, and fine shade trees line well-paved residential streets. The town has been well planned, with attractive parks, flower gardens, and recreation centers. Two thousand children attend the large and spacious red-brick school. A national bank, theater, post office, and hospital have been provided, as well as a police department, a well-equipped fire department, and many business houses, fine churches, and community clubhouses. The company officials take pride in their model town and in administering it in a manner to perpetuate the ideals of the founders.

John Roebling's wire mill has grown into the world's largest plant for manufacturing wire and cables.

When the pioneer established his shop at Trenton, his first product was made by hand in the old ropewalk way. Today the ground where he did it is covered with buildings full of speeding machinery that has little rest—devices that stand in long rows, eating up the strand that unwinds from whirling bobbins, and steadily turning off the completed rope, which passes to "spools," large or small, in proportion to its weight and size.

The machines for the smaller sizes of wire rope are strung out in a long row. The larger ones require elbowroom. For the making of the largest ropes each giant machine has a room to itself and is installed on its own special foundation of steel and concrete. When the mechanism is at work, it somehow suggests the solar rotations. The flying bobbins have a triple motion. On the ends of the radial arms on which they are mounted they travel in a whirling circle around the central column; but, at the same time, each whirling bobbin is unwinding as the strand pays out, and also turns completely end for end, at predetermined intervals.

There is something mysterious and fascinating in the spectacle of these flying reels of steel, or copper, whizzing around and round "like indefatigable moths around a big steel candle, or a dervish round his own spinal column on a spot of ground the size of a dinner plate," and in the sight of the emerging rope, hard, shining, round, packed around its core of hemp or steel, noiselessly gathering all this strength and energy into itself for future use in carrying the loads and burdens of civilization.

8

It is difficult for us to realize how much we are indebted to the invention and development of wire rope. Much of our material progress during recent decades must be credited to the ramifying applications of the family industry that grew from the master builder's original vision.

There are the millions of speeding elevators. The modern skyscraper, even the office building or apartment house of moderate height, was made possible by wire rope. John Roebling made the first elevator rope in 1862. Today millions are in use all over the world.

There is logging in the forests of the Northwest, where the giant trunks, seven, eight, or nine feet in diameter, are slung on cableways, whisked up the sides of mountains, hoisted into the air and deposited on flatcars, to be run down to the river on steep inclines, again operated by steel ropes of great size and strength.

There is quarrying, where wire rope is used in quantity for guying the giant derricks, and for hoisting the blocks of stone out of their beds, and then again to carry the huge blocks on high aerial cableways over long distances to be loaded.

There is underground haulage in tunnel and mine, for which five distinct types of rope are used, enabling the engineer to make light of grades, even with staggering loads; and there are the hoisting ropes going down the vertical shafts to the one-thousand-, or fifteen-hundred-, or two-thousand-foot levels of the subterranean workings, to bring up the cages loaded with men or with precious ore, with everything depending on the strength and safety of the steel rope.

There are the oil fields, where, in the mad search for petroleum to supply the world's shortage, miles of wire rope are used, some of it an inch thick or over, to carry the drills, or the casing and sand lines.

There is shipping—the battleship and the merchantman and the liner; the yacht, the river boat, and the tug—all strung with wire rope from stem to stern, and some of them from truck to keel as well—not to mention mooring lines and towing lines which have their own plan and formula.

There is dredging, for which tough wire rope has supplanted the old cumbrous chain. There is hardly an important harbor in the world where these stout ropes of steel are not busy clearing pathways and anchorages for marine commerce.

The list does not end. There are inclines and funicular railways, in the mountains of America and in foreign countries, enabling young and old to get views from towering peaks which otherwise would have been accessible only to the hardy mountain climber. There are high cableways with which engineers have been able to run cars and trucks and giant buckets on an aerial roadbed of wire over impassable canyons and gorges and morasses, to make fills for railways and other construction, and to place millions of tons of concrete for gigantic dams. There are tramways and traction systems and cable railways and trolleys. And there is the litter of hoisting slings, all over creation, to lift anything from a casting or a crate of cargo to a locomotive or a sunken vessel; for wherever men are doing work of any kind there is a load to lift, and the light but powerful wire rope makes the most convenient sling to lift it.

There are records in the Roebling offices that tell interesting tales of special constructions; and pictures of enormous spools of wire rope, thousands of feet long and in big diameters, running from spool to spool and—since one spool is an ample carload—running from one flatcar to another when loaded for shipment. Such were the huge street-railway cables made by the Roeblings for Australia, for Kansas City, for Chicago and New York.

There is an amusing story of the New York street-railway engineer who insisted that the cable be made in one section, 33,000 feet in length, but who changed his mind about the beauty of it when he got the goods and saw the elephantine spools of packed metal caving in the manholes in the city streets on their way to the point of installation. A gigantic rope machine was built in the Roebling plant to twist this mammoth.

9

In November, 1915, the Roebling plant suffered two fires in one week; the second destroyed a large factory building and caused damage estimated at $1,000,000. These fires were believed to be of incendiary origin. They followed a published threat by Dr. Constantin Dumba, former ambassador from Austria-Hungary, that American plants would be crippled. Two months before these fires he brazenly declared, in a published statement: "We can disorganize and hold up, if not entirely prevent, the manufacture of munitions in America." The Roeblings were producing wire and wire rope for the Allies, and were therefore on the German list of plants to be destroyed.

Immediately after the disastrous fire, a foreign-language newspaper, the Brooklyn *Freie-Presse*, carried the boasting headline: *RENDERED HARMLESS—Factory Building of Roebling Company Reduced to Ashes—Was Used to Produce Wire for the Allies.*

And then, as soon as America entered the war in 1917, the Roeblings and their ten thousand employees accepted the challenge. They threw every ounce of effort into the struggle. They undertook task after task, culminating in one of stupendous proportions and of critical importance in the final victory.

The first big call was for submarine nets, to protect fleet bases and harbors. The Roeblings supplied to the Navy, for this purpose, some three million feet of steel rope. And it was not ordinary rope that was required in this service, for the German submarines had developed a method of slashing through the earlier and lighter nets. For the new type of barrier, the wire rope ranged up to an inch and a half in required thickness. Moreover, the task was not merely a matter of shipping giant reels of rope. Almost all of it had to be cut into specified lengths, and special attachments had to be made and fitted for each length, for these barriers were designed in separate sections to be linked together. And the Army Ordnance Bureau, for its barrier nettings, used another million feet of wire rope, which was shipped to various coast forts.

Then the Quartermaster's Department called for seven million feet of steel rope and the manufacture of three hundred thousand pairs of artillery traces. Although fitting one of the "thimble splices" is ordinarily half an hour's work, the Roebling plant, employing new men previously unskilled, was turning out ten thousand pairs of traces *a day* at the peak of production on this order.

The national government also demanded wire rope for its logging, shipping, mining, and oil-drilling operations. The Spruce Production Board took over eight million feet of steel rope, the Emergency Fleet more than twelve million feet, and the Fuel Administration drew on the rope manufacturers at the rate of thousands of *tons* a month. And all the time the privately owned mines and mills and ordnance plants, and the manufacturers of locomotives and cranes and derricks and freighters, kept demanding—and getting—continually increasing supplies of wire rope to carry on their war-driven work. The total figures were staggering.

But the climax, the call that drew upon all the reserve resources of the wire-rope makers and kept the machines running and the men working and the arc lights burning in the mills for breathless months, was the emergency need for eighty-four million feet of steel rope, and a half-million attached fittings, for the "North Sea Mine Barrage"—the desperately conceived measure that put an end to the menace of the German submarines.

The whole amazing episode of the North Sea Mine Barrage which, with its seventy thousand bottled TNT volcanoes, made that historic body of water a graveyard for German hopes, was hidden from general knowledge. The full story will never be told. While the armies in all the theaters of the war swayed back and forth in the struggle, few people knew that the fate of the world hung virtually by a wire rope. Not until the "moving finger" had written failure across the German plans was any whisper allowed to escape regarding the heroic work that had been done in bottling up the U-boats and making the submarine warfare ineffective.

The Germans had pinned their hopes on the U-boat campaign. Wire rope defeated them.

The actual toll may never be known. The Germans admitted twenty-three U-boats lost there, and other authorities ascribe the fleet's surrender and the final armistice to the defeat of the submarine campaign—a defeat which the North Sea Mine Barrage had forced the enemy to accept.

The swiftness with which the barrage project was executed, the short time allowed for fulfilling of unheard-of demands, the highly specialized equipment and processes that had to be devised and employed in the manufacture—all have become part of the extraordinary story of wire rope.

In January, 1918, the emergency order was received at Trenton. By May 1 it was made and shipped—without a flaw, without a single delay—millions and millions of feet of galvanized rope of the specified lengths and sizes—all of the extra-special quality desired for this work and subject to the most exhaustive tests—to equip and anchor a hundred thousand mines. It was a feverish task, done at a feverish speed, but with outstanding perfection. In three months the Roebling works turned out twenty-seven million feet of this superquality steel rope. In addition the Roebling force prepared and attached a half-million sockets and hooks, at the rate of six thousand a day, besides supplying all of the four-legged lifting slings used in handling the mines.

In the spring of 1918 the Yankee mine-layers were sowing the troubled waters between Norway and the Orkneys with three-hundred-pound cases of death and destruction, at the rate of one every twelve seconds; and twelve hundred feet of fine wire rope went out of sight with every mine. Each mine-laying excursion carried and buried six million feet of steel rope.

Following this stupendous assignment, the Roeblings furnished fifteen million feet more of powerful steel rope for the "Adriatic Sea Mine Barrage." This was an even more ambitious project, since it dealt with a depth of three thousand instead of nine hundred feet. The material was furnished and the barrage was all ready to be laid when it was rendered unnecessary by the German capitulation.

When the fighting stopped, there was a perfectly good mine barrage in the North Sea that had to be taken up and put out of commission. This called for six hundred thousand feet more of wire rope, with special fittings to make it of use. Every mine was canceled without a mishap, and there are now more than eighty million feet of highest grade wire rope reposing at the bottom of the North Sea. But it had done its work, destroying no less than seventeen German submarines in the first week, and more in the succeeding weeks, until this one baffling menace was eliminated and victory for the Allies was assured.

10

In 1931, when the historical society in Mühlhausen proposed an official celebration in honor of Roebling on the one-hundredth anniversary of that memorable day—May 11, 1831—when the town's most famous son bade farewell to his neighbors and departed for America, the city fathers vetoed the proposal on the ground that wire-rope nets manufactured by the Roebling plant at Trenton had captured and destroyed twenty German submarines in the First World War. In the same vein, a strangely conceded tribute to Roebling was expressed by a German biographer, Dr. Wilhelm Auener, who wrote in 1931: "To us it appears tragic fate that this emigrant's cleavage of nationality exerts its effect long after his time. With Roebling the Fatherland not only lost an engineering genius and a great industrialist; but that which he created has worked damagingly against Germany in that war materials in enormous quantities have been produced in the works which Johann August Roebling founded."

Men may fight and destroy for a time, but they can build forever. It is in the constructive arts, in the peacetime progress of humanity, that the products of the new industry created by the inventive genius of the Saxonburg pioneer find their highest and most enduring usefulness.

CHAPTER X

SPANNING NIAGARA

IN ALL the world no place could have been found where the building of a suspension bridge would present a more spectacular accomplishment than over the Niagara gorge, with the Falls thundering a little way upstream and the waters lashing and fuming underneath. No other setting could have provided such a dramatic and impressive background for a slender span symbolizing the genius and courage of man defying the gigantic, furious forces of nature. And to plan such a span—seemingly a frail web in contrast with the overwhelming power of the scene —to sustain the enormous loads of locomotives and freight trains was an unprecedented, breath-taking temerity of the human spirit. The idea of carrying railroad trains over that thundering turmoil of waters on a high and tenuous framework of wire and wood stunned the imagination, and the proposal evoked a storm of protest from nearly all the foremost engineers of the time. But Roebling was a practical man as well as a stubborn one. His one passion in life was to prove his convictions. After all, he was dealing with rock and wire, and he knew what they would do. He built the bridge—thereby disposing of preconceived timidities and establishing an epoch-making new precedent. With that achievement he became one of the world's immortals.

One of the first promoters of the idea of a railroad span across the Niagara gorge was Major Charles B. Stuart, a civil engineer prominent in military and political circles. In 1848 he became the first State Engineer and Surveyor of the state of New York and later he served as Engineer-in-Chief of the United States Navy. He was also author of two technical books on naval subjects; and in 1871—when he had attained the rank of general—he published his valuable volume, *Lives and Works of Civil Engineers of America*, which includes the first published biography of John A. Roebling. But, at the time of this story, he had not yet met the bridgebuilder.

In 1845, while engaged in locating the Great Western Railway of Canada and what later became the Rochester and Niagara Falls branch of the New York Central Railroad, Major Stuart was struck by the possibility of connecting the two lines by a suspension bridge over the Niagara River, between the Falls and the "Whirlpool," where the chasm was about eight hundred feet wide and over two hundred feet deep. As the project was a novel and bold undertaking and generally believed to be impracticable, and as the Major was not himself a bridge engineer, he addressed a circular letter to a number of the leading engineers of America and Europe, asking their opinion of the proposal. Various replies were received, some in open condemnation of the idea, others expressing grave doubts of its practicability and safety at any cost. Of all those canvassed, only four engineers thought the project feasible: Charles Ellet, John A. Roebling, Samuel Keefer, and Edward W. Serrell. And it is a remarkable coincidence that each of these engineers later actually constructed a suspension bridge over the Niagara River, thus justifying their early faith.

Here begins the drama of four men accepting the challenge of Niagara. Four men say they can do the job. Each man in turn wins the opportunity, and each, in his own fashion, tackles the task. The story of the varying adventures of these four aspirants for the honor of spanning Niagara—the way each of them won his big chance and what each man did with the opportunity after he had it—has become part of the epic of bridgebuilding.

2

The reply to Major Stuart's inquiry from Ellet characteristically recorded that brilliant engineer's ready assurance and prompt interest: "A bridge may be built across the Niagara below the Falls," he wrote, "which will be entirely secure and in all respects fitted for railroad uses. It will be safe for the passage of locomotives and freight trains, and adapted for any purpose for which it is likely to be applied. But to be successful it must be judiciously designed, and properly put together; there are no safer bridges than those on the suspension principle if built understandingly, and none more dangerous if constructed with an imperfect knowledge of the principles of their equilibrium." Ellet did not realize the prophetic content of his reply.

John Roebling had also answered eagerly and enthusiastically. An inner awareness of his powers told him that he was the man for this job, and the throbbing call of his ambition told him that this job was the indispensable next stride in his career. In agitated suspense he awaited a reply. Receiving no acknowledgment from the promoter and eager for further information, the bridgebuilder wrote again to Major Stuart at Rochester and, after waiting three more weeks for an answer, he wrote to his friend D. K. Minor, editor of the *Railroad Journal* in New York. Minor promptly responded: "I am glad to learn that you are moving in the matter of the Niagara Bridge—and advise you to be *active*—as Mr. E. is indefatigable in whatever he undertakes. You will do well to make yourself known in Rochester and at the Falls, not only through the press, but also personally. . . . Mr. E. may have got the start of you by offering to take $20,000 of Stock in the Bridge, as I have heard. . . . When Major Stuart comes here, I will have a confab with him and ascertain how the land lays."

Shortly thereafter, Roebling received a belated reply from Major Stuart: "Your letter reached here at a time when I was dangerously ill, and I am just able now to answer it. . . . A company is ready to build the bridge and are now making the necessary applications to the two governments. . . . Chs. Ellet Jr., of Philadelphia, will take the contract, and will probably complete the work the early part of next fall. . . . I hope to visit Pittsburgh this winter to see your bridges and also make your personal acquaintance. As an Engineer, I have long known your reputation, but I believe I have never had the pleasure of meeting you. . . . Excuse my miserable scrawl, as my hand trembles from weakness too much to allow of very intelligible writing."

The Major's excuse that he had been "dangerously ill" was a transparent attempt to cover a more embarrassing reason for his long delay in responding to Roebling's communications. Throughout the body of the letter the writing was firm, and not until the last sentence was the hand of the writer reminded to "tremble from weakness." As the worried and anxious correspondent in Pittsburgh had surmised, his rival Ellet had already enlisted the aid and support of Major Stuart.

This was later revealed, for just about the time that Roebling was vainly waiting for an answer from the Major, the following news release appeared in the Utica *Gazette* and the Philadelphia *Ledger:*

Charles Ellet, Jr., an engineer of Philadelphia, has recently in company with Major C. B. Stuart, of Rochester, inspected the localities in the vicinity of Niagara Falls, with a view of ascertaining the practicability of a suspension bridge across Niagara river. . . . The cost of a hanging bridge at that point, of sufficient strength to sustain the weight of a railroad train, is estimated by Mr. Ellet at $200,000. He offers to construct such a bridge for that sum, and to subscribe $20,000 to its stock.

Roebling, still uncertain as to the status of the project, wrote again to Minor and Stuart. Hastily writing and posting at midnight of Sunday—"that no time may be lost"—Minor replied: "I am expecting Major Stuart here daily and I will then get at the whole matter. . . . You may rest assured that *action, action,* is necessary to have any chance for a fair fight. . . . You must not let your modesty ruin you, nor keep you back. *Puff—puff*—is the order of the day, and if you do not puff, *others will;* and you might as well be out of the world and out of fashion. . . . Ellet is pushing for immediate action. You see he likes to enter the field alone—but we shall see. . . . I said to Stuart that I had a desire to write my name, in connection with his and below it of course, on the eternal rocks of Niagara!— a little 'soft soap' is sometimes useful; and now I desire you to tell me *about* the price you can do it for—leaving a good margin. . . . Unless Ellet has got a long start we will give him a pull."

This letter included a quickly drawn pen sketch of the cross section of the Niagara gorge, with an indicated suspension span swung between two towers above the "more than perpendicular" cliffs. To find current expressions like "confab" and "soft soap" in letters a century old is refreshing; and the advice that "*Puff—puff*—is the order of the day" is seemingly of timeless application.

All this time Roebling was almost frantic with anxiety. Unable to leave his work on the Monongahela Bridge, which was then nearing completion, he accepted Minor's offer to make a trip to Rochester and Niagara on his behalf. He sent the editor $50 for traveling expenses, and wrote him four letters in four weeks, before he received an acknowledgment, in which Minor stated that he had not yet been able to get away from New York.

During this period someone was busy supplying publicity stories to the press, in which Ellet's name was repeatedly coupled with the spectacular enterprise of bridging Niagara. A typical article, printed in the Rochester *American* and other newspapers, briefly mentioned Telford's famous chain bridge in Wales, and Seguin's wire-cable bridge in France, and then described at length Ellet's suspension bridge over the Schuylkill at Philadelphia. In support of the feasibility of the proposed Niagara span, these articles cited the authoritative opinion of "an engineer so competent and justly celebrated as Mr. Ellet."

3

Some time in 1846 a corporation was chartered by the legislature of New York and by the provincial parliament of Ontario to construct an international bridge over the Niagara River, at or near the Falls. After repeated efforts during the ensuing months, enough subscriptions to the capital stock were finally secured to warrant the commencement of the work, and early in 1847 plans and estimates were invited.

Roebling promptly responded, addressing his communication to Major Charles B. Stuart, Commissioner of the Niagara Bridge Company. This letter of the pioneer master builder has become a cornerstone in bridge engineering progress. It was the brave, deliberate, and confident declaration of a man who knew whereof he spoke, and who was only asking for a chance to prove his unshakable convictions:

I have bestowed some time upon this subject, and have matured plans and working details. Although the question of applying the principle of suspension to railroad bridges has been disposed of in the negative by Mr. Robert Stephenson, when discussing the plan of the Britannia Bridge over the Menai, on the Chester and Holyhead Railway, I am bold enough to say that this celebrated Engineer has not at all succeeded in the solution of this problem. . . . It cannot be questioned that wire cables, when well made, offer the safest and most economical means for the support of heavy weights. Any span within fifteen hundred feet can be made perfectly safe for the support of railroad trains. . . . The greater the weight to be supported, the stronger the cables must be, and as this is a matter of unerring calculation, there need be no difficulty on the score of strength. The only question which presents itself is: Can a suspension bridge be made stiff enough, as not to yield and bend under the weight of a railroad train when unequally distributed over it; and can the great vibrations which result from the rapid motion of such trains and which prove so destructive to common bridges, be avoided and counteracted? I answer this question in the affirmative, and maintain that wire cable bridges, properly constructed, will be found hereafter the most durable and economical for railroad bridges. . . . There is not one good suspension bridge in Great Britain, nor will they ever succeed as long as they remain attached to their chains and present mode of construction. . . . The locality of the Niagara Bridge offers the very best opportunity for the application of a system of stays, which will insure all the stiffness requisite for the passage of railroad trains at a rapid rate. The plan I have devised for the structure will convince you that the rigidity will be ample.

To demonstrate his confidence in the design he proposed, in his estimate of cost, and in the success of the enterprise, Roebling offered to construct the bridge on the principles he had outlined, with one railroad track, two roadway lanes for common travel, and two footwalks, for the sum of $180,000; to subscribe $20,000 to the capital stock; and to give security for the complete success of the work in all its parts. In his desperate anxiety to secure the job at any cost, he named a figure which he knew Ellet could not meet. In a letter to Roebling, Minor had confided, "I am quite inclined to believe that Stuart and E. have an understanding and *mutual* interest in the offer of $220,000."

A month later Ellet submitted his proposal to Major Stuart, confirming his prior quotation of $220,000, but offering to build a lighter and narrower bridge—with the railway track and horse-drawn vehicles sharing the same roadway space—for $190,000. "When I made my estimate," he wrote, "I had in view a work of the first order, and as I do not wish to be in any way connected with one of lower grade, I cannot offer to reduce my proposition. But I will now repeat, that a secure, substantial and beautiful edifice—a noble work of art, which will form a safe and sufficient connection between the great Canadian and the New York railways, and stand firm for ages, may be erected over the Niagara river for the latter sum named. . . . With my best wishes for the success of the enterprise in all its magnificence. . . ."

4

For this, the first span to be erected over the Niagara gorge, Ellet was the successful competitor. In November, 1847, the directors of the American and the Canadian Niagara Bridge companies made a contract with him for the construction of a combined railway and carriage bridge over the Niagara River, two miles below the Falls, for the sum of $190,000. The bridge was to be completed by May 1, 1849.

Ellet was high-strung and impulsive. He stopped at nothing to gain his end. His physical courage was already a legend. Above all, he was a born showman, with a brilliant flair for the spectacular.

For the building of a suspension bridge, one of the primary operations is somehow to get a line or wire across the stream. The turbulent whirlpool rapids at the site made a direct crossing of the Niagara impossible. Ellet was ready to start construction early in 1848. Exercising his originality and his instinct for the theatrical, the bridgebuilder enlisted the youngsters of the vicinity by offering a prize of five dollars to the first boy who would fly a kite across to the opposite bank. After many attempts through the day one youngster proudly accomplished the feat and received his reward. The successful lad's name was Homan Walsh; eighty years later, when he was an old man residing at Lincoln, Nebraska, his most precious memory was this exploit of his boyhood—his part in starting the construction of the first span across Niagara.

By means of the kitestring larger cords, and then hemp ropes, small at first but increasing in size, were drawn across, until the first small wire cable, which Ellet had prepared on shore, was pulled across and suspended over the gorge. Ever alert to the advertising value of the spectacular, Ellet promptly had himself hauled across in the first basket ride over the Niagara. On March 13 he wrote to Major Stuart:

"Dear Sir,—I raised my first little cable on Saturday, and anchored it securely both in Canada and New York. To-day (Monday) I tightened it up, and suspended below it an iron basket which I had caused to be prepared for the purpose, and which is attached by pulleys playing along the top of the cable. In this little machine I crossed over to Canada, exchanged salutations with our friends there, and returned again, all in fifteen minutes. The wind was high and the weather cold, but yet the trip was a very interesting one to me—perched up as I was two hundred and forty feet above the Rapids, and viewing from the centre of the river one of the sublimest prospects which nature has prepared on this globe of ours. My little machine did not work as smoothly as I wished, but in the course of this week I will have it so adjusted that anybody may cross in safety."

Ellet thus achieved the distinction of being the first person who ever crossed the chasm alive. Following his sensational trip, others paid for the privilege of duplicating the thrill, and even ladies made the crossing in the suspended basket. A quarter-century later, the famous iron basket was still hanging under the railroad track near the American end of Roebling's bridge, as a historic curiosity for sightseers.

The engineer-contractor proceeded with the erection of a light suspension span, 7½ feet wide, hung on wooden towers, to be used as a service bridge for transporting men and materials across the gorge. So intoxicated was Ellet with pride in his Niagara footbridge that, when it was nearly finished, he mounted his horse and rode across the swaying span, at a height of 240 feet above the rapids and before any side railings had been set up. Women fainted and strong men gasped at the sight of horse and rider moving along the swinging hair line high above the foaming whirlpool. Not until twelve years later was this daring feat surpassed—when Blondin, the famous tight-rope walker, crossed the same gorge by walking on a wire rope, pushing a wheelbarrow and carrying another man on his back.

For ten months after its completion in July, 1848, the temporary span was used as a public footbridge, and the tolls received during this time amounted to nearly $5000, a handsome rate of return on a total cost of $30,000. A difference soon arose between the contractor and the directors concerning the application of the tolls taken on the erection footbridge; the ensuing dispute and litigation were ended by a compromise, by which Ellet relinquished his contract, and on December 27, his connection with the work ended. With some strengthening during the first year the airy temporary structure accommodated ordinary light carriage travel, which crossed the swaying span in its single lane.

With a blaze of advance publicity Ellet had started out to build a railroad suspension bridge across Niagara. When he quit the work, he had built little more than a footbridge. He had enjoyed his brief hour of glory—of spectacular showmanship, of pyrotechnic flourish. He had the shrewd salesmanship to land the job, but he lacked the steadfastness of purpose, the singleness of ambition, to stick to the task and carry it to completion. He had the opportunity for great achievement in his grasp, and he threw it away for some petty immediate profit or pique.

5

The logical inheritor of this challenging opportunity to build a span that would command the attention of the world was John A. Roebling. He had been the first to proclaim the possibility of building suspension bridges to carry railroad loading and, from his pondering on the subject and from his experimental work on other spans, he had more to contribute to a real solution of the problem than any other engineer of the time. In his excited eagerness, he had submitted a proposal that was even more attractive than Ellet's. Now the officials of the Niagara Bridge companies were sorry that they had not dealt with Roebling in the first instance.

For two years the project lay dormant before the officials were stirred to renewed activity. From Saxonburg, from Lackawaxen, from Trenton, from High Falls, and from New York, the eager builder of suspension spans kept in touch with the proposition, and in his spare hours he made studies, calculations, and plans for the great bridge he was hungering to build. Finally the negotiations were concluded, his plans were accepted, and the thrilling opportunity was his.

Freed from Ellet's price-cutting competition, Roebling undertook the work on a new basis: a fixed salary for his engineering and supervision, the owners paying for all labor and materials.

Thus the engineering function was properly separated from the contracting, and a more substantial construction was thereby made possible.

With his string of four suspension aqueducts for the Delaware and Hudson Canal completed, and with his new wire-rope works at Trenton brought to a stage of smooth operation under a trusted superintendent, the master builder was now ready to give his undivided personal attention and energy to the building of the Niagara span. Inspired by the challenge of the task, he concentrated upon it all of the genius and engineering skill with which he was endowed.

John Roebling was a man entirely different in character and temperament from his rival. It was the difference between the profound and the superficial, between the enduring and the spectacular. Ellet's performance was a dazzling fireworks display, quickly burning itself out. Roebling's nature was that of the ponderous granite on which he laid the masterworks of his career. He was a slower man to rouse to any undertaking. But, once convinced that he was on the right track, his tenacity overcame all obstacles.

Bridge engineers abroad regarded Roebling's undertaking as one of amazing daring, for to them suspension construction was associated with undulating and disastrous frailty. A pioneer advocate and builder of suspension structures in England was Sir Samuel Brown; but a whole string of his chain bridges had failed—at Berwick, at Brighton, at Montrose, and at Durham. In 1831 the Broughton Bridge in Lancashire County, a chain bridge only two years old, crashed under a body of marching troops. And, in 1850, a wire bridge at Angers, France, collapsed under a similar load. In the face of this record, how could any sane engineer seriously propose a suspension span, made of wire and wood, to carry the tremendous loads of locomotives and freight trains?

It is difficult nowadays to realize the incredulity which Roebling's proposal aroused. Engineers the world over said it could not be done. The foremost bridgebuilder in England, Robert Stephenson, of Britannia Bridge fame, who had adopted the clumsy and expensive tubular construction for his largest projects because he was convinced that a suspension bridge could not be built to carry railroad loading, said, in a letter to

Roebling: "If your bridge succeeds, then mine have been magnificent blunders." But the Niagara Railway Suspension Bridge proved to be a magnificent success.

6

In 1851 Roebling began his construction operations on the Niagara bridge, and for four years the work continued without interruption, even during the bitter cold of the Canadian winters. The pioneer was primarily an engineer, not a contractor, and his engineering included both planning and execution. His own eye must examine every portion of the structure as it was put together; every step of the erection was planned by him, and closely supervised by him, and nothing beyond the manual labor was entrusted to others. He was not an imitator, nor a duplicator; all his great structures were essentially unique, each planned to meet the special features of the location, and each introducing new improvements in the art and new contributions to engineering progress.

In contrast with Ellet's spectacular exploits and thrills, the building of the great Niagara span by Roebling proceeded without fanfare or flourish. There were no dramatic incidents, no exciting experiences to relate. Once the man had gained the chance for which he had fought, the sole drama was in the courageous tackling of the task and in the final victory. The construction of the great span was not treated as an opportunity to display physical prowess; the exposures of the job were taken as a matter of course. The big test, the real adventure, was that of the mind and of the spirit.

Through the changing seasons, for four years, Roebling gave his concentrated attention to the building of the span, rarely leaving the site of the work for visits to his home or for other professional engagements. His letters to his superintendent at Trenton continued in a constant stream. He ordered various materials for the bridge to be made up by Swan, and instructed him as to the conduct of business affairs at home. Nothing in the business was too petty to escape Roebling's attention. The bulk of the letters related to details of machinery installations at the Trenton plant, orders for wire rope, credit and collections, and other routine instructions, with only an occasional brief reference indicating the progress of the bridge. In July, 1852, he wrote from Suspension Bridge, N. Y.—the name given to the new town springing up at the American end of the span— "Mrs. Roebling wishes me to come home. This I cannot."

There had been additions to Roebling's household. Following the death of her husband Carl Meissner, and finally keeping the hopeful promise she had made twenty years earlier, the bridge-builder's sister Amalia, always devotedly attached to her brother "August," came to America in 1851 with her seven children. They were warmly welcomed and cheerfully taken into the already crowded Roebling home at Trenton. The oldest of the six sons, Fritz Meissner, became a useful representative of the bridgebuilder on some of his enterprises.

In January, 1853, Roebling wrote from Danville, Kentucky, where he had just secured the contract to build another railroad suspension bridge—this one over the Kentucky River, on the projected line of the Lexington and Southern Kentucky Railroad: "Today I have concluded my Contract for the building of the Kentucky Bridge; you may therefore prepare for a heavy business as regards Wire drawing. . . . I will leave here tomorrow but shall remain several days at Lexington and Cincinnati; from there I go to the Falls where I expect to stay about one week only, then I will go to New York by way of Worcester to examine Washburn's wire mill. In New York I will examine C. M. Erickson's Engine. I will be in Trenton about the middle of Febry. . . . Please read this letter to Mrs. Roebling and give my love to the whole family."

In the summer of 1853 a financial depression hit the country: stocks went down, collections became difficult, construction projects were suspended, and orders were canceled. The works at Trenton felt the pinch when payment was suspended for wire ropes made for the Morris Canal inclines. In August, Roebling wrote from Pittsburgh: "Should my Credit at the Bank fall short, then open with Mrs. Roebling my safe and the iron box with the small brass key, and take out of the morocco pocket book my Certificate of Stock of $5000 of the Pennsylvania R.R. Co. and deposit it in Bank if necessary."

Back at Suspension Bridge, he wrote in October: "The progress of the Ky. Bridge is a little doubtful on account of money matters. Several lines have stopped, thousands of men are thrown out of employment in Ohio and wages reduced from 1.25 to 80 cs. p. day. Laborers will be very plenty next winter, and I also think Iron will be a little lower. In regard to the Ky. Bridge, I shall know soon."

The gorge to be crossed by the proposed Kentucky River bridge was both wider and deeper than that at Niagara. Under Roebling's supervision, simultaneously with the progress of the Niagara bridge, the anchorages for the Kentucky span were laid and the stone towers erected, and most of the cable wire and other material for the superstructure were delivered at the site, when suddenly the finances of the railroad company collapsed and the work was indefinitely suspended. The bridgebuilder had invested heavily in this project, as it was to be the longest span in the world and would eclipse even his Niagara crossing. For years the massive stone towers rising above the deep gorge of the Kentucky River stood like monuments in the wilderness, places of pilgrimage from the neighboring settlements. Finally, in 1876, a high railroad viaduct—the first cantilever bridge in America—was built at this crossing, utilizing the towers which Roebling had built in 1854.

7

In December, 1853, the fascinating but exacting work of cable spinning was under way at Niagara, and the bridgebuilder was unable to visit his home and family for the holidays. On December 29 he wrote: "I should like to be at home just now, but must wait, until a whole operation of strand making is gone through. My machinery here works very well and hands are getting practised; the whole operation is very interesting and creates much interest." The next day he wrote again, "We are in spite of the cold & stormy weather getting along so rapidly, that I believe the two first Strands will be completed in three days and that by end of next week the two next Strands will be commenced."

One of the strangest letters ever written—affording a tragic revelation of a man so intensely absorbed in his work that he has almost forgotten his family and his home—was addressed to Swan from Niagara Falls on January 6, 1854: "Your letters of the 2nd and 3rd came to hand. You say in your last, that

Mrs. *Roebling* & *the child* are pretty well. This takes me by *surprise* not having been informed at all of the delivery of Mrs. R. Or what do you mean? Please answer by return of mail.

"I myself was a little doubtful about the sufficiency of a 3 inch Shaft—must try it now. . . . You may also enclose copies of Mr. Cower's or other important letters; let Washington copy. . . . If possible try that flat bar of Habricht's and enclose your report care of Fritz Meissner."

John A. Roebling, master builder, was so engrossed in his work that he was surprised to learn of a recent addition to his family! On January 1, his fourth son, Edmund, had been born —and the absent-minded father learned of it later only from a casual reference in a business letter from his factory superintendent.

8

On May 17, 1854, a news flash by telegraph announced a great suspension-bridge catastrophe. Ellet's bridge over the Ohio at Wheeling had collapsed! In a letter to Swan on May 20 Roebling, busily engaged on his Niagara span, recorded his first reaction upon receipt of this news:

"A telegraphic dispatch from Wheeling states that the Wheeling Suspension bridge broke down on the 17th; the particulars I have not learned. This bridge was not safe against heavy storms.

"On the 18th we had a very severe squall here, which I feared would break down the old footbridge, but it stood the whirl well. I am anxious however to secure the new floor well by Stays. I have no rope on hand to do it with. I can get along with 2000 ft.; but as soon as practicable, send on more."

On May 21 the news was confirmed. Roebling wrote: "The destruction of the Wheeling Bridge is a fact. . . . I wrote to you to send me 2000 ft. more of rope 7 x 7 No. 9, 3 lbs. pr. ft., for Stays. Do not neglect this. I must have this rope very soon to render the new floor safe. If possible send more as fast as you can make it. You can use any bridge wire, for this rope, you have on hand; if it is a little larger, does not matter."

In 1847 a company had been formed to build a toll span over the Ohio River at Wheeling. Ellet and Roebling, both described in the local press as "young engineers of ability," submitted bids for the construction of the bridge. Ellet advocated a single, long span of 1,010 feet from bank to bank; and Roebling, in view of the narrow width of the structure, urged a central span not to exceed 600 feet, flanked by side spans and stiffened by *inclined stays*. The decision was left to the local board of directors, a group of businessmen who knew little or nothing about bridge design. A vote was taken, with the result that Ellet's plan was approved and he was given the job. The span was light and narrow for its length, carrying only a seventeen-foot wagon way and a 3½-foot sidewalk. Most important of all, the span had no stiffening construction.

In 1849 Charles Ellet completed the Wheeling Suspension Bridge—his greatest work and, at the time, the longest span in the world. Five years later, that proud record-breaking structure was destroyed by the wind. Technical publications recorded the disaster merely as another bridge wrecked by a storm, and the lesson was lost to the profession. We find the complete story of the disaster, however, in an eyewitness account by a reporter, printed the following day in the Wheeling *Intelligencer* and reprinted four days later in the New York *Times*—it took four days then for a news story to reach New York. A remarkable anticipatory parallel to the 1940 Tacoma Narrows bridge catastrophe is revealed by the following excerpts from the original newspaper account:

With feelings of unutterable sorrow, we announce that the noble and world-renowned structure, the Wheeling Suspension Bridge, has been swept from its strongholds by a terrific storm, and now lies a mass of ruins. Yesterday morning thousands beheld this stupendous structure, a mighty pathway spanning the beautiful waters of the Ohio, and looked upon it as one of the proudest monuments of the enterprise of our citizens. Now, nothing remains of it but the dismantled towers looming above the sorrowful wreck that lies beneath them.

About 3 o'clock yesterday we walked toward the Suspension Bridge and went upon it, as we have frequently done, enjoying the cool breeze and the undulating motion of the bridge. . . . We had been off the flooring only two minutes, and were on Main street when we saw persons running toward the river bank; we followed just in time to see the whole structure heaving and dashing with tremendous force.

For a few moments we watched it with breathless anxiety, lunging like a ship in a storm; at one time it rose to nearly the height of the tower, then fell, and twisted and writhed, and was dashed almost bottom upward. At last there seemed to be a determined twist along the entire span, about one half of the flooring being nearly reversed, and down went the immense structure from its dizzy height to the stream below, with an appalling crash and roar.

For a mechanical solution of the unexpected fall of this stupendous structure, we must await further developments. We witnessed the terrific scene. The great body of the flooring and the suspenders, forming something like a basket swung between the towers, was swayed to and fro like the motion of a pendulum. Each vibration giving it increased momentum, the cables, which sustained the whole structure, were unable to resist a force operating on them in so many different directions, and were literally twisted and wrenched from their fastenings. . . .

We believe the enterprise and public spirit of our citizens will repair the loss as speedily as any community could possibly do. It is a source of gratulation that no lives were lost by the disaster.

The newspaper man who wrote the foregoing dramatic account unknowingly summarized the crux of the aerodynamic phenomenon he had observed when he used the significant phrase: *"Each vibration giving it increased momentum."* And when he stated that the mechanical solution of the failure "must await further developments," he wrote better than he knew. In those days bridgebuilders were not thinking in terms of aerodynamics, and the profession had to wait nearly a hundred years for the further developments that finally gave them an understanding and mastery of the problem.

The only engineer of the time who grasped the full significance of the Wheeling bridge disaster was John Roebling. With intuitive genius, a century ahead of his time, he realized the need of bracing and stiffening suspension spans against cumulative undulations that may be started by the action of the wind. From his very first bridge, the Monongahela, he provided his spans with *ample stiffening* and with *inclined stays*. These were incorporated from the start, in 1850, in his design for the Niagara bridge; and immediately upon receipt of word of the Wheeling failure, Roebling rushed an order to Trenton for additional wire rope to supplement his system of bracing on the Niagara span by adding inclined stays below the floor of the bridge *to resist wind uplift*. The Wheeling catastrophe ended Ellet's spectacular but brief bridgebuilding career, but it gave added strength and confidence to Roebling and his work. It confirmed the practical soundness of his original and prophetic thinking on the subject of the security and stability of suspension bridges.

During the disastrous gale in 1854 the Wheeling span had overturned, the south group of cables was wrenched and broken, and the whole structure, except the north cables, was blown into the Ohio—a wrecked and twisted mass of wire and timber. Six years later the bridge was rebuilt by Roebling, at a cost of $42,000. Since then the Wheeling bridge, in its stronger reincarnation, has been performing valiant service. During the Civil War the newly rebuilt span was used extensively for the movement of troops. Half a century later the bridge was used for truck transport trains during the First World War, and then again, three decades later, in the Second World War. Today, eighty-five years after its reconstruction by Roebling, it is a heavily traversed bridge, often loaded with autos and trucks from end to end. It is now the oldest existing suspension bridge in America; in fact, except possibly for some old stone arches and covered timber bridges, it is the oldest existing span of any type in the New World.

9

In June, 1854, the cables and suspenders for the Niagara span had been erected and Roebling recorded the start of work on hanging the roadway beams, high above the whirling rapids: "We commenced suspending beams last Saturday & will finish today—all done in three days. I have the satisfaction of seeing my Suspenders come out perfectly correct, so much so, that it will be hardly necessary to touch a Screw."

Elated at the successful completion of the most critical phases of the construction, the builder, in a letter two days later, told how well all of the parts were going together as the erection entered its final stages: "My Bridge is the admiration of everybody; the directors are delighted. The woodwork goes together in the best manner. The suspenders require scarcely any adjustment at all."

While construction of the bridge was proceeding in July, an epidemic of cholera broke out in the vicinity. The quickly spreading plague, especially virulent in the narrow flats on both banks of the Niagara River, found many victims among the bridge workers. Patients went into a collapse immediately after being attacked, and many succumbed within a few hours. Over sixty persons—men, women, and children—died during the first week. A panic seized the inhabitants. The doctors were helpless, few nurses could be found, and many of the stricken unfortunates were abandoned in terror and left to their fate.

The medical science of the time seemed powerless to stop the raging epidemic or to help the victims. All forms of treatment were tried, from allopathy to homeopathy, but all alike failed.

Roebling did his best to minister to the bridge employees. Since conventional medical treatment had proved futile, he tried his own favorite remedy for all ills, hydropathy. He was a great admirer and personal friend of Priessnitz, the leading German specialist on the subject—fully believing in his theory. When the plague was at its height, Roebling improvised a hospital and had eight patients subjected to hydropathic treatment, but they all died before night. He persisted, and in the succeeding days he was rewarded with some apparent cures—two of his patients somehow recovered.

Roebling offered two stout Germans $200 to go and set fire to a house in which the plague had broken out in a very malignant form. They went, and both caught the contagion, of which they died.

The bridgebuilder himself, in his battle against the Niagara epidemic, was constantly exposed to the infection. He *determined* not to have it. At one time he walked his room all night, fighting with his will power against symptoms which threatened to make him a victim. A man who later related the incident said it was the most striking exhibition of the power of mind over disease that he had ever witnessed.

In a letter written July 29, Roebling reported the waning of the epidemic and told of his treatment: "There are no more new cases of sickness. I have saved two by water treatment; all the rest are in the grave. The Doctors had bad luck. . . . Keep off *fear*—this is the great secret. Whoever is afraid of cholera will be attacked, and no treatment can save him. . . . Should the cholera appear in Trenton, keep up your courage and do not fear, and the disease should not attack you. Best not to think of it at all."

However strange Roebling's philosophy of physical remedies and preventive treatments may seem to us, his personal application of that philosophy provides a superlative illustration of the power of the mind in combating the ills of the body, particularly when the mental armor takes the form of an iron will. On his ocean voyage to America in 1831 the pioneer, determined not to be seasick, had more than once walked the deck all night in his unyielding fight to ward off an attack.

On August 2, 1854, the bridgebuilder recorded the end of the epidemic: "No more cholera cases here; all well, myself included; say this to Mrs. R. . . . I have again commenced work. . . ."

10

During Roebling's absorption in his work at Niagara, the wire business at Trenton had some hard sledding, with the financial depression and the consequent shrinkage of orders, collections, and funds. The bridge contract was a lifesaver during this period. In September Roebling wrote from Suspension Bridge: "The Co. has paid me $17,000 already on Wire, Rope, & Salary; all this is gone into the Trenton business, in the shape of drafts and Cash, which I brought home on various occasions. My balance sheet of the Trenton business next New Year will be no brag, but very little if any over *minus*. It would show a loss perhaps without the bridge orders, which have greatly helped, and which have been very promptly paid."

The following month the bridgebuilder recorded an unfortunate accident on the work. His principal carpenter, Mr. Kenzie, was working on the Cascades side, removing the wooden towers that Ellet had built, when suddenly one of them fell down with a crash, and a timber struck the carpenter across the face. "His recovery will take time and is doubtful under the bad medical treatment which he has," Roebling noted. "I am now in a bad fix, and am compelled to complete the work without him."

Late in the fall of 1854 more financial difficulties loomed on the horizon. Word reached Roebling that Peter Cooper's business—the Trenton Iron Company—was forced to close down, and that the local banks were threatened. The bridgebuilder wrote to his superintendent at Trenton: "You had better draw out to within a few $100 and keep the Notes in the Safe, and if there is danger, send Philip or Bernhard at various times to draw out gold for their bills. But I do not see that those Banks can be much in trouble unless it was on account of their connection with the Trenton Iron Co. . . . Next winter will be a very hard one for great numbers of laboring men, on account of scarcity of work and dearness of provisions. . . . The Philadelphia Banks are, I believe, considered safe and in good condition. . . . My Assistant in Ky. has a good deal of trouble in raising money, but he is managing remarkably well; he is about completing the masonry. . . . Keep me well informed, and keep a good lookout for breakers among our customers, particularly the small and suspicious Coal concerns. These may soon meet with tighter times, at least those who have no capital."

The country weathered the financial storm, and in January, 1855, the uncompleted Niagara span successfully weathered a physical storm. Roebling wrote from Suspension Bridge: "We had a tremendous gale for the last 12 hours; my bridge didn't move a muscle. I expect an order of 2000 to 3000 ft. 7 x 7 No. 9 for Stays for the Lewiston Susp. Bridge, which needs guying badly."

The reference to the Lewiston Suspension Bridge related to a light highway span built over the Niagara River between Lewiston and Queenston in 1851 by Serrell, one of the four courageous engineers who had replied favorably to Major Stuart's original canvass concerning the feasibility of a railroad suspension bridge. Constructed without the benefit of Roebling's intuitive—or deeply rational—understanding of the need of bracing, Serrell's span proved shaky in the January gale and had to be promptly equipped with wire-rope guys to make it safe.

In the spring of 1855 Roebling decided to build a new home for his family at Trenton, and this time it was to be of brick. The first modest frame dwelling was crowded far beyond its capacity, with eight or nine Roeblings, seven Meissners, and some ten workmen quartered there, filling every corner of the single story and attic. Early in March the bridgebuilder communicated his instructions to his faithful factotum: "I shall want about 290,000 brick for the new house. I wish you to look around and to make a conditional bargain and write to me. . . . You can also enquire about lime and sand, and masons to build the cellar walls and to lay the brick. I do not wish to make a long job of it, but to get all under roof this season." This new home must have been for its owner a symbol of success; two months later, in his final official report on the Niagara span, he found a way of casually mentioning "my brick dwelling at Trenton."

Even after completing the epoch-making Niagara span, following his building of six prior notable structures, and after five years of operation of the Trenton plant, Roebling was still in restricted financial circumstances. A picture of the modest and crowded living conditions of the time is disclosed in a letter from the bridgebuilder to Swan, written in April, 1855: "I think it would be well that you should board Mr. Bulkley. He can sleep for the present in the bathroom, and I confine myself with the other little room, until the new house is finished, then he can get a room in my house. If you do not like to board him, perhaps I will board him myself. He being a young man yet, it is advisable and certainly better for the interest of the business, to keep him near the mill. If he and you are satisfied with this arrangement, Mrs. Roebling can furnish the bedding and keep his room in order."

Although funds were scarce, Roebling was mindful of the personal needs of his employees and tried to help them. He communicated his wishes to Swan: "Fritz Buttendorf wrote to me and enquires if I would not advance him $100 or guarantee the purchase of $100 of furniture to furnish his home. He wants to marry now and has got no money to do it with. I think it will be safe to trust him. I therefore request you to offer him either the cash in advance or my credit."

II

In March, 1855, the Niagara bridge was completed. History had been made. John A. Roebling had built the first successful railway suspension bridge in the world. Many engineers in both hemispheres had challenged the feasibility of the proposed span, and its failure had been freely predicted.

On March 16, 1855, the first train passed over the completed structure—*the first train in history to cross a bridge suspended from wire cables*. And to the amazement of the public the span did not collapse.

In another sense bridge-engineering history had been made. The Niagara span was the first suspension bridge ever built with *stiffening trusses*. This was a major contribution to bridge engineering—the first radical improvement in the suspension-bridge art in two thousand years. From the most primitive swinging spans of twisted vines and fibers, there had of course been successive improvements in materials and in details of construction, but the full potentialities of the suspension type could not be realized as long as it continued to be represented by swaying, undulating structures. As Roebling expressed it: "Suspension bridges have generally been looked upon as loose fabrics hung up in the air, as if for the very purpose of swinging. Repeated failures of such works have strengthened this belief." Even in his earliest spans, the master builder grasped the fundamental importance of stiffened construction. His success in the construction of suspended aqueducts—"as rigid as stone or cast iron aqueducts"—demonstrated the truth of his thesis that cable spans could be built as stiff as desired. In those canal structures, his solidly braced flumes supplied the stiffening element. In his first highway structure, the Monongahela bridge, he made the outside railings or "parapets" in the form of latticed trusses, which served to some extent as stiffening elements for the suspended spans—forerunners of real stiffening trusses. But it was in the Niagara bridge that this new concept received its first full expression—the first use of stiffening trusses in all the history of bridgebuilding. Filling the vertical height between the upper and lower decks of the span, this trussed bracing was eighteen feet deep. The amazing success of this radical innovation established it thereafter as a recognized essential feature in the design of suspension bridges.

During the week preceding the formal opening Roebling had tested the rigidity of the bridge by running locomotives of various weights over the span and measuring the deflections.

SPANNING NIAGARA

The results splendidly confirmed the efficacy of his stiffening girders and trusses.

The opening of the bridge to traffic was described by the builder in a letter: "Last Sunday I opened the Bridge for the regular passage of trains. The first one was the heaviest freight train that will ever pass, and was made up on purpose to test the bridge. With an Engine of 28 tons we pushed over from Canada to New York twenty double loaded freight cars, making a gross weight of 368 Tons; this train very nearly covered the whole length of floor between the towers. . . . No vibrations whatever. Less noise and movement than in a common truss bridge. . . . Yesterday the first Passenger train from the East with three cars, crowded inside & on top, went over in fine style. Altogether we passed about twenty trains across within the last 24 hours. No one is afraid to cross. The passage of trains is a great sight, worth seeing it."

At the end of the letter, the proud bridgebuilder added: "Please communicate the above to Mrs. R.; this will save me writing to her."

12

On May 1, 1855, Roebling presented to his clients, the directors of the Niagara Falls Suspension Bridge Company and the Niagara Falls International Bridge Company, his final report on the work. This document has become a classic in engineering literature. It was a masterly summary of the concepts underlying the success of a courageous and epochal undertaking, the world's first successful railroad suspension span:

It gives me great pleasure to be enabled to report the Niagara Railway Suspension Bridge complete in all its parts. The success of this work may now be considered an established fact. The trains of the New York Central, and of the Great Western Railroad in Canada, have been crossing regularly since the 18th of March, averaging over thirty trips per day.

One single observation of the passage of a train over the Niagara Bridge will convince the most sceptical, that the practicability of Suspended Railway Bridges, so much doubted heretofore, has been successfully demonstrated.

The work which you did me the honor to entrust to my charge, has cost less than $400,000. The same object accomplished in Europe would have cost four millions without serving a better purpose, or insuring greater safety. . . .

Sitting upon a saddle on top of one of the towers of the Niagara Bridge during the passage of a train, I feel less vibration than I do in my brick dwelling at Trenton during the passage of an Express Train over the New Jersey R.R., which passes my door within a distance of two hundred feet.

Professional and public opinion having been adverse to Suspended Railway Bridges, the question now turns up: What means have been used in the Niagara Bridge, to make it answer for Railway traffic? The means employed are *Weight, Girders, Trusses* and *Stays.* With these any degree of stiffness can be insured, to resist either the action of trains, or the violence of storms, or even hurricanes. And I will here observe, that no Suspension Bridge is safe without some of these appliances. The catalogue of disastrous failures is now large enough to warn against light fabrics, suspended to be blown down, as it were, in defiance of the elements. A number of such fairy creations are still hovering about the country, only waiting for a rough blow to be demolished.

Without adding much to the weight of the structure, a surprising degree of stiffness has been obtained by the united action of the girders and trusses. They have fully realized my expectations. The pressure of an Engine and of a whole train of cars is so much distributed, that the depression caused is not readily observed. . . . The next means of stiffness I have applied are wire-rope stays, above as well as below the floors. . . . Their principal duty is to guard against the forces of winds, but at the same time they contribute materially to preserve the equilibrium of the structure during the passage of trains.

Bridges of half a mile span, for common or Railway travel, may be built, using iron for the cables, with entire safety. But by substituting the best quality of *steel* wire, we may nearly double the span, and afford the same degree of security.

In the last two sentences the master builder showed a keen anticipation of future developments. His bold prognostication of the feasibility of longer spans has already been largely fulfilled —with the Golden Gate span of 4,200 feet accomplished in 1937 and still greater spans projected.

The Niagara bridge carried the railroad track on the upper deck and the highway below. This double-deck span, 821 feet long and weighing a thousand tons, was suspended from four ten-inch cables of wrought iron wire. The arrangement of the cables was unique, two of them being connected by wire-rope suspenders directly to the upper deck of the bridge and the other two similarly to the lower deck. The greater part of the wire for the cables was manufactured in England. This was supplemented by the wire removed from the cables of Ellet's temporary span, erected in 1848 and taken down in 1854.

Conviction of rightness and pride of workmanship were apparent in Roebling's discussion of the results of his method of cable building: "The appearance of the cables is not only pleasing, but their massive proportions are also well calculated to inspire confidence in their strength. The reflecting man however will naturally inquire: Is this mass of wire put together so that the different strands bear all alike? Does each individual wire perform its duty, so that when exposed to a great strain, they will resist with united strength to the last? I can answer this question in the affirmative, and can assure you that the tension of these 3640 wires, composing each cable, is so nearly uniform that I feel justified in using the term perfect." And then, in justification of his assurance, the builder described in detail the construction of the cables: the preliminary oiling, straightening, splicing, and reeling of the wire; the perfected procedure, with his patented machinery, for stringing the fifteen thousand wires across the gorge; the careful adjustment of every wire for identical sag and consequent uniform tension; the grouping and anchorage of the wires to form twenty-eight strands; the consolidating and compacting of the strands to form the four massive cables; and the final application, by the use of his patented wrapping machine, of a tight, continuous wire wrapping around each finished cable. The bridgebuilder knew that the cables he had built would defy time and load: "What I consider of most importance to the durability of the cables," he said, "is the fact that their strength is nearly six times as great as their ordinary working tension, and equally important is the fact that their strength will never be impaired by vibration."

The Wheeling bridge disaster, occurring only a year before the opening of the Niagara span, naturally left many questions in the public mind. Most engineers were mystified, and bridge-owners were deeply concerned. Were all cable bridges equally subject to sudden destruction by the wind? Was the great new span over Niagara liable to be overtaken by the same untimely fate? What defects of design made the Wheeling bridge so peculiarly vulnerable? What were the safeguards in the Niagara design to prevent a duplication of the catastrophe on an even more tragic scale?

In his report to his clients Roebling felt it his duty to address himself to these questions in order to clarify the mystery and set all doubts at rest. His lucid exposition of the action of the wind in building up dangerous undulations in an unstiffened span showed his unique grasp of the problem. Containing, as it did, reasoning and conclusions that engineers of a later century tragically failed to perceive, Roebling's analysis of the problem was, in itself, proof of his genius as a master builder:

The destruction of the Wheeling bridge by a high wind on the 17th of May last year, the greatest disaster of the kind on record, has naturally given rise to doubts as to the safety of suspension bridges generally. . . . Although I would much prefer to leave this subject alone, I cannot conscientiously do so. It is my duty to establish the safety of the Niagara Bridge, which has already, and in the brief space of one month, become one of the greatest thoroughfares on this continent. I cannot do so without drawing a comparison with other works, and without pointing out the defects which caused the destruction of the Wheeling bridge, and on the other hand explaining the means of safety which have been employed in the Niagara Bridge.

The Wheeling bridge formed a span of 1010 feet from centre to centre of towers; its weight, including cables was about 440 tons. The number of cables was 12. . . . There was no provision in the whole structure which could have had an effect in checking vibrations. . . . A competent eyewitness stated that the waves of the floor, caused by the wind, rose to a height of over 20 feet. This may have been an exaggeration, but no ordinary strength of cables can resist the momentum produced by such a weight falling even 15 feet. . . .

Weight is a most essential condition, provided it is properly used in connection with other means. If relied upon alone, as was the case in the plan of the Wheeling bridge, it may become the very means of its destruction. That bridge was destroyed by the momentum acquired by its own dead weight, when swayed up and down by the force of the wind. . . . A high wind, acting upon a suspended floor devoid of inherent stiffness, will produce a series of undulations. . . . These undulations will increase by their own effect until, by a steady wind, a momentum may be produced that may prove stronger than the cables. ... Weight should be simply an attending element to a still more important condition, viz: *stiffness.*

The destruction of the Wheeling bridge was clearly owing to a want of stability, and not to a want of strength. This want of stiffness could have been supplied by *over-floor stays, truss railings, under-floor stays,* or *cable stays.* ... In the Niagara span most ample provisions for stability have been made. The superstructure, forming a hollow box or beam, twenty-four feet wide by twenty feet deep, with solid girders of five feet depth, and *effective trusses,* possesses enough of stiffness to resist the action of any gale. There are 64 diagonal stays of 1⅜-inch diameter rope, above the floors, equally distributed among the four cables. To be prepared, however, for the greatest emergency, there are 56 wire rope stays or guys attached to the lower floor, which are firmly anchored to the solid rock of the cliffs. Each of these ropes has an ultimate strength of 30 tons.

Not the slightest motion from high winds was ever noticed. The work has been frequently tested by the strongest gales that blow in this vicinity. I am also convinced that it will be proof against a hurricane.

The report of the master builder to the owners of the bridge concluded with an eloquent and fitting expression of appreciation of the faith and trust they had reposed in him:

In reporting to you the final and successful completion of the bridge, I would be doing injustice to my own feelings as a man, if I did not avail myself of this opportunity, to thank you publicly for the unwavering confidence which you have always placed in my professional ability. When Engineers of acknowledged talent and reputation freely expressed their doubts as to the success of this work, a wavering of confidence on your part would have been but natural. But I am happy to state here, that in all my operations I have always met with a cordial support. It is a great satisfaction to me that this work has turned out equal to my promise, and also to know that, on taking leave of you, the mutual confidence that exists will not undergo any change.

Respectfully submitting the above, I remain
    Gentlemen,
        Your obedient servant,
            JOHN A. ROEBLING.

The bridgebuilder also prepared, for the profession, a "Memoir of the Niagara Falls Suspension and International Bridge." Illustrated with his own drawings, it was published in

1856 in London in a volume entitled *Papers and Practical Illustrations of Public Works of Recent Construction, Both British and American*. In this scientific report Roebling referred to the earlier discussions in England that had led to the adoption of the tubular plan for crossing the Menai Straits on the mistaken conclusion that the suspension type was unsuitable for railroad bridges. He admitted that a suspension span will deflect under load; but so will all other bridge types. It was simply a matter of degree, and this was subject to control—a fact not previously realized. With keen insight he posed the basic question: How rigid do spans need to be made?—a question that has not yet been fully answered. And he proclaimed and proved a principle unknown to his predecessors and unrealized by most of his successors—namely, that suspension bridges can be built of any desired degree of rigidity.

As Robert Stephenson had generously offered to concede, the success of the Niagara span proved that the earlier choice of the heavy and costly tubular type for the British structures had been a mistake. Thereafter that form of bridge construction was abandoned. True, new span types were subsequently developed—the cantilever truss and the continuous truss held sway from 1880 to 1920—but finally the suspension form came into its own, and became the dominant type for the longest and greatest spans in the world. On the foundations laid by Roebling's genius and prevision the monumental achievements of a later century were reared.

In the Niagara Railway Suspension Bridge, Roebling made an outstanding and enduring contribution to the fundamental principles of suspension-bridge design. He taught the profession the importance of stiffening suspension spans, and he showed the world the effectiveness of such stiffening. His predecessors and his rivals had failed because they did not realize the importance of this feature and had omitted it in their structures. The necessity of providing an adequate stiffening system was a new concept to the profession. For nearly a century thereafter this basic idea was recognized and governed suspension-bridge design, producing structures of a new order of stability and longevity until, in the swing of the pendulum, engineers began to forget or to discount the principle that Roebling's genius had evolved and demonstrated. With more mathematics than the pioneer master builder had available in his time, but with less native intuition and constructive perception, a later generation of designers carried the reversed trend to a climax until they were awakened by the impact of catastrophe—the destruction of the Tacoma Narrows span, by aerodynamic oscillations, in a mild gale, in 1940.

In the Niagara achievement, guided by insight, intuition, and genius, and profiting by his own earlier successes and by the disastrous experiences of his contemporaries, Roebling developed and introduced features of suspension-bridge design that assured strength, rigidity, and stability, not only against the useful loads to be carried but also against the potentially destructive forces of wind and storm. The pioneer gave these principles to the profession—in his published account of the work and in his printed reports—but the profession forgot his teachings, only to rediscover them three generations later at great cost.

13

The location of the bridge was in a superlatively picturesque setting. With its shapely stone towers rising from the high cliffs of the Niagara gorge, and with the long graceful sweep of its cables, the span seemed almost a natural part of the scenery. The thundering Falls in the distance and the Whirlpool Rapids beneath lent an enhanced charm to the airy appearance of the bridge itself.

The Prince of Wales, during his American tour in 1860, made a special point of visiting Roebling's bridge. He had his train stopped on the span while he walked over the famous structure, examining its construction. It had become the most talked-about bridge in the world. Other royal visitors and distinguished personages from all parts of the globe were attracted by the fame of the span and had to see it with their own eyes.

In the same year the eminent English railway and bridge engineer, Peter W. Barlow, C.E., F.R.S., F.G.S., visited America, in order to see for himself the notable bridge that Roebling had built. On his return to London, Barlow published an enthusiastic account in book form under the title, *Observations on the Niagara Railway Suspension Bridge, Made During a Recent Tour in America.* He recorded his conviction that, when the Britannia tubular bridge was under construction, the suspension principle had been prematurely condemned; and then he paid tribute to Roebling's achievement in showing the world how to build a suspension bridge economically, efficiently, and safely, to carry the heaviest loadings.

In his discussion he compared Roebling's span with the famous Britannia Bridge built by Stephenson. The comparison between the two bridges was striking. The one represented the proudest achievement of the foremost bridgebuilder in England; the other was the work of a hitherto unknown engineer in America. There were the two contrasting solutions of the problem of spanning wide gaps to carry railroad traffic. Robert Stephenson, with comparatively unlimited resources at his disposal, had built the longest-span railroad bridge in Great Britain; but his span length was only 460 feet, and he required 3,000 tons of iron girders to carry the load in each span. John Roebling, with his wire cables, to which he gave longitudinal stiffening by means of timber trussing—because he could not afford iron trusses—had bridged a gap of 821 feet in a single span! Moreover the Niagara bridge had to carry the heavier loading. With a span length approaching double that of the Britannia Bridge, and with a greater moving burden to be sustained, the genius of Roebling had produced a rigid railroad span while using only about one-fourth as much material.

After discussing his actual observations of the rigidity of the Niagara span under heavy loading and his calculations of the efficiency of the design of that bridge compared with other railway structures, Barlow presented his conclusion: "The preceding investigation leads to the inevitable conclusion, viz., that the Niagara Bridge is the safest and most durable railway bridge of large span which has been constructed."

Another prominent English engineer, John Provis, had published a paper in the *Transactions of the Institute of Civil Engineers* in which he described the effects of wind on the most famous suspension bridge in Great Britain, Telford's Menai Straits Bridge. In 1839 the thirteen-year-old Menai bridge had been seriously injured by a gale. "This injurious action of the wind," said Provis, "arises from the undulation of the platform." He repaired Telford's span and corrected the weakness by stiffening the suspended roadway. In a previous gale, in

1836, the amount of the wave in the span, as observed by the bridgekeeper, was sixteen feet. This subjected the bridge, in each undulation, to the shock of its weight of four hundred tons falling sixteen feet. To the British engineer the rigidity of Roebling's span at Niagara was amazing.

Three years after the bridge was opened to traffic, Buffalo papers expressed an apprehension—evidently emanating from some rival engineers—that the wires in the cables would "crystallize" under the vibrations of passing trains and that the bridge would fall down. Roebling, with the assistance of his son Washington, thereupon made an exhaustive series of experiments on the subject of "crystallization" of iron under all conditions of strain and temperature. He refuted the whole fallacious theory in a pamphlet published at the time.

In 1860 Roebling inspected the Niagara bridge and wrote, for the owners, a report on the excellent condition of the structure after five years of service. After discussing train speeds, deflections, strength and rigidity of the structure, and resistance of the materials to time and the elements, he stated in conclusion: "I will close this report by repeating, once more, that the cables of the Niagara Bridge are made of a superior quality of material; that they are free from vibration; that they are well preserved; and consequently that they may safely be trusted for a long series of years."

Roebling's Niagara span was the only railroad suspension bridge in the world to stand the test of time. Five years later, a smaller railroad span, suspended from wrought-iron chains, was built in Vienna; but it proved to be much too light and showed many defects, with progressive weakening and consequent increase in deflections under load, so that it had to be taken down.

Following its completion in 1855, Roebling's bridge carried the progressively increasing weights of trains and locomotives for more than four decades. After twenty-six years of service and revenue, the wooden suspended structure—which Roebling had been constrained to use by the original limitation of funds —was replaced by iron and steel trusses. After thirty years of service, it was found that the saddle rollers at the tops of the towers lacked the necessary freedom of motion and that the resulting strains had affected the stonework. A most interesting and successful feat was performed in the substitution of iron towers for the masonry towers, without interrupting traffic across the bridge. After this operation—which Roebling would have enjoyed planning and executing had he lived—the bridge continued in excellent service condition. Occasionally, in the succeeding years, chord members of the new metallic stiffening truss broke under the ever-growing locomotive loads, but the trains were permitted to travel uninterruptedly over the structure—with the truss members fractured—pending and during the repair operations. No other type of bridge would have remained standing under those conditions.

Finally, in 1897, a combination of circumstances made it expedient to retire the veteran structure after its long record of distinguished service. Three considerations governed this decision. The actual sizes and weights of locomotives and trains had increased far beyond anything imagined in the early period when the bridge was planned and built. Freight cars had doubled in tonnage capacity and locomotives had quadrupled in weight. In consequence the bridge in its later years was actually loaded to three times the capacity for which it was designed. As the second impelling consideration the tremendous increase that had taken place in the volume of rail traffic between the United States and Canada made it imperative to have a two-track bridge, whereas in Roebling's time the single track he was commissioned to provide was ample. The final consideration was that new developments in the immediate locality made it necessary to provide, in addition to highway and railroad facilities, for an electric trolley line—a totally unexpected item in the march of progress since Roebling's day.

So in 1897 the world-famous span which Roebling had built to carry the railroad trains and highway traffic of his day over the Niagara gorge was finally taken down, to be replaced by a wider and heavier structure to accommodate the new demands of more modern transportation facilities and loads. The town adjoining the crossing retains its historic name of Suspension Bridge, New York.

When the old bridge was taken apart, the wire cables constructed by Roebling were found to be as sound and strong as when they were originally erected. The service and strain of forty-two years had not impaired them. After sustaining the burden of carrying some two million locomotives and trains across the gorge, the wire had retained all its elasticity and even coiled back into the original ring form in which it had been drawn, proving that it had never been overstrained; and although it was ungalvanized, its original protection had proved effective. In the possession of the writer is a section of the original cable of Roebling's Niagara span; and, despite its age since its erection in 1853, it is remarkably free from any rust, and retains to this day the original springy curvature of the wires.

14

Shortly before Roebling was engaged to build a span over Niagara, Edward W. Serrell landed his opportunity. He secured the commission to construct a highway suspension bridge across the Niagara River between Lewiston, New York, and Queenston, Ontario, near Brock's monument, seven miles below the Falls. The light structure was quickly completed, at a cost of only $50,000, and in March, 1851, it was opened to traffic. The span was 1,043 feet long, and it was suspended from ten small cables, each containing only 250 wires. This type of construction, following the French system, was not conducive to enduring stability.

In January, 1855, Serrell's span was severely shaken in a gale that visited the locality. Roebling, who was then completing his own bridge near by, came to the rescue by promptly supplying a system of wire-rope guys, which the span sorely needed to make it safe. He installed these hold-down guys under the floor of the bridge, anchoring them to the shores on either side. These rope stays served to hold the span against uplift by the strong winds sweeping up the gorge.

Early in 1864 a great ice jam piled up about the rocks to which the guys were fastened; and, in order that the ice should not damage them, the ropes were loosened. During the pleasant spring weather that followed, the bridge attendants overlooked the loose guys and failed to refasten them. On April 16 a high wind came up, and the unsecured bridge fell an easy prey to its force. The gale broke the roadway, severed the suspenders, and left the cables dangling uselessly in the air. For thirty-four years the damaged cables were left hanging there, with no attempt to rebuild the span. There hung the cables as the sole remains of the wreck, a silent reminder of nature's power and man's forgetfulness.

The amazing and dramatic developments in bridge construction in America were being followed with the keenest interest by the engineering profession on the other side of the ocean. The new span lengths tackled represented a boldness beyond precedent. The failures simply paralleled familiar experiences with smaller structures in England and on the Continent. But the successes, especially the outstanding achievement of Roebling, made the engineers abroad blink their eyes and readjust their earlier prejudices and judgments. Henceforth the suspension principle must command respect. The deficiencies and failures must no longer be charged to the type, but only to the examples unscientifically designed or improperly constructed.

15

The success of Roebling's railway suspension bridge over the Niagara rapids created a demand for a bridge two miles farther upstream, close to the Falls, where the scenic attraction was even more impelling. The promoter and builder was Samuel Keefer, a civil engineer of Ottawa—the fourth and last of the engineers who had testified to the feasibility of Major Stuart's proposal in 1845. After much opposition a charter was obtained, and in the winter of 1867–68 a rope was carried across the river on the natural ice bridge at the site of the proposed span; thus physical connection was made between the cliffs at this point for the first "Honeymoon Bridge"—a crossing that was to be the scene of many dramatic incidents in bridge construction and bridge destruction. The span was completed and opened to the public on January 2, 1869. It was a wooden structure suspended from cables of iron wire, with the record-breaking span length of 1,268 feet.

The roadway was only ten feet wide, and carriages were unable to pass one another on the span. This led to long waits at either end; lines of carriages in caravan formation moved in one direction across the bridge, while many others were waiting at the opposite end for the line to pass before they could have the right of way. Those were the days when the Niagara hackmen were in their prosperous prime, and the electric trolley had not yet been introduced. The bridge was profitable. In 1872, only three years after its completion, the span was rebuilt, with steel replacing wood in the stiffening chords; and in 1884 the wooden towers, in which sight-seeing elevators were operated on the Canadian side, gave way to towers of steel. In October, 1887, the work of widening the bridge was begun, and it was completed the following June, with new steel cables, new anchorages, and a virtually new steel structure from bank to bank. The reconstructed span was the admiration of all who visited the Falls. But, within seven months, it was doomed to an untimely fate.

On the night of January 9, 1889, the Niagara was visited by a terrific hurricane, and when daylight came in the morning not a single inch of the suspended structure remained. The span had been torn away from the cliffs as though cut out by a giant knife, and the entire crumpled mass of steel and wire lay bottom up in the gorge below. On the slopes of the bank on each side of the river the ends of the fallen mass were visible, while beneath the deep, turbulent waters of the river the greater portion of the wreck was hidden; and there it remains to this day.

While the public mourned the loss of the bridge, the controlling companies were equal to the occasion, and at once ordered the span to be duplicated. The rebuilding of the bridge was a feat of surprising rapidity; as the iron mills still had all the blueprints and patterns, the steel parts were quickly at hand. On March 22 the duplicate bridge was started, and in a month and a half, on May 7, the new span was opened for travel. The structure had a width of 17½ feet, and when it was built the men behind the enterprise believed they were building for all time.

But in 1889 they did not anticipate future developments. They little realized that the ensuing decade would bring such great changes in the Niagara region as to demand a voluntary destruction of the proud structure they had built, in order that it might give way to a better and more modern bridge. The hacks and the horsecars had to give way to electric trolleys. The new electric cars were heavy, and the existing bridge was not strong enough to carry them. It was therefore determined to replace the suspension bridge with a new span; and since the suspension span did not seem rigid enough under the loads, the engineers of the time decided to adopt the arch type for the new construction. When taken down, the eight-year-old suspension bridge was rebuilt on the site of Serrell's old Lewiston Bridge, which had been destroyed by wind in 1864.

The new highway span at the Falls was built in 1897, and was hailed as the greatest steel arch in the world. It was the fourth bridge built on the site, counting those built in 1888 and 1889 as the second and third. On January 27, 1938, this fourth "Honeymoon Bridge" was wrecked by a high ice jam, which crushed the steelwork at the springing points of the arch. To replace it, a new arch span—the "Rainbow Bridge"—was built in 1940-41.

16

In 1845, in response to Major Stuart's circular inquiry, four engineers had declared their readiness to bridge Niagara. In the years that followed, each in turn secured an opportunity to build a span across the famous gorge. There is drama in this chance competition of four bridgebuilders. The widely differing measures of success they attained afford an illuminating contrast of personalities and capabilities.

First there was Ellet. Cocksure and confident, he had started out in a spectacular burst of showmanship. He was going to build a great railroad suspension bridge. He got as far as erecting a light, temporary span, 7½ feet wide, and then he quit. His shaky suspended plankway remained up only six years, and was then torn down to make way for Roebling's bridge at the same site.

Next came Serrell. He built the Lewiston-Queenston span in 1851—a flimsy highway bridge, dancing in the wind. Anchor guys had to be added to hold it down. It lasted but thirteen years, then it was wrecked by a gale.

Keefer also built a light highway span—the Niagara-Clifton bridge—in 1869. It had the most spectacular location, nearest to the Falls. Three times in its brief life it was strengthened or reconstructed, until a hurricane swept it completely out of existence.

And then there was John Roebling. Of the four bridgebuilders who accepted the challenge of Niagara, Roebling was the only one to start a railroad suspension bridge and to *finish* it—not a shaky footbridge, not a light, swaying, highway span, but a stiffened suspension bridge, with powerful consolidated cables, to carry the heaviest known trains and locomotives, in addition to affording a wide roadway on a separate deck for heavy highway traffic. It was a real bridge, an enduring structure, the work of a master builder. Through winds and gales and hurricanes the bridge remained staunch and unshaken. Loads increased—longer and heavier trains of freight cars, more ponderous locomotives thundering over the span—until the loading was three times as heavy as any anticipated when the span was projected, and still the bridge firmly carried its steadily augmenting burden. Decade after decade, for forty-two years, it remained in service, carrying the constantly increasing load of international traffic on both rail and highway decks. At a time when other railroad bridges, even those of the most substantial construction in conventional types, had a customary service life of but twenty years before obsolescence and replacement, Roebling's more daring Niagara span continued in powerful usefulness for twice that period. And it would be standing yet, as far as any destructive forces of nature are concerned. Only the voluntary decision of man determined its final replacement —not for any flaw in the structure, not for any inefficiency in the design, but to make room for a more modern span proportioned to new weights and dimensions and capacities that transcended prior conception.

That is the whole story. Four men had started out to span Niagara. One built a footbridge, and quit; two built light, flexible, highway spans, at the mercy of the wind; but only Roebling met the original challenge—he built a rigid, railroad suspension bridge, the first successful railroad suspension span in the world. And when we consider the smallness of the sum that was available to build the structure in the first place and the lack of guiding precedents and of design theory at the time, we are stirred by the genius of the man who, defying the doubts of his contemporaries, eagerly accepted the challenge and built a bridge that changed all preconceived notions and shattered all prior conventions.

With his invention of wire rope John Roebling was started on his road to success. With his achievement of the epoch-making Niagara Bridge the master builder was started on the road to immortality.

CHAPTER XI

THE MIND AND THE HEART

IN LATER life Washington Roebling invariably spoke with the highest regard of his father's genius and professional achievements. But in his reminiscent moods, he would also recall, without reserve and even with a certain apparent gusto, the unrestrained punishments which he had so often endured at the hands of an irascible parent. The severe castigations, the almost unremitting scoldings to which the boy was subjected during his growing years left their mark upon mind and character. Neither the rod nor the child was spared. Fortunately these punishments, though often unmerited, did not embitter, but were recalled in riper years with a philosophic calm mingled with an appreciative sense of their seriocomic aspect.

Of his mother, who was evidently devoted in every way to her children, he always spoke with reverent affection. And it was doubtless from her that he inherited the milder qualities of his personality—gentleness of heart, serenity of spirit, and patient resignation under adversity and affliction.

His strongest childhood attachment was for Julius Riedel, the itinerant tutor who gave him his first instruction in "the three R's in German." And when this beloved teacher and preacher had to leave Saxonburg to seek a new post in Ohio, the sobbing child "was dumped into the same wagon and dropped at Pittsburgh to live with a German teacher named Henne."

After the family moved to Trenton in the autumn of 1849, the youngster was enrolled in the Trenton Academy, which was one of the distinguished private schools of the time. The academy catalogue gave this significant warning: "If there is necessity for corporal punishment it is promptly and judiciously administered." This phrasing must have caused a wry smile to appear on Washington's face, for "corporal punishment—*judiciously* administered" would be for him a new and rare experience. Here the youth secured the classical Latin education which the father had missed. And here the son of the immigrant had as his fellow students the sons of American aristocracy—politicians, judges, senators, educators, and authors.

During the years of Washington's attendance at the academy his father was away from home much of the time, completing his suspended aqueducts at High Falls and Neversink, and then designing and building the Niagara bridge. In the bridge engineer's absences his faithful steward, Charles Swan, looked after affairs at the mill, aided by young Roebling in such time as he could spare from his studies. Even before he reached his teens, the boy had proved useful as an assistant to his father and had already demonstrated his aptitude in things mechanical.

Four years of study at the Trenton Academy found Washington prepared to enter the Rensselaer Polytechnic Institute at Troy, New York, then the foremost engineering school in America.

2

When young Roebling entered the institute, the students still wore a prescribed uniform; it consisted of a dark-green frock coat, with a black velvet collar, and a cap with a gold symbol on the band in front. He doubtless purchased an outfit upon his arrival at Troy.

There were no dormitories for students in those days. The printed catalogue supplied information and advice on living expenses:

Members of the Institute find Board and Lodgings with respectable private families in the City. . . . The total living expenses, which include board, furnished lodgings, laundry, fires, lights and attendance, vary from about $150 to $300 for the scholastic year, according to the habits and inclinations of the student. . . . There is little necessity for spending much money during the student's life at the Institute; and the supply of any more money than what is sufficient is very apt to be *worse than useless*. . . .

Washington was not subjected to the hazard of having "too free command of pocket money." The Roebling children had to account for every penny they expended, and even had to write home to Mr. Swan to send them a few postage stamps when needed.

Unfortunately the scholastic records of young Roebling's college career during his three years' attendance are not available—they were lost, together with his graduation thesis, in one of the fires suffered by the institute. The only records of his school days that have survived are two or three letters written by himself during his first year at Troy and addressed to the family's factotum at Trenton.

In one of these letters, dated in March, 1854, he asked Mr. Swan to forward to him "a collection of logarithmic tables by Baron de la Vega, published in Vienna and written on the left hand in Latin and the other in French." He told Swan exactly where to look for the book at home and asked him to send it to his Troy address, 97 Ferry Street.

In another letter, written in May, Washington said he was writing "just after having swum across the Hudson" and was therefore tired out. He continued: "There is no want to me of having plenty to do. It comes of itself, if you only want it, and sometimes there is a little too much of it. As my circle of acquaintances increases, paying even an occasional visit to them takes time. But I will take care of that." Evidently his social popularity was becoming onerous, as it was cutting too much into his hours required for study.

In the same letter he expressed concern over the "rolling" troubles Swan was having at the mill, and reported that, as requested, he had tried to find a skilled hand to go to Trenton to work at that operation, but could find none. He also stated that he had heard from his brother "Ferdie" that the Trenton dam had broken away again, and observed "had they taken my advice that would never have happened."

To engineering graduates of today, and their teachers, a glimpse of the curriculum of a century ago, when engineering education was in its infancy, would be a fascinating and startling revelation. In the list of nearly a hundred subjects Washington Roebling had to master in three years of concentrated application were such diverse and advanced courses as Analytical Geometry of Three Dimensions, Differential and Integral Calculus, Calculus of Variations, Qualitative and Quantitative Analysis, Determinative Mineralogy, Higher Geodesy, Logical and Rhetorical Criticism, French Composition and Literature, Orthographic and Spherical Projections, Gnomonics, Stereot-

THE NIAGARA RAILWAY SUSPENSION BRIDGE

ROEBLING'S PITTSBURGH-ALLEGHENY BRIDGE

omy, Acoustics, Optics, Thermotics, Geology of Mining, Paleontology, Rational Mechanics of Solids and Fluids, Spherical Astronomy, Kinematics, Machine Design, Hydraulic Motors, Steam Engines, Stability of Structures, Engineering and Architectural Design and Construction, and Intellectual and Ethical Philosophy. It was a stiff schedule. Under such a curriculum the average college boy of today would be left reeling and staggering. In that earlier era, before colleges embarked upon mass production, engineering education was a real test and training, an intensive intellectual discipline and professional equipment for a most exacting lifework. Only the ablest and the most ambitious could stand the pace and survive the ordeal.

Washington A. Roebling was graduated in July, 1857, with the degree of Civil Engineer. For his Senior Thesis, always an important requirement for graduation, he chose as his subject, "Design for a Suspension Aqueduct."

3

Colonel Roebling has left on record his conviction that seventeen, at which age he entered the engineering school, was at least a year too soon for what he called "that terrible treadmill of forcing an avalanche of figures and facts into young brains not qualified to assimilate them as yet." He declared that the director of the institution was the hardest taskmaster that ever lived, and that "the boys were ground down and crammed with knowledge and mathematics that their poor young brains could not make use of. . . . I am still busy trying to forget the heterogeneous mass of unusable knowledge that I could only memorize, not really digest. When a class starts with 65 and only graduates twelve it is proof of the terrible grind." He added that "the few who graduated left the school as mental wrecks."

In all probability the Colonel in his memories exaggerated somewhat the strenuous conditions of his student life, and his delight in forceful verbal utterances carried him beyond the boundaries of sober fact. In any event he never appeared to suffer very much in later life from the evils which he depicted, for obviously he never became one of the "mental wrecks" which he accused the institute of producing in the educational grinding mill of his descriptive metaphor.

His class numbered only twelve at their graduation in 1857, but all twelve men subsequently gave evidence of the character of the training they had received. Four of them made names for themselves as bridge engineers. Charles Macdonald built the railroad cantilever bridge over the Hudson at Poughkeepsie and the Hawkesbury River Bridge in Australia. Francisco Trujillo laid out the modern city of Havana. Other classmates achieved success in related fields, especially in railroad engineering. Two became engineering teachers of eminence. Four of the twelve were Lieutenant-Colonels in the Civil War, including young Roebling himself.

During his three years at Troy Washington established lasting friendships not only with his classmates but also with fellow students in the other classes entering or graduating while he was there. Among these were a number who achieved distinguished careers. Two attained the rank of Chief Engineer of the United States Navy. One became Engineer of Bridges of the Pennsylvania Railroad, and another, A. J. Cassatt, became President of that railroad. Another contemporary student was Theodore Cooper, whose brilliant career as one of America's leading bridge engineers was tragically terminated by the collapse of the ill-fated Quebec Bridge during its erection in 1907.

Five fellow-alumni later worked loyally under Washington Roebling on the design and construction of the Brooklyn Bridge, notably Charles C. Martin as Principal Assistant Engineer and Francis Collingwood as Assistant Engineer. Lefferts L. Buck, who graduated after the Civil War, began his engineering career as a draftsman on this work and, following this apprenticeship, he became the builder of other bridges, including two spans over Niagara and the second bridge over the East River, the Williamsburg Bridge.

At an alumni reunion in 1881, two years before the completion of the Brooklyn Bridge, Francis Collingwood paid a tribute to his absent chief, Colonel Roebling, who "despite a nervous shock, from which he never recovered, has never lost his hold on the work"—and then the speaker gave due credit to the school that had trained them for the task: "The men who have come from the Institute to the bridge have come to stay; and this seems to be true on whatever enterprise they are employed; they seem to have a wonderful sticking power. . . . Higher praise can hardly be accorded the Institute than this; for it shows that her men are all equipped for their work, that their worth is everywhere recognized; and that, with experience, they are fitted for the highest engineering positions."

Despite his vivid criticism of the scholastic procedure of the time, the Colonel retained a grateful and loyal regard for his alma mater. In after years his benefactions to the Rensselaer Polytechnic Institute aggregated over $100,000; in addition he left a bequest of $50,000 to the institution in his will. His son, John A. Roebling, 2nd, and his brother, Charles G. Roebling, also graduates, added another $100,000 to these testimonials of loyalty and affection.

About 1906, after a disastrous fire at the institute, Washington Roebling sent a check for $10,000. In thanking him, Director Palmer C. Ricketts explained the needs of the school and asked the Colonel for his advice. On receipt of this letter Roebling replied: "Dear Ricketts, What you evidently need is not advice, but money." Attached to this brief note was another check for $10,000.

Recalling his expressed belief that the school, in his time, had failed to provide properly for the physical and social needs of the students, the Colonel was one of the first alumni to establish a fund for the erection of dormitories. The "Roebling Dormitory," the finest and largest unit on the campus, was erected by gifts from the three members of the family who were graduates of the institute.

About 1917, when "wireless," as it was called, was still a plaything, Director Ricketts wrote to the Colonel to ask him whether he would like to build a pair of radio towers on the campus, which would have both educational and publicity value for the institute. A favorable answer and a check for $5,000 led to a serious study of this new development in science. As a result the Roebling family constructed Station WHAZ, for broadcasting and research. It was the first radio station in any institution of learning.

Like the Brooklyn Bridge, his alma mater still bears visible testimony to the devotion of Colonel Roebling.

4

The human part of Washington Roebling's education was in his own family circle. Admiring respect for his father, loving tenderness for his mother, humorous appreciation of his brother Ferdinand, and a deep affection for his younger brother Charles —all these went into the shaping of the man's responses to life.

Between him and his brother "Ferdie," only five years his junior, a relationship of whimsical and tolerant understanding was established, for there was a wide divergence between their personalities. Ferdinand was not an engineer; he was of the hearty salesman type—the advertising man, the merchandising expert, the business getter. He left the engineering problems, the expansion of the plant, and the mechanical operation to his brothers, while he busied himself with promotional activities, with public relations, with social and political affiliations. He had none of the shyness and reserve of his brothers; he was not afraid to meet people, and he knew how to make them like him at the first meeting. Perhaps for these reasons—because he showed these very differences of talents and traits—he became the father's favorite. Ferdinand took life more lightly than the others; he had no serious ambition apart from his impulse toward business achievement and financial success. For this brother, to whom he quizzically referred in one of his early letters as "that lazy chap Ferd," Washington entertained a humorous and tolerant regard.

But for his second brother, Charles, twelve years younger than himself, Washington felt throughout his life a deep and tender affection. This youngest of the three brothers was a rare personality—thoughtful, modest, shy, impulsive, considerate of others, and generous to a fault. These were traits that endeared him to the eldest brother, for Washington had early learned to recognize human values and to prize the things that are precious and rare. Only to the notes that were beautiful and true did the chords in his own heart respond.

A rare and heart-warming book could be written on the life and personality of Charles G. Roebling. Where other men achieved by sacrificing the finer instincts of life or by trampling over the feelings and aspirations of others, he kept his gentleness of character throughout his career and yet left behind him monuments of achievement. Engineer, scholar, musician, student of metaphysics and science, with a keen appreciation of art and a passion for botanical and floral forms, with sympathy for suffering and unquestioned loyalty to friends, reticent and avoiding publicity, he was little known to the general public; yet his varied engineering accomplishments—the model industrial town of Roebling which he planned and created, the manifold enlargement and expansion of the family plant, mills, and factories which he engineered and executed, the bridges he built and helped to build—are monuments to his extraordinary talents.

Fortunately we can obtain an intimate picture of the man as he was through the eyes of the one who knew him and understood him best—his brother, Washington Roebling. And in this reflected image, we can also perhaps glimpse some of the hue and quality of the heart that mirrored it.

Upon the death of Charles in 1918, his brother wrote this tribute:

IN MEMORIAM
Charles G. Roebling
WRITTEN BY HIS ELDER BROTHER, WASHINGTON A. ROEBLING

Born December 9, 1849

Died October 5, 1918.

There may be some people whose sole activities can be confined to two lines such as these, others who have done a little more than living and breathing.

But at rare intervals, once in a century, a man appears who towers so high above his fellows that his activities and accomplishments surpass a hundred-, yes, a thousand-fold those of the average man. Such a man was Charles G. Roebling.

Then Washington, unconsciously and indirectly, paid a tribute to the personality of his father, when he wrote that Charles, whom he lovingly idealized, "was his father over again":

I am a strong believer in heredity. He was his father over again, to a far greater degree than any of the other children. He inherited his temperament, his constitution, the concentrated energy which drives one to work and be doing something all the time. It might be argued that if a man inherits everything he deserves no credit for what he is. That would be so in a life of universal monotony, but with each generation in turn totally different conditions and environment arise. These have to be met by the new individual who must develop his own powers to adapt himself to them, to overcome them and use them as his tools.

In the next passage, sketching the childhood of his younger brother, Washington revealed a glimpse of his own early years:

The early childhood of Charles passed along rather peacefully. The father was away much of the time at Niagara, Pittsburgh and Cincinnati, so the boy was spared the educational experiments which were applied so disastrously to the older boys.

The reference to his father's "educational experiments" was the Colonel's euphemistic allusion to the overworked corrective measures of scolding and cowhide.

Then came a dramatic illustration of Charles's innate warm-hearted sympathy and generosity, manifested in childhood and forming the pattern of his character through life:

As he grew older the inherent impulsive traits began to show themselves as this little incident will show. While standing at the door one day, a beggar woman came along asking for old clothes. Charles rushed upstairs to his mother's closet, seized an armful of costly silk dresses, and bestowed them on the astonished mendicant, who marched off with them.

In telling how Charles selected his own middle name, the elder brother also recorded a glimpse of the family's religious life:

The boy was not christened as a child; his father waited until the two younger brothers were ready to march with him to the First Presbyterian Church, where a pew was maintained to enhance the family respectability. My mother was a Lutheran and father more of a Universalist, until he evolved a religion of his own, a thousand written pages of which are still mouldering in a trunk in the garret, forgotten and unread. Charles was baptized without a middle name. A little later he noticed that his older brothers rejoiced in one, so he selected the middle name of Gustavus of his own accord.

The reminiscences went on to sketch Charles's education:

Charles' older sister, Laura, had married a German school teacher from Mühlhausen named Methfessel, who established a boarding-school and pedagogium in Stapleton, Staten Island. Thither the boy was sent when he was twelve or fourteen; he thrived under his sister's motherly care and followed the usual academic course for three or four years. He developed great ability as a piano player, and became a brilliant performer, by far the best in Trenton. (His father before him had been a virtuoso on the flute and piano until his left hand was maimed.)

When he was eighteen he was sent to the Rensselaer Polytechnic Institute at Troy, where I had graduated in 1857. . . . When my father died and I undertook the Brooklyn Bridge, he still had two years to serve at Troy. . . . I have at hand an ambrotype taken when he was twenty-one, wearing a high stovepipe hat, with a perfectly smooth face and still capable of doing boyish tricks. He was a fine draughtsman and fair at mathematics, and learned more about machinery than I did. . . . Being an engineer, he naturally took to the manufacturing end of the business. . . . The history of Charles G. Roebling is practically the history of the Roebling Company from now on. . . . Thus we find him down at the mill every day in the year, early and late. That was his pleasure and delight. An occasional business trip gave him all the relaxation he needed. . . . Charles had one very strong point—he never copied. Every task was an education for him. . . . Charles did not like to write. I have never received a letter from him. . . . Even about such important subjects as the cables of the Williamsburg bridge he never wrote a line. He might at least have written a short monograph so as to get credit. He loved his work; it was his life, his happiness. . . .

By the death of Charles, I not only lose a friend and brother, but Trenton loses its greatest citizen and the country at large its foremost mechanical engineer.

The diverse facets of John A. Roebling's character and talents were distributed among these three sons. Washington inherited his engineering skill, analytical power, high courage, thoroughness, steadfastness, and articulateness, if not his brilliance of creative invention. Ferdinand inherited his ability to handle people, his salesmanship, and his excellent business sense; and later, as secretary and treasurer of the company, he was largely responsible for the financial success of the family industry which the pioneer had founded. Charles inherited his idealism and impulsiveness, his musical gifts, and, above all, a passion for building, which he applied to the founding of a model town and to the physical expansion of the Roebling plants; but he had none of his father's faculty of expression—he hardly ever wrote even a letter.

When we know what a man best loves and admires, we have a measure of him. In the traits of the brother whom he idolized, we find a clue to Washington's own character and ideals.

Finally, we cannot have an object of admiring and reverent regard without being influenced by it. In the case of Washington Roebling, respect, tenderness, appreciation, and affection for the members of his family constituted an essential and significant part of his education. Yes, the deepest part—the heart of the man.

5

Much is said of heredity and environment, the influence of the parents in shaping the character and lives of their children. But may not the process work both ways, with an interaction of characters within the family circle, so that something rare and gentle in the children may soften the hardness in a parent?

Whatever it was—whether the family relations, or the new influence of Emerson upon his philosophy of life, or the natural mellowing of the advancing years—something changed the inner life of John Roebling in his later years. He became gentler, kindlier, more considerate, with a new perception of the values he had missed. The younger children were spared the harshness and the severity with which the eldest had been raised, and long-suppressed emotions of tenderness began to show themselves, involuntarily, through the man's protective shell of reserve.

Following this change in outlook, he also began in his later years to give generously of his means—shyly, privately, unostentatiously. Even his own family little knew of whom he helped and how much he gave. The hardness and the penury that marked the earlier period of struggle—the initial striving for dollar profits and material achievement in the building up of his wire-rope business—gave way ultimately to a feeling of warm obligation and responsibility toward his workmen, for the welfare of all who were dependent upon him, and for the needs of the destitute and the helpless among his fellow townsmen and fellow countrymen.

CHAPTER XII

"THE BRIDGE WILL BE BEAUTIFUL"

WASHINGTON ROEBLING'S preparation for his lifework—home discipline, schooling, professional education, apprenticeship in the shop and in the field—closely followed his father's concept, his father's plan. Even John Roebling's impatience and severity toward the growing boy were manifestations of an impelling urge to mold the youth to the exacting pattern of a foreshadowed destiny. As the bridgebuilder designed a bridge, and forged and hewed the members to fit, so he planned his son's career and obstinately endeavored to shape his mind and character to fit into a preconceived design.

Following his graduation at Troy, the young engineer returned to his home at Trenton and took up work in the mill. As the senior Roebling and the plant superintendent, Charles Swan, were away for a part of this time, Washington was left in charge of the works—a heavy responsibility for a youth of twenty. He wrote regularly to his father—addressing him as "Dear Sir"—giving him full reports on the business of the plant, and particularly on the matter of credits, which were then causing much concern. Collections were slow, and substantial amounts due were lost, for 1857 was the great panic year when widespread failures of banks and business enterprises played havoc with the nation's financial structure.

In that year John Roebling received a commission to build still another bridge at Pittsburgh—this one to cross the Allegheny at Sixth Street. It was to take the place of an old covered bridge that had served for nearly four decades before it had finally become unsafe. The aged structure consisted of a string of six wooden truss spans, reinforced with timber arches. The roof was flat, and had a railing at the sides and steps at each end. This convenient railed platform covering the bridge was used as a promenade by the belles and beaux of days gone by.

The removal of the old structure and the construction of the new, including the building of the piers and the anchorages, required three years. The ambitious builder was determined to make this the finest bridge he had thus far built, to embody in it all the improvements in his art since his first structure erected over the same river twelve years earlier. For the first time he was able to adopt iron, instead of timber, for the suspended construction.

To withstand the pull of the cables, heavy anchorages of masonry were built on the banks of the river. Embedded at the bottom of these, under hundreds of tons of rock knit together with cement, were the large anchor plates to which the cables were later connected by wrought-iron links.

2

In the spring of 1858 Washington joined his father at Pittsburgh to assist him in erecting the new suspension bridge over the Allegheny. At last John Roebling's dream had come true—father and son were working together on the building of a bridge. In a letter to the Trenton office written from the Monongahela House, the father briefly mentioned the presence of his new assistant: "Glad to hear that orders are coming in. Here business is generally reviving; a good deal of life about the Wharf since the river got up. I am getting along well here. Washington is about the work."

During the two years that followed, both father and son wrote many letters to Swan, affording a sketchy record of the progress of the work and some glimpses of life in this growing city at the head of the Ohio. John Roebling's letters were principally confined to business matters, including inquiries concerning affairs at home and orders for supplies needed for the bridge construction. The younger engineer's correspondence was more breezy and personal, revealing a permeating sense of humor and a wide range of interests. Happy on his first bridge job, he was in high spirits. He wrote informatively and entertainingly on community progress and local lawlessness, weather and sunspots, music and chess, Christian morality and Democratic politicians. Although he was working diligently and strenuously, his letters gave but little inkling of the fact.

In his first letter Washington gave his new impressions of the place, which he had not visited since his Saxonburg days. Even at that early date Pittsburgh was already a smoky city, a locally welcome sign of flourishing industrial growth. The river was crowded with steamboats and rafts, and frequent collisions put the pumping boat out of commission, delaying the foundation work for the bridge piers. One anchorage was nearly completed. The young engineer was living at a "very pleasant boarding house" where he hungrily enjoyed his meals, and where "we have no female boarders, fortunately."

In the following months, the father had to make frequent trips to Cincinnati, and during these absences young Roebling was left in charge of the work at Pittsburgh.

In June John Roebling reported high water holding up the construction. By the end of the month the river level had fallen, permitting the foundations to be started: "The weather has been intensely hot for the last week," he wrote. "The rivers are getting low. I am preparing for laying foundations for the Piers. . . . Washington will send you a Pittsburgh Gazette with my article on *Steamboat Explosions*, which matter I mean to keep stirred up until the proper legislation has been obtained." In deference to both weather and patriotic spirit, Roebling gave his construction crew "a long 4th July to enjoy themselves during the dog days."

For this contract the Roeblings, father and son, had to build three new granite piers in the river, in addition to the two massive anchorages on the banks. The tall piers were built with round ends, to deflect current, drift, and ice, and were neatly finished with copings of large slabs of granite.

In a typical letter, written from the Monongahela House in July, Roebling ordered a ton of wire to be shipped to Iowa, inquired about the account of the Pennsylvania Coal Company, referred to a new harness, some notes to be found in Washington's desk at home, a new lubricating mixture for the machinery, and some trees injured in a recent storm, and then he added: "Please say to Ferdy he should write to me, he has never written yet."

A new note of tenderness is revealed here. The man of iron had a soft spot in his armor. The bridgebuilder was wistfully hoping for a letter from his son Ferdinand.

3

In August, John Roebling reported the successful conclusion of the financing for the Allegheny bridge: "Our new Bridge Stock was all sold the other day at auction at par, surprising everybody. I have not got as much as I wanted. I first wanted to see the result of my foundations. . . . People have great confidence in this Bridge."

As the construction advanced, the enthusiasm of the bridge company officials grew in proportion; and John Roebling was given a free hand to do as he saw fit. Here was the opportunity he had long been seeking—to produce the finest suspension bridge, both structurally and artistically, that had ever been built. He submitted his ideas to the owners as he went along, and they were all approved. Following the public's generous response to the offering of bridge stock, the purse strings were freely open. For the first and only time in his career the builder had ample funds made available to him to build a bridge according to his best ideas of substantial construction and to finish it in a manner that would impress the traveling public of the day. His clients knew that this would be a profitable investment.

Unable to leave the work for a visit to his home at Trenton, the bridgebuilder had asked his wife to join him and was anxiously awaiting her arrival. His eagerness was revealed in a letter to Swan: "Do not forget to telegraph me about Mrs. Roebling's departure or non-departure, so that I know whether she is coming or not. I have been running to the depot to no purpose." Whether Mrs. Roebling finally came is not known.

In the spring of 1859 Mr. Roebling again referred to "Ferdy," from whom he was still hoping to receive a letter: "I am pleased with your account of Ferdy. He has not written to me yet. We have commenced unroofing the old bridge."

During the Easter season Washington made a brief vacation trip to spend a few days with the family at Trenton. During his absence his father wrote from Pittsburgh: "I want Washington to come here next Friday or Saturday. My plans for the Cincinnati Bridge have been accepted and I am going on with the work." The bridgebuilder had to leave for Cincinnati. This was Washington's opportunity. He promptly came back to take charge of the job, and his next letter to Swan revealed his enthusiasm: "In one week I bet you will see me jumping about here; you had better come out and see how it is done."

On each granite pier two handsome iron towers were erected to carry the cable saddles. Each tower consisted of four cylindrical columns of cast iron, slightly inclined inward to form a neatly tapering structure. The spaces between these corner columns were filled with artistically effective latticed ironwork for bracing, with openings left above the sidewalks to form tapered portals. Above the cable saddles each tower was finished with ornamental spires, one at each corner and a higher one in the center. The resulting effect of this finial construction might have been tawdry if unskillfully handled; but in Roebling's planning, with careful design and innate sense of artistic proportions, a pleasing composition was obtained, accentuating the graceful lines of the spans.

In June Washington reported progress on the bridge: "By the middle of this week the cast-iron towers will be up in place. . . . We put up two columns a day; they are awkward things to handle because they can't stand up alone. . . . About the 4th of July the masonry will all be finished—Thank the Lord."

In erection and adjustment operations the building of a suspension bridge demands the nicest calculations and the widest range of engineering ability. The labor involved in the construction is far greater than the casual observer would imagine. The heart of a suspension bridge is in the quality and strength of the cables. The making of the cables, stringing them wire by wire, requires so much time that the public is often impatient with the delay in the progress of the undertaking, only to be surprised by the rapidity with which the rest of the work is completed after the cables are finished.

The traveling wheel, hung from an endless moving rope, had to make the complete trip over the towers 4,200 times from shore to shore in the stringing of the two main cables; and then the spinning of the two outside cables, which were smaller, meant 1,400 additional trips from one anchorage to the other. In October Washington happily recorded completion of the stringing of the main cables: "Yesterday the last strand of the big cables was finished; tomorrow we commence wrapping."

4

Washington Roebling was enjoying the strenuous life of a bridgebuilder, comprising involved office computations alternating with active physical exertion out on the work. On anchorages and piers, on high towers and swaying catwalk, stringing the cables and hanging the ironwork, he was there with the bridgemen—planning, helping, watching, inspecting, and supervising—with eyes and ears open to learn and to make improvements, and with a heavy burden of responsibility on his young shoulders.

One day, during the wrapping of the main cables and the hanging of the suspenders, the young bridgebuilder met with an accident. While he was out on a suspended "buggy," a falling wrench struck him on the head, knocking him unconscious. He described the experience in a postscript to Swan: "I was delayed in putting this letter into the post office by having had a hole knocked into my head accidentally, but I am happy to inform you that it is healed up again. I am all right, however. I experienced for the first time the delightful sensation of being knocked insensible."

In November Washington reported gratifying progress on the construction: "We have got along well on the bridge; the big cables are wrapped, all but about 50 feet at each end when we ran out of wire— You will have to make some more. The suspenders are all hung up in the main spans and in a day or two we hang on beams. Wrapping goes very well; no trouble after once getting started fairly."

In a postscript he added: "Tell that lazy chap Ferd to write."

Later in the month he wrote to Swan concerning morality in Pittsburgh, incidentally revealing his own capacity for indignation at hypocrisy: "I send you two papers containing an account of Rev. Mr. R.'s trial. You will see that although Pittsburgh claims to be the most orthodox Christian city in the United States, there are still some very rotten persons in it."

Very few of Johanna Roebling's letters can be found. In one of these, written a week before the Christmas of 1859, she was hoping that her two bridgebuilders would come home for the holidays. She wrote in German:

My dear Washington—

It is a long time since I received your letter. Today I will write you only a few lines, for next Sunday is the long-awaited Christmas festival, and we all look forward with pleasure to seeing you here again. If you should bring various small remembrances with you, they will be welcome. Your sisters have also remembered you, each one according to her circumstances. . . .

Fritz Meissner offered me a few dozen stereoscope pictures, at 10½¢ each—various views from here as well as Europe. I want to get some. He sent me one to try out and it fits in our stereoscope.

With the hope of seeing both of you here, in good health, next Saturday evening, I remain,

Your loving Mother,

J. Roebling.

5

The stringing of the two smaller cables was in progress when a violent February storm almost wrecked the footbridge. Young Roebling described the experience: "Yesterday we had a furious storm. All hands had to quit; they could not stand up, and I expected to see our footbridge go to the dickens, every moment; she jumped and pitched and reared like a wild horse; the new bridge shook a great deal sideways also because no planks are on yet to make her stiff in that direction. Yesterday morning we had 65 degrees heat and in the night it snowed; that explains the storm."

He wrote of the anticipated time of completion of the work, also of the continued opposition of the people on the Allegheny side, who were vexed over the proposed schedule of tolls: "I am afraid that the contemplated opening of the bridge by April 1 won't be exactly accomplished; it will probably be the 15th. All the stays have to be cut off yet and put on before we can open, and you know that those are not a few by the bill of rope you sent us. . . . The wrath of the Alleghenians has now reached its full height and they have determined to build a free bridge right next to ours; that is to say, a petition has been circulated for a charter, and since that don't cost anything, it was fully signed and was nine feet long according to last accounts. The whole move is only intended to frighten our people to put the toll down to nothing."

During the early spring the erection of the floor system was under way, and the work proceeded without a hitch. The spans of the new structure were stiffened with wrought-iron stiffening trusses extending the full length of the bridge. These were effectively proportioned and neatly designed. For additional stiffening, a system of inclined stays was installed. These wire ropes ran over the saddles and radiated downward from the tower tops for a distance of a hundred feet on each side of the tower. Where these diagonal stays crossed the vertical suspenders, they were lashed together with small wire.

In March John Roebling wrote of the large force of men engaged to speed the completion of the job: "The weather has been very cold for a few days, but is now fine. We have 42 Carpenters on the Bridge, some 80 men in all & are driving as hard as we can. . . . No more orders come in?"

In another of the rare letters preserved from the mother's correspondence, written to her son in Pittsburgh, she inquired repeatedly when the bridge would be opened and hoped she could see it when it was finished: "Today is Sunday and that is also usually letter writing day. . . . This morning we had confirmation in the German Church. In the absence of an organist, Laura substituted. . . . When will the opening of the bridge take place? I wish I could see it when it is finished and before it is dedicated. . . . Mr. Swan complains about the hard times. Not many orders are coming in. . . . Will Papa go to Kentucky to the railroad convention? How long will it be before the bridge will be finally completed? . . . All is well here. Willie wasn't very well for a time, but is now all right again. Willie says he is tired, and the others are studying. . . . My best greetings to Papa, and I hope that you both are well."

The reference to William, the last child born to the Roeblings, was pathetic. Sickly in his fourth year, he died soon after this letter was written.

An April freshet, flooding lower Pittsburgh and doing much damage on the Monongahela, was described by Washington: "Last Monday we had a rousing freshet. . . . The water commenced running into all the streets; all the Point rats were drowned out as usual. . . . Beyond the filling of cellars and damaging of floors there was not much damage done, except in the Monongahela, where 17 coal boats were sunk. The Monon-

"THE BRIDGE WILL BE BEAUTIFUL" 213

gahela bridge will have about $4000 damages to pay for coal boats sunk against its piers. The company has to pay those damages every year and the effect will be to break it up sooner or later. . . . Tomorrow we start to take down the small footbridge, which will take a week; the main floor will not be opened before the first week in May. We have quite a pile of men at work now; the check roll runs up to $600 per week. In the next *Frank Leslie's* you will see a picture of the bridge as she appeared during the flood. I am getting along quite well putting up the stays; it takes some force to handle those heavy ropes. . . . A Russian government engineer has been here for a few days; Father says he knows more than any other engineer he has ever met; he is travelling in this country to see the public works and will call at Trenton next summer. . . . This is the day of the prize fight, I believe. Hurrah! for the one that beats."

In his letters to Swan, John Roebling had been repeatedly and anxiously inquiring whether any new orders were coming in for wire rope. On May 1 the bridgebuilder recorded his happy surprise at news that large orders had been received, and at the same time he announced that the Allegheny bridge would be opened for wagon traffic on the following day: "You have quite surprised me by your letter. Those orders amount to 59,130 feet and will weigh about 256,000 lb. How will you be off for wire and iron to fill them? This is a six months business in place of three months. To fill all these orders will keep you busy all Summer. It is well you have got through with our improvements & machinery for the present. Always good policy to be prepared for such heavy orders. . . . Tomorrow I shall open the bridge for teams. All well."

Later in the month he made a brief trip to Kentucky and Cincinnati. Following his return in June, he reported the fine behavior and appearance of the new Allegheny bridge and wrote of the finishing touches still required before he could leave.

That year John Roebling had secured a patent on his invention of a railroad passenger car made entirely of iron, for greater strength and safety compared with the wooden cars previously used. The inventor promptly arranged to have a number of his new cars constructed in a near-by plant at New Brighton. Following a visit of father and son to see the cars in the process of manufacture, Washington wrote to Swan, reporting enthusiastically on the successful operation of the finished cars. In describing the trip, he mentioned the "misfortune" of running into the Western delegations to a Democratic convention—"an awful hard set in the way of looks."

A few days later Roebling wrote that he and Washington would soon be able to leave Pittsburgh: "I expect I & Washington can leave here on the 1st or 2d July. . . . We are now finishing Tollhouses and the ornamental parts of the Work, which is a slow business. The Bridge will be beautiful, when entirely completed."

The thrill of the artist is revealed in the last sentence, "*The Bridge will be beautiful.*" The man had planned and wrought and, as he saw his work approaching completion, his heart leapt at the realization that his dream had taken shape as a thing of beauty. The builder's pride was justified when others—artists, engineers, and laymen— proclaimed this bridge to be one of the most graceful structures in the land.

In his next letter Swan sent word that he had been injured in an accident at the mill. Washington had to return to Trenton to take over the work at the plant, while his father remained at Pittsburgh to look after the final details which would complete the bridge.

6

The new Allegheny bridge was a beautiful example of high technical skill in design and of the finest workmanship in execution. It differed from Roebling's other later masterworks—the Niagara, Ohio, and Brooklyn bridges—in consisting of four comparatively short spans instead of having one long main span. Although lacking the grandeur of larger proportions, it had a peculiar beauty of its own, due to the recurring curves formed by the cables, and the pleasing appearance of the minaret towers over which the cables were suspended. The new bridge was not a massive monument but, in its own way, it was an artistic gem.

The bridge, when it was finished in 1860, was generally regarded as the finest in the world. John Roebling had not then acquired the European celebrity which came later with his building of the great Cincinnati span; but the fame of the new bridge at Pittsburgh soon reached across the ocean.

The newly completed structure evoked the admiration of the Prince of Wales and his suite when they passed through on their American tour. They were astonished to find such a work of art in what was then a rather small provincial town. The Duke of Newcastle, who was a "practical man," examined it closely and informed the Prince that it was the best bridge he had ever seen.

Through the years following its completion, Roebling's Pittsburgh-Allegheny bridge elicited the admiration of competent authorities. "A structure of importance in the history of bridges" . . . "considered the finest in the world," said the editors of *Engineering News*. "Long noted as the most beautiful and graceful in its lines," said W. G. Wilkins, in 1895, in a technical paper describing the bridge in the *Proceedings* of the Engineers' Society of Western Pennsylvania.

Roebling, when asked what the strength of the structure was, said that he had ascertained the power of the greatest hurricane in the history of Pittsburgh, and had multiplied this by nine, which represented the strength of the bridge. This assurance gave the owners such confidence in the power of the bridge to resist pressure that no insurance against flood damage was considered necessary, especially as the spans had been put five feet above the highest recorded flood mark, that of 1832. With its iron trusses and with its solid floor construction of heavy timber on iron beams, it was also considered almost impossible to set the bridge on fire. For these reasons the owners carried no insurance of any kind on the structure; and for twenty-one years their supreme confidence remained apparently fully justified.

In 1881 a fire of mysterious origin occurred on the bridge. Most of the superstructure was of iron; but the timber floor suffered serious damage. The fire aroused consternation in view of the importance of the structure, and incendiarism was suspected. In an account printed the following day, the Pittsburgh *Dispatch* voiced the current suspicion: "The bridge was set afire by some evil designed person; and Old Peter, the night man at the Allegheny Toll House, said last night there was no doubt in his mind that it was set on fire."

But soon the mystery was solved. The detective work was done by John Harper, president of the bridge company; and the culprits were found to be "those feathered English pests," the sparrows, who had built their nests in large numbers on the under side of the spans. In making their nests, the birds had used quantities of hay, straw, and other combustible materials. On the day of the fire several steamboats had passed under the bridge and, at the high stage of water prevailing at the time, their smokestacks came close to the under floor. Sparks or flames from the stacks had set fire to the nests and started the blaze. The owners of the toll bridge not only suffered the loss of several weeks' wagon travel, but also had to repair the damage themselves "out of their accumulations" from past earnings.

Roebling's graceful suspension bridge over the Allegheny at Sixth Street carried a mounting load for thirty-two years. In 1891, in order to take care of the rapidly increasing volume of travel and to allow electric cars to cross without reducing their speed, it was decided to build a new structure—heavy steel truss spans designed by Theodore Cooper. In 1928 the bridge was replaced once more, this time with a graceful self-anchored chain bridge of prize-winning design.

Four bridges have successively occupied the Sixth Street site, one preceding Roebling's structure and two after it. Their combined chronicles give a cross section of four generations or eras of bridge evolution, from timber through iron to steel and finally to supersteel. The second and the fourth were suspension bridges and were unquestionably the most artistically effective of the series for grace of line and proportions. And of this sequence of structures, the one that represented the greatest engineering and architectural advance for its time was undoubtedly the beautiful cable bridge completed in 1860 by John Roebling and his son.

CHAPTER XIII

A RIVER DIVIDES *

F ROM a myriad of crystal springs hidden in the emerald twilight of the Appalachian forests, down the rock-ribbed slopes of the ancient time-defying mountains, through the deep shadowed valleys into the sunlit foothills, across the soft rolling plains and vast wind-swept prairies, the waters from a quarter of a million square miles of North America unite to form the mighty Ohio, flowing majestically on its thousand-mile course to add its flood to the great Father of Waters. For thousands of years, fed from the ageless hills by winter snow and summer rain, the resistless river has followed its self-made channel westward across the land.

With the coming of the white man the river soon gained a new significance. It became an indispensable artery of transportation, a mighty waterway, carrying the growing stream of men and civilization from the East into the wilderness. But it was also a barrier, an obstacle in the path of overland travel and commerce, a wall dividing North from South. The barrier of water called for a bridge.

Coming across the mountains of Virginia, blazing trails through the timber, planting corn in the clearings, the first pioneers to the West began to fill the fertile Bluegrass country of Kentucky while, on the other side of the Ohio, bands of painted Indians freely roamed through the forests and often, with their light canoes, made sorties across the river to attack the white settlers. After a few years venturesome pioneers began moving into the land north of the Ohio. Floating downstream from Brownsville and Pittsburgh in flatboats until they

* In Chapters XIII and XV, the author has based his account of the building of the Ohio River bridge upon the material in Harry R. Stevens' *The Ohio Bridge*, Cincinnati, Reuter Press, 1939.

came to some pleasant cove, they would land and establish a farm or a settlement. In 1788 one of these parties landed at a little inlet on the northern shore and built the first cabins of the future town of Cincinnati.

Separated from the nearest important white settlement up the Ohio by hundreds of miles of wilderness, with wolves and bears among the least of the dangers which the dark forests concealed, pioneer Cincinnati had to look elsewhere for human contact and for trade. Across the Ohio River, only eighty miles to the south, was Lexington, already one of the "oldest" and "wealthiest" cities in this new western country. In Lexington the Ohio farmers purchased their tools and seed and the Cincinnati merchants bought their supplies; and during the Indian wars, from 1790 to 1815, thousands of Kentucky militiamen crossed the river at Cincinnati to fight the savages in the north. For trade, for travel, and for defense, ready communication across the Ohio had become a necessity.

In 1792—three years after the first log cabins were built at Cincinnati—the first ferry was established there. This provided the needed transportation across the Ohio, but its usefulness was subject to limitations. Winter ice blocked the river; floating timber and debris damaged or sank the boats; spring floods swept them away; and the rates of toll were excessive.

During those pioneer years the small country town grew steadily. But little boys still picked blackberries in a thicket in the heart of town and looked for stray cattle in the willow bottoms along Mill Creek. On the Fourth of July a brass band paraded up and down the dusty main street while the local orators aired their patriotic eloquence. In winter the men sat around the stove in the general store, chewing black tobacco, telling of Indian raids, and discussing the latest news of Napoleon.

Then in 1811 came a new stirring symbol of progress—the first steamboat puffed around a bend of the river! And presently a brand-new steam cotton mill was put up—the town had its first real factory! The industrial future of the city of Cincinnati was born.

By 1815 Cincinnati, with a throbbing awareness of growth, had developed a fever for great things. The pioneer riflemen had defeated the marauding Indians in the War of 1812, and the surrounding countryside was at last safe. Farmers began to move up into the Miami valley to raise wheat and corn. Across the river the town of Covington was laid out, and the first houses were built there. Apples, cheese, pork, and brandy were carted to the Cincinnati landings, loaded on the steamboats, and shipped down the river to the new southern markets at Memphis and Vicksburg, Natchez and New Orleans.

With the coming of the steamboat the dusty, bedraggled little town of Cincinnati had suddenly boomed. Thousands of young men, poor in money but rich in energy and ambition, came pouring in from the East, to see an inland empire springing up before their eyes. It was going to be vaster, more wealthy, and more momentous than any the world had ever known. No wonder the young men had visions of great things. They talked of bridges, a bridge across the Ohio—across a river a quarter of a mile wide! Nowhere in the world was there such a bridge. These people in the raw frontier were thinking of things the greatest engineers in Europe scarcely dreamed of.

2

But building bridges calls for more than vision. It calls for capital, and capital was scarce in the new country of the West.

The difficulties of financing such improvements, even on a small scale, had been painfully learned early in the history of the frontier town. In 1798 a citizen's committee headed by William Henry Harrison—who later became the ninth President of the United States—undertook to raise funds for a bridge across Mill Creek. They were barely able to raise $290, and the creek had to remain unbridged for another eight years. Finally a Mr. Parker built the bridge, using the yellow poplar that grew along the banks. A spring freshet came; a boat that happened to be under the span rose and lifted the bridge on its back, and together the boat and the span drifted off down the Ohio. With the bridge went Parker's investment of $700. The next bridge over Mill Creek was built by Ethan Stone. The great flood of 1832 carried it away; the whole span floated off downstream and came to rest on an island above Louisville. So precious was the bridge that it was towed all the way back to Cincinnati and reinstalled.

The dream of a bridge across the Ohio continued to be a topic of live interest. The plans which someone drew up at that early date for the proposed crossing are amusing. The wooden bridge would be carried on eight piers in the river, with a draw span at mid-channel to allow boats to pass. The estimates of cost ran from $100,000 to $200,000, and capital from the East would be sought to finance the construction, to be repaid by tolls.

For another ten years little was done with the project. Money was invested heavily in steamboats and factories and turnpikes—especially the new "macadam" roads—but capital was not available for a bridge.

Half a century had passed since the first pioneers had come to Cincinnati. The little cluster of mud-plastered log cabins had become a flourishing city. Steamboats plied up and down the river, and commerce from half the nation crowded the public wharf. The need for a bridge was greater than ever before, yet the project had not progressed beyond mere talk.

Finally action came—but not from Covington or Cincinnati. By an odd turn of fortune the real movement for a bridge at this crossing came from an inland city eighty miles away. During the 1830's proud, aristocratic Lexington discovered that she was falling far behind in the race for glory. The course of commerce had changed, cutting directly across its former north and south route. The steamboat had suddenly made the Ohio River the great highway to wealth; and Lexington was not on the river. Economic survival now depended upon establishing direct commercial connections with Cincinnati. Early in the spring of 1839 the people of Lexington held a public meeting "with reference to the erection of a bridge across the Ohio River, from Covington to Cincinnati." Other cities and towns along the line took up the idea and joined in with enthusiasm.

The news was eagerly received in Cincinnati, and plans were made for a regional convention. Finally, on Friday evening, April 5, "a highly respectable portion of the citizens" met in the hall of the old Cincinnati College to hear the combined delegations. Eloquent speeches were made by the most prominent civic leaders; but soon the steamboat men, insurance men, and politicians took control of the meeting. The bridge project was adroitly sidetracked, and a committee was appointed instead "for the purpose of obtaining subscriptions from the citizens of

Cincinnati for the stock of the Lexington, Georgetown and Covington Turnpike Company." In time the money to improve the road from Lexington to Covington was raised, and the road was improved, to the advantage of both cities. But the proposals for a bridge, which had started the campaign, dropped into the background.

The dream of a bridge across the Ohio seemed more remote than ever before. The plans which had been suggested in 1826 were now impossible, for the steamboat traffic had grown so heavy that no such bridge could be permitted—it would block the navigation on the river. The need for a bridge was greater than before, but the new problem created by the heavy steamboat commerce appeared to defy solution. As the urgency grew, the difficulty increased. There seemed to be no way out.

3

At this very moment, when the high hopes for a bridge seemed doomed to disappointment, the path out of the impasse was being cleared. Two men were working on plans which would revolutionize bridgebuilding. One of them, Charles Ellet, lived in Philadelphia. The other, John A. Roebling, lived even closer by—along the headwaters of the Ohio near Pittsburgh. These two men were each on the brink of a solution which would make the Ohio bridge a possibility.

In the fall and winter of 1841 Ellet, as engineer and contractor, built his first suspension bridge, the Fairmount bridge over the Schuylkill River, the structure which Roebling had longed to build. With the success of his first accomplished work Ellet's self-confident ambitions continued to grow. In the glow of his triumph he was soon talking of spanning the Ohio. In a pamphlet which he published about this time, he suggested that a bridge would be possible at several points along the river— he named Louisville, Maysville, Marietta, and Wheeling— without a single pier in the stream; and he wrote at length of the beauty, safety, strength, and utility of suspension bridges. The thought took hold in the minds of Cincinnati businessmen.

With the idea of suspension bridges in the air, local constructors soon had to try it out, and in 1843 they proceeded to build one in Cincinnati. Spanning the Miami Canal at Race Street, it was constructed in four months at a cost of only $1,650. It was a little span, only sixty feet long, but it was the first suspension bridge west of the mountains.

By the end of another year the movement had gone so far that Colonel Long, of Kentucky, declared that the Ohio River bridge could be built for $150,000, "and the architect of the wire bridge, now in course of erection over the Allegheny River at Pittsburgh, himself a deserved celebrity as a builder of bridges, we understand is willing to build it for $100,000."

The anonymous master builder of suspension bridges, the "deserved celebrity" to whom Colonel Long referred, was none other than John Roebling.

No longer was the Ohio River bridge merely a romantic dream. It had become a practical proposal. Bridgebuilders had discovered how to build a long span high above a river. A suspension bridge was the solution; and Roebling was ready to build it.

The old difficulty still remained the chief obstacle: there was not enough capital. The eastern part of the country had been hit by a financial panic in 1837, and this reached the West in 1839. Shops and factories shut down; merchants went bankrupt and banks failed; men lost their savings, their homes, their jobs. Cincinnati had to establish food kitchens and bread lines.

After seven such lean years the country seemed to turn a corner in 1845. A new phrase was invented, America's *Manifest Destiny*. The future held promise of great things. It was America's "Manifest Destiny" to become the greatest nation on earth.

With a brighter outlook businessmen began to make plans and to talk more optimistically about the future. The people of northern Kentucky, like those elsewhere, felt that the country was again on the upgrade. Talk went around of holding a convention of businessmen and civic leaders to discuss the possibility of a bridge. The local newspapers up and down the Licking River promoted the idea. Thus a great project was born.

On the last Thursday in October, 1845, the convention assembled in Covington. A resolution was introduced by Robert Wallace, of Falmouth, calling for "a bridge across the Ohio, at Covington." It was received with acclaim, for it was in the mind of every delegate present.

From that time on, the idea of an Ohio bridge was not allowed to drop. Colonel Long, who had first suggested Roebling as the potential designer, had declared that the structure could be built for $150,000. At that price the bridge stock would all be subscribed in a single month—people would scramble for it! In February, 1846, the Kentucky General Assembly met in Frankfort; the Covington lobbyists were on hand, and soon the long-sought charter was enacted. By its terms a group of fifteen leading citizens were named as the incorporators of "the Covington and Cincinnati Bridge Company." The company was officially created and launched. But this was only the beginning of a twenty-year battle against all sorts of difficulties, obstacles, and delays.

4

Some of the leaders in the group wrote to Roebling to ask about a bridge plan. In his reply the bridgebuilder briefly outlined his ideas. His plan called for a single span of the unprecedented length of 1,200 feet! The middle of the arching span would be a hundred feet above the river! Such an elevation, he pointed out, would permit steamboats to clear even at a high stage of water.

Anti-bridge interests became alarmed when they saw how popular the proposal was becoming. Ferry owners and steamboat men, commission merchants and insurance brokers, boiler makers and engine builders, and all others whose business depended upon the river traffic joined forces to oppose the building of the bridge. They advanced every conceivable objection and pictured dire consequences if the scheme should go through. "Was there anywhere in the world so long a bridge without intermediate piers?" That proved it was impossible to build one. "If piers were placed in the river, water would be dammed up and the banks and wharves would be flooded." "Steamboats could not pass under the bridge, and that would destroy the trade and prosperity of Cincinnati." "Commercial and manufacturing houses would rush to cheaper sites across the river, and Cincinnati would become a deserted city." Thus the arguments ran—the same pattern of objections that has so often been used, even in more recent times, to obstruct bridge projects.

Meanwhile the Ohio legislature was considering the question. Here the opposition knew how to make its influence felt. The committee of the State Senate presented a brief report: The bridge would depreciate the value of real property in Cincinnati; and it would obstruct the navigation of the Ohio. The legislature refused to grant the charter, and the opposition celebrated a day of victory.

The bridge company was prepared for a long battle and was by no means ready to surrender at the first setback. Three months later, in May, 1846, they had Roebling himself in Cincinnati. It was the first time he had ever come there. He surveyed the line across the river, measured the banks, and examined the geological structure of the river bed. Having completed his notes, he returned to Pittsburgh.

In four months Roebling finished his plan and drew up a report, which was immediately published in Cincinnati. It was a thirty-six-page pamphlet containing sections and elevations of the proposed bridge, a lucid and convincing technical discussion, and a brief but brilliant analysis of the commercial problems involved.

Addressing himself first to the major premise of the opposition, namely that the free and uninterrupted navigation of the waterway was a national object, paramount to any individual or state interest, Roebling admitted that this, as a general proposition, was quite correct. "On the other hand," he pointed out, "it appears but justice that the States should be allowed the right of forming communications across the river for the promotion of commerce and intercourse, provided such communications do not impede the navigation."

"The idea of bridging the large rivers of the West," he continued, "could not be entertained before the system of suspension bridges was fairly introduced. . . . The construction of suspension bridges is now so well understood that no competent builder will hesitate to resort to spans of fifteen hundred feet and more, where localities may require it, and where the object will justify the expense."

On the moot question of the required height of the bridge Roebling collected a mass of evidence. The largest packet on the Ohio River was the *Pike No. 7*, whose smokestacks rose sixty-nine feet above the water line. Roebling had heard the captain of a steamer say he would consider any bridge an obstruction even if it were two hundred feet high. "Such objections," he declared in his report, "are, of course, beyond the reach of argument and not deserving of any notice."

This defense of the bridge Roebling followed up with a vigorous offensive on the steamboats. The difficulty came from the unnecessary and wasteful height of the smokestacks. "High chimneys are the strongest proofs of defective arrangements," he wrote. "They will disappear from our rivers sooner or later, as improvements advance. They are objectionable in every respect; if they could be forced from our rivers by some low bridge, it would be the greatest service which would be rendered to the navigation of our western waters."

As there were a number of boats with high smokestacks still running, however, and as the fashion seemed likely to continue, he said, "we shall be obliged to adopt in the construction of the proposed work an elevation unnecessarily high." He added that there was an easy method of raising and lowering smokestacks which was known and practiced on ocean steamships.

All of this was a pioneer, trail-blazing discussion. No bridge had been built across the Ohio at any point, and the answers to these questions of navigation clearance were vital to the future of bridgebuilding in America. The same arguments that Roebling originated and mustered in so clear and convincing a fashion have had to be used by succeeding generations of bridge engineers to this day. That his proposed underclearance of 121 feet was "unnecessarily high," as he maintained, is confirmed by the fact that the official standards more recently agreed upon and established for Ohio River bridges require a height of only ninety feet above low water.

Roebling dismissed the subject of smokestacks with a parting shot in lighter vein to puncture the seriousness of this chief obstacle of the opposition. "Apart from the passage of bridges, or waste of fuel," he concluded, "high chimneys prove objectionable in other respects. They increase the surface exposed to head-winds, make the boat top-heavy, and are in danger of being knocked off, or of becoming entangled among the branches of trees, when the boat is running near the shore."

As the charter granted by the Kentucky legislature left the location of the bridge entirely at the option of the company,

Roebling strongly recommended that the same provision should be included in the charter to be obtained from the legislature of Ohio. "It will be to the interest of the company to select the best location," he said, "and as their interests and those of the community are identical, the public will be best served if the company is left entirely free in its choice. Where the business of the community can be best promoted—wherever the greatest intercourse is likely to be, and where the approaches of the bridge do the least injury and offer the greatest accommodations —there will be the most advantageous site."

Such broad-gauge analyses of governing conditions reveal Roebling to have been a real engineer, in the higher sense of the term, and not merely a trained technician or technologist. Contrary to the popular conception, the calculation of stresses and sections, the designing of members and details, and the supervision of their erection are not the only things a bridge engineer has to do. In addition he has to conceive a project, develop it, verify and demonstrate its economic justification, battle opposition and secure authorization, wrestle with the problems of legislation and financing, determine the best location from both engineering and city-planning considerations, determine the best type and layout for economy, efficiency, and durability, plan approaches and connections for maximum convenience and safety of traffic, design the form, lines, and proportions of the structure for beauty as well as for strength, select and specify the best materials for the work, prepare construction contracts and specifications, deal with public officials, property owners, clients, contractors, and material men, present recommendations and reports in a clear and convincing style, and, above all, he has to have a sound and comprehensive grasp of the civic and social import and influence of his plans. Few men in the history of the profession have measured up to this mark. Roebling was one of the few.

As soon as Roebling's report was issued, opponents of the bridge lost no time in starting their attack, for they saw that the bridge company was proceeding in all earnestness. The new attack was effective. The 1847 session of the Ohio legislature met in Columbus, but it passed no charter for the company.

The Mexican War began, and the young men of Covington and Cincinnati marched gaily off to fight. General Scott took

Vera Cruz, and the war came to an end. But still there was no charter.

Up the Ohio River, at Wheeling, in what was then a part of Virginia, another bridge company was more successful. It had secured a charter, and Charles Ellet was at work there on the largest bridge he had yet tried to build, with a record-breaking span length of 1,010 feet.

With the example of a bridge almost finished at Wheeling, the Ohio legislature changed its mind and, in March, 1849, passed an act confirming the charter granted three years before by Kentucky to the Covington and Cincinnati Bridge Company. But, to satisfy the opponents of the project, the Ohio legislature wrote into the act a number of arbitrary and almost prohibitive conditions fixing the arrangement, span, and height of the bridge. Piers offshore were barred, the span was fixed at the excessive length of 1,400 feet, and the naturally desirable locations were prohibited. These restrictions—a case of engineering by legislative fiat—seriously handicapped the company and the engineer.

When the city of Covington had been planned in 1815, the streets were so laid out as to lie directly opposite the main streets of Cincinnati. It had been the hope of the sponsors that the bridge would form the link in one of these prepared lines of communication. Now the company was forbidden to build directly in the line of any of the Cincinnati streets running down to the river. Years later, when the bridge was opened, Roebling observed that everyone, even those who had put the amendment through, regretted the shortsightedness of the move. It deprived Cincinnati forever of one of "the finest and most magnificent avenues on this continent."

5

Before the sponsors of the project could make a start under all these difficulties, another disaster overtook the company. At the beginning of the agitation for a suspension bridge many enthusiasts had recommended the building of a small structure on the Kentucky side to connect Newport with Covington across the Licking River. Those who endorsed it did so with the hope that, when its success was demonstrated, the greater project, to span the Ohio, would find readier approval.

In January, 1854, after many delays and false starts, the Licking River Suspension Bridge, from Covington to Newport, was at last completed and opened as a toll structure. The cost was $80,000, but the value of the bridge seemed likely to exceed even such a large outlay. For two weeks after it was opened, all went well. Then one day, while a drove of cattle was passing over the span, *the bridge suddenly collapsed!* In a few terrible minutes the hopes of many investors were shattered—and the faith that had been placed in the strength and security of suspension bridges was destroyed. After this disaster only the hardiest adventurers had any heart left to push the idea of a suspension bridge across the Ohio.

The failure of the small pioneer span at Covington left a lasting impression on Roebling: he feared the effects of trotting cattle on a suspension span more than he did any strains or vibrations that might be produced by heavy locomotives or speeding trains.

Following the Licking River bridge catastrophe, financing of the proposed Ohio River span seemed hopeless. Yet the times were favorable. Ever since 1849, with the precious tide of gold from California flowing like new blood into the arteries of trade, the country had been growing more prosperous. By 1854 and 1855, there was the first great railroad boom. Every investor in the country foresaw a fortune for himself in railroad stock. A mere bridge seemed to lack the appeal that made the railroads so attractive. Money poured into the offices of the railroad companies; but the Covington and Cincinnati Bridge Company had a hard time selling any stock.

6

Ten years had passed since the company had been formed. It had secured its charters from both states, Kentucky and Ohio. It had employed an engineer, John A. Roebling, to draw up a plan for the bridge. And there it had stopped. Ten years of struggle and discouragement, and nothing to show for it! Not a spadeful of earth turned, not a stone placed! The bridge was still a dream.

But early in 1856 the company turned another corner. A new man was elected to the board of directors—Amos Shinkle. From that day a new spirit of energy and determination entered the company, and a different fortune waited on the bridge.

The story of Amos Shinkle is a typical picture of individual courage and initiative in American pioneer life a century ago. Born in 1818 on a farm in the narrow, winding valley of White Oak Creek in Brown County, Ohio, Shinkle came to Cincinnati at the age of eighteen, with seventy-five cents in his pocket. There he found work as a cook on a flatboat, which took him down the river to New Orleans. Again and again he made the long trip, and each time he had to find his way back to Cincinnati. When he came back by land, he rode with a party, for there were long stretches of wilderness in Louisiana, Mississippi, and Tennessee still infested with Indians, and with thieves! Cash was not abundant, but from each trip the youth managed to save a little. Finally he bought a flatboat of his own. He loaded lumber in eastern Kentucky and took it down to New Orleans, where it found a ready market. Then he bought a steamboat, and ran it up the river from Cincinnati; this business was profitable, and he continued it until 1846. And then a new venture occurred to him. Steamboats needed coal! Wharf space in Cincinnati was too valuable to be used for loading fuel. Across the river was Covington, a new city, growing rapidly. There he set up as a coal dealer. In the fifteen years before the Civil War, he made a fortune by supplying coal to steamboats. By 1861 he had three large boats of his own.

Through his business interests and his earlier experiences Shinkle recognized the paramount importance of transportation. He invested money in the building of turnpikes. The efforts to build a bridge to Cincinnati touched his imagination. Finally he bought stock in the bridge company, and was elected a director.

The day Shinkle was elected to the board was a turning point in the history of the project. Immediately things began to hum under the new spirit of directed energy. A committee was appointed to draw up by-laws for the company. Within three weeks the Kentucky legislature passed an amendment to the charter authorizing an increase of the capital stock to $700,000. The city of Covington was permitted to subscribe $100,000 of this amount, to be paid for by the issue of municipal bonds. A month later the Ohio legislature passed a similar amendment. It authorized the same increase in the capital stock, and, in addition, allowed the reduction of the prescribed span length from 1,400 feet to any span not less than 1,000 feet.

No time was lost in going after new capital. Two days after Shinkle's election the board appointed a committee to solicit subscriptions. Greenwood and Shinkle were appointed to estimate the prospective traffic for the bridge. For one week they stationed men to count the wagons, drays, carriages, buggies, foot passengers, cattle, sheep, and hogs that crossed the river on the Covington and Cincinnati ferries, as well as on the Newport ferry. At the end of the week the committee made its report.

7

The president of the company, Richard Ransom, wrote to Roebling, inviting him to come to Cincinnati at the company's expense. The bridgebuilder arrived in Cincinnati in April, after an absence of almost ten years. He met with a committee of the board, and presented a new report on the *Practicability and Probable Cost of a Suspension Bridge across the Ohio River*. One thousand copies were printed, together with "an appeal to the people for material aid for the bridge," and were distributed in Cincinnati and Covington.

In July the board received Roebling's proposition. While negotiations upon it were under way, the company called for bids on furnishing the stone and timber required for the foundations of the towers. At the same time a committee was appointed to buy the right of way required for the bridge on the Ohio side.

Then the company called on the stock subscribers, and money began to come in. During August further arrangements were made for obtaining stone, timber, sand, and cement for construction work.

Negotiations with Roebling were going along slowly. His reputation was now established, and he no longer needed to give away his services and energy for inadequate compensation. Since the failure of the Wheeling bridge in 1854 Ellet was eliminated from the picture, and Roebling no longer had to meet his rival's terms and methods of competition in order to get a start.

Roebling was now working on a bridge at Waterloo, Iowa. As he had "other work to do at Chicago and Niagara Falls," he suggested coming to Cincinnati about September 1. The bridge sponsors were by this time too enthusiastic to take a chance on any further delay. President Ransom himself went to Iowa to see Roebling and complete the arrangements. This was much quicker than correspondence. On August 18, 1856, the contract was entered into between the engineer and the company. After more than ten years the company had at last secured the services of the man who could build the bridge they wanted. At the same time Roebling secured the opportunity for his next great achievement, a bridge larger and more difficult than any he or any other engineer had thus far tackled, a span that would be the longest in the world.

After four decades of talking and hoping, human heads and hearts and hands were ready to bring the dream to reality. Under the terms of the contract both Roebling and the company expected that the bridge would be completed in three years, or four at most. Neither of the parties had any inkling of the long siege of ordeals and tribulations ahead of them.

8

At last the stage was set to begin work on the great structure to span the Ohio. At the end of August, 1856, Roebling was in Cincinnati ready to start construction.

All of the conditions were favorable. Stock in the amount of $314,000 had been subscribed for. A long, dry summer had left the river lower than it had been for seventeen years. But so quickly had success come to the reorganized bridge company that it was caught unprepared to take advantage of the opportunity. There were no materials on hand; and, much to Roebling's disgust, no machinery had been provided and no effective pumps were available.

At the meeting of the board on September 2 more business had to be attended to than ever before. There was much to be done and no time to be lost. President Ransom and Mr. Roebling were delegated to visit Dayton and other places to examine stone that might be used for the construction. Messrs. Greenwood and Roebling were authorized to buy engines, pumps, derricks, and hoisting equipment. Messrs. Shinkle and Roebling were directed to arrange for the purchase of the stone at Six Mile Lock on the Licking River. And finally, the president was authorized to draw on the treasurer for the first expenditures, $2,500, to pay off the men who were already at work excavating for the tower foundations.

The contract for the foundations was promptly awarded. Engineering examination of the Covington shore indicated that the builders would have little difficulty in finding a good foundation there. Below the visible river bank of clay a heavy bed of coarse sand mixed with gravel was found, and below this was the bedrock of blue limestone.

The excavation on the Covington side proceeded easily and quickly; within less than three weeks it was finished. The first timbers for the foundation grillage were laid down on the bottom of the excavation; heavy oak logs were hewed square and laid so as to form a solid platform. A second layer was then placed crosswise over the first, and the two were fastened together with iron bolts. On these, additional timbers were laid and bolted to make thirteen courses in all. The spaces between the timbers were filled with concrete.

On the Cincinnati side a quite different situation was encountered. The original river-bottom clay covering the bank had been largely washed away and had been replaced by loose dirt fill from cellar excavations in the city. A strong oak sheet piling was first driven into the mud along the river front. Together with the earth embankment, this proved an effective cofferdam. In the low stage of the river the excavation gave no trouble at all, and in a few days the workmen had dug down to river level.

But a new problem arose. By the great depth and extent of the excavation, most of the wells along the rising ground back of Cincinnati were drained dry, the ground water finding its way into the foundation pit. "We drained their supplies," Roebling explained, "and had to pump them out; and this copious influx came from a quarter totally unexpected."

By mid-September the workmen on the Cincinnati side had dug as deep as low-water mark, which was substantially below the river level even in its reduced stage. Water from the river, seeping through the bank, began to flow into the excavation, seriously interfering with the work. Engineer Gower, one of Roebling's assistants, placed an engine in position to operate a pump, which he expected would remedy the problem. Meanwhile contracts were made for limestone, oak and pine timber, and another pump.

At the end of September the flow of water into the excavation for the Cincinnati tower still remained the chief difficulty. Pumps worked by steam engines were kept going constantly; but though they seemed to throw out tremendous quantities of water, they were inadequate. Water kept flowing in as fast as it could be pumped out. Roebling saw that the want of powerful pumps was a serious drawback, threatening to defeat the enterprise at the very start.

Optimistic reports were given to the newspapers, for the public was curious; but it was hard to hide the fact that the work was almost at a standstill.

9

Two months went by. The constant flow of water into the excavation prevented further progress. Confronted with this apparently insuperable difficulty, the contractors were almost ready to give up in despair. The city was full of steamboat pumps, but they were so small and of such an unsuitable type of construction that they could not cope with the conditions on this job.

"Raising clean water is an easy process," Roebling said, "but to raise large masses of soft mud and sand is not so easy. After experimenting and losing a few precious weeks in an endeavor to work some patent rotary pumps, which utterly failed, we came very near to a complete halt. There was no time to get proper steam pumps of large dimensions constructed at any of the shops in Cincinnati, nor could we expect them in time from the East; every day's loss was irreparable—and so we were thrown back upon our own resources."

In November the fall rains were at last beginning, and the river was gradually creeping back up the shores. Every day, indeed every hour, counted now as Roebling cast about for some way to meet the situation. Finally he determined to design and build the necessary pumps himself.

It was a critical situation, a race against the rising river. There was no time to make elaborate drawings, patterns, and castings; iron or other metal construction was out of the question, as the necessary machining would consume weeks of precious time.

Roebling sent for some large pine planks, three inches thick. He called the carpenters together and made a quick sketch to show them what he wanted. Enthusiastically they set to work under his direction, and within forty-eight hours he had constructed four large square pumps, ingeniously and simply fashioned of the materials at hand. Before another day had passed, he had them set up in pairs, one pair to work at a time, the other pair to be held in reserve. Now, where could he obtain the power to work his pumps? Roebling had the answer ready. Right in front of them was a source of power, but nobody else had thought of it. Close to the landing lay the *Champion No. 1*, a powerful tugboat owned by Amos Shinkle. Roebling had the tug moored alongside, and connected the new pumps by chains to one of the engines. The novel solution was a success. The wooden pumps worked, and never once failed. They lifted and threw out mud and sand as effectively as pure water, and they discharged forty gallons at each lift. Before the day was done, tons of water and soft mud had been ejected from the excavation by the home-made pumps. No difficulty was now experienced from the flow of water, and the excavation proceeded rapidly, day and night, until all the clay and sand were removed. The bridgebuilder's resourcefulness had saved the day.

10

Then, much to everyone's relief, the rains stopped. The river had risen four feet in the first week of November; now it began to recede.

With the surface layer of clay and sand removed, a hard layer of coarse sand and gravel was now exposed in the Cincinnati excavation. Soundings made by driving down long iron bars showed that the limestone shale was twelve feet farther down. A decision had to be made whether to go down to the rock, or to drive piling, or to lay down a solid timber platform on the gravel bottom.

Roebling made the decision. "A compact bed of gravel," he explained, "if left undisturbed, and protected against undermining and washing, stands next to a solid rock foundation, provided unequal settling is guarded against." He therefore decided "to stop at the gravel, and build a solid timber foundation up to low water mark, thence to commence the masonry." At less risk and less cost, he declared, this kind of foundation was much more certain to succeed than any other.

There the enormous hole gaped, waiting for the foundation to be built. It was not empty for long. Within a few days the sand and gravel bottom was hidden by a flooring of oak timbers, just as that on the Covington side had been. A crowd of spectators constantly surrounded the excavation. Eighty men were now continuously at work, trying to bring the Cincinnati construction up to schedule. The huge improvised pumps connected with the engines of Shinkle's *Champion No. 1* kept the excavation clear of water, and the work progressed rapidly. By the end of November the last of the ten courses of timber was bolted in place, and the grillage foundation was grouted and leveled off, ready to receive the first course of masonry.

On both sides of the river the foundation platforms had been placed safely below low-water level, so that the timbers would be insured against any possibility of ever becoming dry. Their indefinite preservation was thus a certainty.

Up to this time the foundation contractors had been miraculously favored by the unprecedented duration of low water in the river. Navigation was interrupted. The ferries had stopped; the *Belle*, running to Covington, went aground on a shoal, and thereafter passengers had to be rowed across the river in light skiffs. Deckhands, cooks, and chambermaids who worked on the steamboats were in distress for want of work. But for the bridgebuilders to count upon the low water continuing would have been hazardous. The season had now arrived when one week's steady rain would result in a sudden rise of ten to twenty feet.

Materials had therefore been assembled and prepared to protect the work against high water. Cofferdams were quickly built, twenty feet high, around the top course of the foundation, well braced, and embanked all around. The planks were all jointed and calked, to form a watertight box surrounding each tower base.

Now a rise of twenty feet in the river could not interfere with the work. Sure enough, on the day after Thanksgiving the late fall rains began. In a few hours the river started to rise. The rains continued, the water kept rising, and in a few more days river traffic had picked up. Early in December the continued rise in the river began to slow up the work somewhat, but Roebling's foresight in providing the cofferdams was rewarded. The builders were in excellent spirits; they were making good progress in laying the masonry within the cofferdams, and they were confident that, within the next few weeks, they could bring the work above all danger from rising water.

The early and middle part of December proved to be mild and fair. Hundreds of citizens gathered daily to watch the work on the towers. By the time the building season came to a close, just before Christmas, the masonry had been laid to a sufficient height above the foundation timbers to take the pier work out of the reach of danger—safe against the elements until construction could be resumed in the spring.

The company now concluded new arrangements for handling the work during the coming year. The contractors agreed to furnish the stone, cut it, and lay it, for the erection of the two towers, and the company was to supply all the cement and sand needed for the masonry, as well as all the steam power and machinery necessary for hoisting and laying the stone. Work under this revised contract was to commence in the spring, as soon as the weather and the stage of the river would permit—the entire masonry work to be completed in two years, by December, 1858. The contract was to be subject to modification by "unavoidable calamities, such as epidemics and uncontrollable contingencies, such as the state of the weather or of the river." Thus the company and the contractors tried to peer into the future and anticipate contingencies that might affect the progress of the work. But they never suspected the calamity which was to come eight months later.

11

The winter of 1856–57 was one of terrible severity and suffering, and long remembered in Cincinnati. A week after New Year's Day the Ohio River froze over. Intense cold continued, week after week, without breaking. Boats of any sort found it impossible to reach the city. All commerce was cut off except the little carried on by the new railroad lines. In time coal could not be had at any price, and the food supply was dwindling. Great was the distress of the poor, with the shortage of food and with no fuel for their stoves in the freezing weather, and even those with money shared in the general suffering.

Then came a sudden change—rains, flood, and thaw. The river ice began to grind and shift, rocking uneasily. It started to move. On Thursday morning, February 5, the ice broke. Then followed terrible scenes which brought tragic losses to the population. Boats which had been frozen fast in the ice were splintered or torn apart. Some were forced down under the floes; others were raised high over the jammed ice, only to plunge crashing into the jagged mass. During this whole day of warm, gentle winds and springlike freshness the people of the river cities stood helplessly along the shores, watching the awful spectacle of destruction. Floods followed, only prolonging the disaster.

As winter passed into spring, the high water continued, from March to April, from April to May, from May to June. On May 19 Roebling arrived at Cincinnati to resume the construction, but wrote home: "Spring is no farther advanced here than at Trenton. The river is two feet over our caissons—can do nothing now."

Not until July did the stage of the river at Cincinnati fall low enough to permit work on the tower masonry to recommence. The remaining months of 1857 were given almost entirely to building up the great cable piers.

The stone used at the core of these massive towers came from far up the Ohio River. Two hundred miles above Cincinnati, between the Muskingum and the Scioto, great sandstone cliffs overlooked the river on either side. Roebling had early decided to use this Buena Vista sandstone. "I preferred for the lower portions of the masonry," he said, "the petroleum variety of the Buena Vista quarries, because of their greater strength and impregnation with petroleum, which latter condition promised a greater resistance to the action of the water."

The first twenty-five feet of the towers above the foundation would be constantly exposed to the action of the current. Roebling therefore decided to have this lower section of the piers faced with a good quality of limestone. Various stones were tried. A few boatloads of shell marble and other limestone from Ohio and Indiana quarries were examined, and finally the principal supply was taken from Dayton, Ohio. Dayton limestone had been used in Cincinnati since 1842, when the old St. Peter's Cathedral on Plum Street was built of it.

Above the first offsets the towers were wholly of sandstone. The faces of the blocks were left rough, just as they came out of the quarry. A heavy draft was cut around the margin of each stone before it was moved into position. Thus the faces of the towers had, as Roebling observed, "a massive look quite suitable to their function."

Never before had the people in that part of the country seen such enormous masses of stone raised and set at such great heights. It was a new problem, and one that exercised the skill of the engineers. Animal power could not handle it, so two large steam engines had been set up, one on each side of the river. The raising and setting of the stones was done by high-masted derricks, secured in their position by wire-rope guys.

12

In the meantime the company was running into more serious difficulties than high water or cold weather. Almost from the outset difficulty was encountered in collecting the pledged payments on the stock subscriptions. Finally, in January, 1857, the president of the company was directed to institute suits against delinquents.

It soon became apparent that additional stock sales would be necessary to meet current demands for cash. In March President Ransom packed his suitcases and set off for New York to do what he could. He returned from the East in April—empty-handed.

Spring had already come. The river might at any moment fall to a low stage, and the company should be ready to begin work again at a moment's notice. Money or credit was urgently needed.

The bridge company was not the only business faced with this trouble. Money was scarce in this new, raw West, where no one had yet had time to accumulate any capital reserves. The eight-year upswing to prosperity was approaching a climax. By 1857 the shortage of actual cash was pressing hard on the heels of investors. Scores of business concerns were in straits endeavoring to make collections.

People had faith in the enterprise. But it takes more than faith to build a bridge. Money had to be found somewhere to pay contractors and laborers for material and work. Bridge bonds had been issued by the city of Covington to pay for her subscription to the bridge but had never been sold. Now the bridge company was willing to take the bonds instead of money, and to try to sell them. President Ransom met the Covington City Council and completed the transaction. Once more he went to New York to see what could be done with the Covington bonds. Only worse discouragement than before awaited him. When he returned to Cincinnati to report to the board in mid-July, he had not been able to dispose of a single bond. Another call for payment on stock subscriptions was made, but with disheartening results. At that meeting the secretary of the board resigned. Five weeks later, while construction was still going on but with no funds on hand to pay for the work, the treasurer resigned.

13

The storm had broken! On the afternoon of August 24, 1857, news came by telegraph of a bank failure: the Ohio Life Insurance and Trust Company, with its office in New York and its main branch in Cincinnati, had closed its doors. The failure was totally unexpected; only two days before, the stock of the company had sold at ninety-eight. But things had now gone so far that nothing could stop a general collapse.

In a few brief weeks banks, insurance companies, and brokers, first in Cincinnati, then in New York and Philadelphia, and soon all over the world, were closing their doors. Like the panic just twenty years before, the great panic of 1857 swept across the country with hurricane violence. East and west, north and south, railroads went into bankruptcy, factories were shut down, and nothing but despondent gloom remained. The prosperity from 1850 to 1856 had been just an illusion, it seemed, and now that illusion had vanished.

Facing personal financial ruin, the master builder remained cool and collected in the emergency. As soon as the crash came, Roebling started sending detailed instructions to his son Washington and his superintendent Swan at Trenton, cautioning them to restrict credit to doubtful customers and to watch the banks in which the funds of the business were held. On September 28 he wrote from Pittsburgh: "The New York Banks are good yet, but I fear there is more trouble ahead. We must make safe all we can, while it is time. If you have not made a special deposit yet at the Trenton Bank, do not wait any longer—that Bank is not safe in times like these. . . . The old Pittsburgh Bank has a hard run but keeps up well, paying specie. . . . Before confidence is restored there will be more failures, particularly in Philadelphia and Pennsylvania. In the West it is only commencing now. . . . All checks we get now must be drawn out in gold. . . . I am going to Cincinnati tomorrow."

As the financial situation grew more critical, Roebling wrote to his superintendent at Trenton: "Unless the replies of the Pennsylvania Canal Company and from Talcott are quite satisfactory, we must discharge a portion of our hands at once and reduce them to a *winter force*, only keep our good old hands, also reduce their wages for the winter season. We must prepare for *the worst*. There will be *no* demand for Wire Rope next year, at least such is my present opinion. . . . The number of failures is increasing every day. Where this will end, nobody knows. Only those who are without any debts can weather this storm."

In order to salvage some of his funds tied up in a Philadelphia bank, Roebling wrote to Swan on October 11, instructing him to stock up with provisions to be purchased in Philadelphia: "You should go to Pha. and purchase with Pha. money a lot of provisions, say $300 worth—flour, bacon, ham, coffee, sugar, molasses, rice, &c. for our own use, your household, and all the hands in our employment. Let all the hands understand that they will have to take one-half their wages in provisions & groceries &

coal. You can convert the south part of the cellar of the new house into a store room. . . . This will not only be a saving all around, but also a great service to our hands. We may be able by this means to employ our old good hands all winter & make rope ahead, and also wire. Mr. Riedel & Washington can manage this store business, & keep a book of entries in the store room, to be squared up every week. Say to Mrs. Roebling that she must not buy any more provisions and groceries in Trenton —all must come from Pha. at wholesale prices, bought for bills of the Pha. bank. Perhaps it will be well to order an additional boatload of Stove Coal from the Schuylkill region (to be paid with Pha. money) for the use of the hands and our own. We must not touch our Specie funds in Trenton and N. Y. The Pha. funds will take us through the winter by pursuing the above Store Keeping plan."

Two days later Roebling wrote from Pittsburgh: "Tonight we get the news that 8 or 10 N. Y. Banks have suspended. Things begin to look worse and worse." And the very next day: "N. Y. Banks all suspended. This is the last news."

A week later he instructed Swan to withdraw the funds held in a special deposit in a New York bank: "I think it would be advisable to withdraw the Special Deposit from the L. M. Bank of N. Y. and put it away in my Safe at Trenton & let Washington sleep in that room. . . . You might take Ferdy along & a good bag to bring the gold home. Take Fritz Meissner along with you to the Bk. to assist in counting. Should the times continue so severe, then we shall want *all our funds*. In that case I shall *refuse* to make any more rope for Morris Canal, next Spring. We will have to look out for Penna. Canal & Delaware & Hudson. They will be good, but slow in paying. A good deal will depend upon consumption of Coal next Winter."

As the situation rapidly grew more desperate, with many thousands of men thrown out of employment, inflammatory meetings and rioting began to take place in the principal cities. On November 1 Roebling wrote to Trenton: "Those inflammatory meetings, lately held in Pha., with such exclamations: *Fight or Bread*—indicate what may be expected next winter in these large cities. . . . We must economise as much as possible, but keep a *small force* at work, if it can be done at all."

14

For seven months of winter and spring work on the Ohio bridge was suspended. The partly completed towers on either bank were mute witnesses of a resolve to build a great span. While prosperity might disappear overnight, the stone of the great bridge towers was lasting stuff. Through frost or summer heat, through ice or flood, through prosperity or panic, those were towers which would stand for centuries.

Like the stone of the bridge towers the spirit of the builders was of unyielding stuff. One misfortune after another barred their progress and challenged their strength, but they were not men to give up without a fight. Banks might fail and railroads go into receivership; New York investors and brokers might refuse to take bridge stock and city bonds; subscribers might fail by the carload to pay the calls on their pledged subscriptions; but as long as there was a fighting chance, the bridge company was resolved to carry the work forward.

At the beginning of 1858 the company had only a desperately small cash balance on hand. The last two calls for payment on stock had been made in the very teeth of the panic. The president and directors tried threats, persuasion, and negotiation—all to no avail.

In desperation the company voted to dispose of the Covington City Bonds for whatever they might yield. A few of the bonds were accepted by creditors who saw no other way of getting early payment. But more than half, fortunately, were disposed of to contractors in exchange for promises of materials or labor in the future, as might be required.

Because of high water overtopping the cofferdam, the workmen were unable to reach the base of the towers until early in the summer. About July 1 the machinery was moved into place, and operations were started. Since the company's resources were limited, the contractors agreed to confine work to one side of the river at a time.

Toward the end of November the company suffered a severe loss in the sudden death of President Ransom. He had carried the burden and the strain for almost three years.

Early in 1859 the Covington and Lexington Railroad was

sold, a failure; and the city of Covington, which had helped to finance it, defaulted on the interest payment on the bridge bonds. The dwindling assets and credit of the company were thus wiped out.

With both cash and credit gone, the project was at a standstill. The year 1859 had come and gone, and nothing had been done on the bridge.

Early in 1860 the directors drew up their annual report. After recounting the disappointments and failures of the preceding fourteen months, they recorded their continuing confidence in the project for an Ohio bridge as basically sound, necessary, and wise. With courageous optimism, they still held hopes for the future. It was more than a year since the last stone had been laid in place on the towers and the machinery had been taken away; but the next spring, perhaps, would see another turn of events and work might be started once more.

The spring passed into summer, and the directors viewed the river front with discouragement. Two years ago it had been the scene of hopeful activity. Now the bridge was so far forgotten that a Covington builder, Mr. Beatten, was preparing to put up a building in the middle of the proposed entrance to the bridge. Still clinging to hope, the directors went to court and prevented the obstruction.

As summer ripened into fall, a greater tragedy caught the bridge company in its grip—no mere business depression, no financial panic, but a mightier crisis of far-reaching sweep and proportions. The nation itself was caught in the throes of a cataclysm; it was being torn asunder.

For almost half a century the two great sections, North and South, had been growing more self-conscious, and drifting further apart. With differences in viewpoint growing into belligerent defiance, the ties that held them together were becoming dangerously strained. In a few brief weeks at the end of 1860 the last bond between them was suddenly broken—the bond of allegiance to one government and one flag. If Lincoln were elected, the Southern states announced, they would secede from the Union. In November, 1860, Lincoln was elected. Instantly a wild spirit of patriotic defiance swept across the land where cotton was king. The rest of the country looked on in bewildered amazement, paralyzed by the imminent catastrophe.

With the future so mysterious, so unpredictable, there was not a chance in the world for any company to raise new funds. The board of directors met to survey the situation and to do what they could in the crisis. The contractors were demanding to be paid. The accounts were long overdue, but there was no money in the safe. The only alternative to bankruptcy and foreclosure was to mortgage the project. With heavy hearts, the directors agreed to recommend the issue of bonds. On December 20 the stockholders met to hear the news. John W. Finnell, who had succeeded Ransom as president, told them how the contractors were "importuning, nay demanding, the bonds in conformity with the contracts." The stockholders were as helpless to deal with the catastrophe as the directors. They adjourned to celebrate the holidays as best they might.

When the stockholders met again, two days after Christmas, they had worked out a compromise plan for issuing preferred stock instead of bonds. This solution had been suggested by Roebling. The proposed remedy was a drastic one, but no other plan even came near to protecting the property and rights of the company. The bridge had to be saved in order that it might some day be finished. Prices had risen since 1858, and it was now apparent that the bridge could not be completed for less than $1,000,000. A preferred-stock issue for half of that amount was approved.

A number of economy moves were adopted. The treasurer was to receive no compensation, since his duties now were "nominal." The secretary's salary was reduced to $100 per year. The offices of the company were to be given up, and the records were to be moved "to a place of safety."

During this critical period the company was waiting—not for a financial depression to be shaken off, not for the attractions of their project to interest investors; they were waiting to learn whether America was to be one nation or two, whether, when the bridge was built, there would be a gap separating the North and the South even wider than that which the bridge was designed to span.

The weeks of terrible suspense lengthened into months, the old year passed into the new. The two great sections of the nation, already divided, stood tensely expectant. Allegiance to a common flag was gone. Links between the sections were wanting. The mighty unfinished bridge piers stood facing each other from opposite shores of the Ohio, ready to bind North and South, yet not connected by a single strand.

15

It was a warm spring evening just before the middle of April. The hedges were clothed in fresh green, though the locust trees were still black and naked. Rattling over the telegraph wires came the fateful news: Southern troops had shelled a fort of the Federal Government! Three days later President Lincoln called for 75,000 volunteers, to serve three months in suppressing the rebellion.

Along the Ohio, fear and anxiety held sway. Kentucky, by location and by nature, was physically a Southern state; she lay below the Mason-Dixon line; her farmers grew a Southern crop —tobacco—with slave labor. Would she join the Confederacy? Yet her ties with Ohio and the North were strong. Perhaps they would prove too strong to be broken, after all. It was a critical hour.

The Confederacy had been crystallized by January, 1861. No more states seemed likely to secede. Months passed, and no move was made on either side. But when Lincoln called for troops, a new wave of Southern solidarity passed over Dixie. Virginia, North Carolina, and Tennessee withdrew from the Union. Kentucky was the child of old Virginia. Where would she go?

Less than a week after the firing on Fort Sumter, the citizens of Covington, Cincinnati, and Newport met in Covington to form a home guard for local protection. By the early part of May the shotgun company of home guards thus formed in Covington elected officers: four corporals, four sergeants, three lieutenants, and a captain—Amos Shinkle.

In the approaching election in Kentucky there was only one real issue, union or secession; and the president of the bridge company was the Union candidate for the state legislature. When the returns were made known, President Finnell had been elected.

With the president away at the state capital and Captain Shinkle busy with the Kenton County home guards, bridge activities were necessarily suspended. In the ensuing months almost all of the directors were occupied with work relating to wartime needs.

Before long the Government called on Shinkle for assistance. The coal dealer owned three boats, which he turned over to the army for the transportation of soldiers and supplies down the Ohio; and his coal office at the corner of Fourth and Scott streets was taken over for a recruiting station. With Captain Shinkle in charge of the work, embankments were thrown up on the hills overlooking the city of Covington. By October the earthworks were completed and ready to receive siege guns.

With the coming of winter the problem of caring for the needy became increasingly acute. Amos Shinkle called a public meeting for the relief of the poor. Under a committee composed largely of directors of the bridge company, a soup house was opened where free meals were given, and fuel, clothing, and food were distributed.

As the thoughts and energies of the builders were absorbed in the more pressing emergencies of the national crisis, the unfinished bridge was almost forgotten. Through 1861 and 1862 the project lay dormant. It was waiting for another dramatic turn of events to give it new life.

CHAPTER XIV

THE SOLDIER

WHEN the Civil War descended upon the nation, few families escaped its wounds and its tragedies, its sufferings and its upheavals. The Roeblings felt its impact and its tension, and shared in the heartache and the heroism of the struggle.

Upon completing the Allegheny bridge at Pittsburgh, in the autumn of 1860, Washington Roebling had returned to Trenton and had taken up his work in the mill.

During the next few months the national crisis was rapidly coming to a head. Lincoln was already feeling the weight of the impending tragedy. On February 21, while on the way to Washington for his inauguration, the President-elect stopped at Trenton and made a short speech. The Roeblings, father and son, heard the address, eloquent in its simple sincerity. Their earlier sympathies were strengthened; and from that day they were resolved, should the need arise, to support the President in his efforts to preserve the Union, even if civil war were the price that had to be paid. But each kept his thoughts to himself.

On April 12 came the startling news that Southern troops had shelled Fort Sumter, and three days later came President Lincoln's call for volunteers.

That evening an unspoken tension pervaded the Roebling household. Washington's mind was filled with thoughts of enlisting, but he hesitated to give voice to his half-formed resolve—he did not know how his parents would take it. But John Roebling was not the man to spare himself—nor even his eldest son, who was his right hand—in a cause in which he felt that great principles were involved. Suddenly, at the supper table, the father turned to the son and said to him, with characteristic harsh abruptness, "Don't you think you have stretched your legs under my mahogany about long enough?" According to the story that has come down in the family,

Washington was eating a hot potato at the moment, and thus was unable to make an immediate reply, even if he would. As soon as he could recover himself, he dropped the half-consumed potato, rose from the table without a word, and walked out of the house. The next morning he enlisted—never to return to his parental roof. For four years thereafter he neither saw his father nor communicated with him.

2

It was on April 16, 1861, the day after President Lincoln had issued his call for 75,000 volunteers, that Washington A. Roebling, then twenty-four, was enrolled as a private in Company A of the New Jersey State Militia. On the same day Company A, composed of the young men of Trenton, reported for duty in defense of the Union.

For two months young Roebling remained at Trenton with Company A, which was performing all the duties of a military garrison for defense, including the guarding of bridges, and was at the same time rendering important service in the work of arming and equipping seven regiments of troops for the field. But mere garrison duty in his home town did not satisfy the impatient youth. Looking for more active service, he secured a discharge from the New Jersey Militia and immediately enlisted again as a private in Company K, Ninth Regiment, of the New York State Militia. With that unit he soon saw action at the front, acquiring some artillery experience, and there he was not long in obtaining the promotions for which his natural abilities and his engineering training qualified him. In four months he was made a sergeant, and in another four months a second lieutenant.

In the meantime John Roebling was seeking some way to make his contribution. He saw young men bravely volunteering for the service, and he observed that no provision had been made by the state to arm and equip them. He at once proposed that a sum should be raised for the purpose by public subscription. He himself started the list with a liberal amount, and then he called on other men of means, soliciting and securing their subscriptions to the fund. So deeply interested was he that he traveled to the national capital and personally visited government officials, urging them to take decided, vigorous measures. It was criminal, he declared, to send young men to the battlefront ill-equipped and unprepared. He did not mention his own son.

The young soldier's letters from the front, addressed as usual to the family's faithful confidant and correspondent, Charles Swan, supply a fragmentary picture of experiences and impressions in camp and on the battlefield. In the summer of 1861 Private Roebling wrote from Camp Cameron at Washington, indicating his impatience to get into action: "Things progress very slowly here. Loafing in the camp seems to be the principal occupation. In the cool of the morning and evening we practice sabre exercise; probably we will know something about it in six months from now. . . . We would have had our horses before now and the whole lot was ready for us, when Major Sherman stepped in and picked out all the good ones for his battery. . . . We got a wagon and 2 horses yesterday from the Maine boys—that is something, at any rate. If they keep on this way, the war will last 10 years."

After a few months Company K became the Sixth New York Independent Battery of Horse Artillery. In September Corporal Roebling wrote from Poolesville, Maryland, near the Potomac River: "In addition to being commissary, I am now chief gunner and corporal of one of the new rifled cannon, which arrived last week. It is the place which I like best, because the gunner has all the sighting and pointing to do and can have the pleasure of popping over the rebels. . . . The river bank on the opposite side fairly swarms with Secessioners. They have thrown up heavy earthworks, mounted with cannon, at every place where there is a chance to ford in low water. . . . Skirmishes take place daily." His military experience was becoming more real.

In December the unit moved some forty miles southward to a new location along the Potomac. "The roads are very bad," wrote Sergeant Roebling, "the ground being mainly wet and swampy. . . . Our next day's march was to a swamp of ten miles length, the trees being principally cedar, pine and green holly full of red berries. That night we encamped within hearing distance of the guns of the rebel battery. . . . I have examined their battery pretty well with my glass. It is quite a heavy work, and protected by the high river bluffs behind. . . . They fire every day at the passing vessels of which they have hit none, besides sending a few shells across to our camp. . . . The only white woman I have seen around here is Mrs. *General* Sickles, who rides around in full uniform, attended by a bodyguard of cavalry. . . . Our slumbers were disturbed this morning about four o'clock by a terrific discharge from the rebel battery. Their shell burst half a mile to the right of us behind a blacksmith shop. . . . The rebels are miserable shots. They fired thirty times at a schooner and succeeded in putting only one shot through the mainsail."

That winter the Sixth New York Battery remained stationed at Budd's Ferry, some fifteen miles below Washington. Its duty was to protect shipping, going up and down the Potomac, from the Confederate batteries on the west side of the river. The camp was about a mile from the river, hidden by woods from the view of the enemy on the Virginia side. Some sixty-five years later, one of Washington Roebling's tent mates, George H. Brown, wrote of his reminiscences in a letter to Mrs. Roebling. "The tent," he recalled, "housed ten choice 'sports.' . . . Nearly everyone could, or thought he could, play some kind of musical instrument, but none of them could come within a mile of your late husband. *He could make a violin talk*."

In January Lieutenant Roebling wrote from Budd's Ferry: "The other night a party of us gave a musical soiree at the widow Mason's house, down on the river bank. The music consisted of singing, piano, guitar, and violin, your humble servant performing on the latter. . . . Most of the furniture is removed from the house on account of the range from the battery. We had scarcely finished supper when two shells fell in the yard alongside the house, fortunately without bursting. The rebels have mounted a very large mortar. . . . One of the gunboats here is covered with a sort of iron crating, intended probably to keep out fragments of bursting shell. . . . This is a splendid Sunday, clear and warm, reminding one of spring which will soon come. The bluebirds stay here all winter, as also the mocking birds. . . . I believe I have not thanked you yet for the cigars and do so now. They are the best I ever smoked. That jelly was also excellent; it came safe and sound."

About this time the young officer received word from home

that he could have the rank of major in the Tenth New Jersey Regiment of Infantry, Volunteers, which was then being formed under Colonel William A. Murphy, of Trenton. Happy in his artillery assignment and proud of his unit, Washington wrote to Swan: "I don't think I would accept the offer, because my present place suits me excellently well. This company is the best artillery company in this division. . . . The officers in our battery would be very sorry to lose me and are very much opposed to my leaving. I have forgotten almost all the little infantry drill I used to know; the artillery is so much different. . . . General Berry will send down two Whitworth guns in a short time for the purpose of dismounting the guns of the rebels' battery opposite, piece by piece. Our new captain, Bramshall, will have charge of them and he has promised me the superintendency of them. . . . I fired a few shots across the other day which hit very well. They replied without doing any damage. Roll-call, and I must close. Give my love to all."

On March 9, the day of the famous battle between the *Monitor* and the *Merrimac*, Lieutenant Roebling wrote from his camp at Budd's Ferry, describing the evacuation and burning of the Confederate forts and batteries across the river: "This Sunday has been the most eventful since we have been here. We were all sitting quietly in the sun, after dinner, when suddenly large quantities of smoke were visible on the other side. Immediately all available tree tops were manned by an eager crowd. . . . In a short time, all their camps, to the number of six or more, appeared one mass of smoke extending far up the Quantico. Presently flames began to shoot forth from their steamer *Page* and also from three schooners which she had captured. . . . By this time the flames had reached some of the magazines, resulting in some tremendous explosions. Some of their guns had been left in position, double shotted, and as the flames reached them they went off, the shots striking the bank on this side. Apparently no attempt had been made to remove the guns or any of their commissary or ordnance stores, the size and color of the flames indicating the burning of flour, forage and provisions; and even now, nine o'clock, the sound of exploding shells falls on the ear."

The following day, in the rain, the young officer crossed the river to make a reconnaissance and to examine the abandoned batteries. "I have just returned," he wrote to Swan, "from a visit to the sacred soil of Secession. . . . The battery is an immensely strong work, with plenty of bomb-proof shelters. . . . It must have been the fear of having their rear cut off which made them decamp so suddenly. . . . Their camps are very filthy concerns. . . . Most of their guns were burst and all but two magazines. . . . A large number of letters have been found; their uniform tenor is hard times, and harder times coming. . . . At Dumfries only women remain, many of them very pretty. . . . Several masked batteries were discovered in unlooked for places. On the whole we may congratulate ourselves that the rebels evacuated them; it would have been a bloody battle to take them."

The lower Potomac was now safe and the mission of the Sixth New York Battery at Budd's Ferry was terminated. On April 30 Lieutenant Roebling wrote: "General McClellan was here last night, stopping to see General Hooker. . . . Day after tomorrow we leave from here for certain; our destination being, most likely, the York River. . . . I expect we go in schooners towed by steamer. The guns and limbers will have to be taken apart, and lowered in the hold, while the horses will stand on deck. They are sea-going schooners with high bulwarks so that the horses can't jump overboard as they did in crossing the river at Ball's Bluff. . . . I wish the paymaster would come around; we haven't been paid since shortly after New Year's."

3

In the spring of 1862 Lieutenant Roebling was transferred to the staff of General McDowell and assigned to various engineering duties, notably planning and supervising the construction of suspension bridges for military use. He had to prepare a number of standard designs that could be used for unknown crossings; and he had to assemble the material and equipment. Without guiding precedents, this kind of bridge designing called for special skill and resourcefulness, since the conditions were quite different from those of peacetime construction.

The quantity of wire rope that might be required, he ordered from the family works at Trenton. As the shipments arrived at

Washington, he had the rope trucked across the river to Alexandria where he assembled it on the dock of the Manassas Depot, so that it would be ready for prompt transportation by rail or boat to any point where it might be needed.

This first military handling of the large spools of wire rope revealed deficiencies both in the wooden reels and in the army trucks. "In moving the ropes to Alexandria," he wrote to Swan, "the reels were pretty well smashed and I have a carpenter at work putting in slats; all slats having no backing under them are broken. The reels, on which the 400 feet lengths are, had better be wider and lower; they fall over on very slight provocation. The bottoms of all the army wagons used for hauling them were broken through."

At the same time he assembled at the depot all the necessary tools, blocks and tackle, a full stock of iron parts, and four hundred bridge bolts. The bolts and fittings he specified were made for him at the government shops in Alexandria.

The young bridgebuilder prepared complete drawings and detailed instructions to facilitate erection of the suspension bridges by officers and men who knew very little about bridge construction. He sent the plans and descriptions to General Meigs, who ordered five hundred copies printed. Lieutenant Roebling wrote home: "The written instructions are being printed very nicely. The drawings are being engraved in Philadelphia. I have received several proof sheets already for correction. The engraving is done very well—500 copies are to be struck off. I wish they were done; it would save me a heap of trouble in explaining to folks."

A few days later he went to Front Royal, Virginia, to erect a small suspension span across the Shenandoah, but the wire rope he had intended to use for it failed to arrive. Anxious about the rope, as he was charged with military responsibility for the material, he returned to Washington to see what had become of it. There he found that, during his absence, General Meigs had taken the rope and had sent it to Harpers Ferry to take the place of a ferry rope which had been ordered from Trenton. Everything was explained, and the young staff officer's anxiety was relieved.

During this brief visit to Washington, Lieutenant Roebling received orders to go at once to Fredericksburg to rebuild the road bridge across the Rappahannock, which had been swept away by a flood. Upon his arrival he at once prepared a plan for the proposed bridge. The length of the structure was over a thousand feet, divided into some fourteen short spans, so that he was able to get along with three wire ropes on each side to serve as the suspension cables. On account of the length of the bridge he did not have quite enough rope and had to order more from Trenton. Planking for the bridge floor was so scarce around Fredericksburg that he had to have it sent down from Alexandria.

While waiting for this material to arrive, the young engineer wrote to Swan: "I am looking for a horse now, if I can find a cheap one. . . . Tell Ferdie to send me a box of Bower's two-cent Plantation cigars, direct by Adams Express to Fredericksburg. . . . It is hot here."

In the meantime he was busy with the construction of the anchorages and the towers, to have everything ready for erecting the cables.

This was Washington Roebling's first suspension bridge—the first bridge independently designed and built by him. Time did not permit the spinning of wire cables, so he used prefabricated wire ropes instead. He was thrown completely on his own resources, acting as designer, computer, engineer, superintendent, foreman, and master mechanic. He had to break in green men, teach them to perform highly specialized operations such as connecting and anchoring wire ropes, and train them to work on the swinging cables.

He found some soldiers who could do rough carpentry work. Two of the old piers were gone, and he had to replace these with timber trestle construction. Then, on each pier, he erected a tower of bolted timber framing. On either shore he built an improvised anchorage of concrete and stone. He was then ready to erect the cables. With one end of a wire rope secured to the anchorage, the reel was loaded on a flatboat and towed across the river, allowing the rope, as it was paid out, to lie on the bottom of the stream; the other end was then secured to the anchorage on the opposite shore. Men climbed to the tops of the towers and, by means of attached manila ropes, lifted the cable to the iron saddles that had been provided on top of the timber frames. Each of the wire ropes was erected in the same manner, three on each side of the bridge. The anchored ends were adjusted until all six ropes had identical sags in each of the fourteen spans.

With Lieutenant Roebling setting the example, the amateur riggers then went out along the cables in suspended slings, lashing the ropes together and attaching the suspenders, which were also made of wire rope. To the dangling ends of the suspenders timber crossbeams were secured. Upon these crossbeams the floor planking was laid, forming a continuous wagon roadway from shore to shore. Lighter wire ropes were strung to form the railings.

Although the bridge was not built under fire, there was no time to be lost. The soldiers enjoyed the novelty and excitement of the work. They had to do some swimming and diving in the course of operations, as well as some climbing and tightrope walking. There were few mishaps—an occasional tumble into the Rappahannock was all in the day's fun. Swinging out on the cables or walking the narrow beams held no fears for men who had been dodging shells and bullets.

After the wire ropes arrived, it was only a matter of a week or two before the thousand-foot bridge was finished.

4

During the summer of 1862 Lieutenant Roebling served on the staff of General John Pope and took part in the campaign which ended in the Second Battle of Bull Run. He was also in the fighting at Antietam and South Mountain.

In October he was building another suspension bridge, this one over the Shenandoah River at Harpers Ferry. He was handicapped by the lack of mechanics and skilled workers. "The only men I can obtain," he wrote, "are prisoners, and a very lazy set they are." He asked the Trenton office to send David Rhule, who had worked as foreman on a number of the elder Roebling's bridges, commencing with the Monongahela bridge in 1845; but Rhule failed to show up.

"Hurry up the strands," Washington impatiently wrote to Swan. "It takes ten days to reach me." A fortnight later he wrote that the announced shipment had not arrived, owing to some carelessness on the part of the Baltimore and Ohio Railroad.

He made a hurried trip to Baltimore to look after his bridge timber, which had also failed to arrive. There he found that the army clerks had made some "horrible" blunders in transmitting his order. The floor planking, which he had ordered 2 inches by 12 inches, they had carelessly changed to 12 inches by 12 inches. Instead of one carload, five cars were standing loaded with these big logs for planking the bridge. The loss fell on the Government.

The spans had to be longer than those at Fredericksburg. Some method of stiffening was required, since the bridge was to carry cavalry and possibly artillery. To provide the structure with framed trusses would involve too much delay. The bridge-builder therefore decided to install a system of inclined rope stays to stiffen the spans. In addition he braced the structure against wind and lateral sway by adding rope stays in the horizontal plane under the floor. He thereby secured a bridge of unusual security and rigidity for this type of quick, light, temporary construction.

The erection operations were rendered more difficult by the raw, wintry weather. Cold rains alternating with chilling winds were not conducive to rapid progress, especially when the work was done by unwilling prisoners. The weather was so bad that the men could scarcely work half the time.

David Rhule finally showed up—too late to be of much help, as the most difficult parts of the work had already been done; moreover, he promptly fell sick from the drinking water at the camp. Despite various delays and difficulties, the bridge was completed in December, less than two months after young Roebling had received the assignment. He wrote to Swan; "The bridge has turned out more solid and substantial than I at first anticipated; it is very stiff, even without a truss railing, and has been pretty severely tested by cavalry and by heavy winds."

The bridge at Harpers Ferry was later captured by the Confederate forces. To prevent its use by the pursuing Union Army, they completely destroyed the flooring system. While it was still in the hands of the enemy, Lieutenant Roebling was commissioned to reconnoiter it in order to ascertain the damage. When he found what was needed, he prepared a complete new flooring at a distance from the crossing; and when the Union forces reoccupied the position, he was able to restore the bridge without loss of time.

Once, on his way to headquarters with some important papers, Lieutenant Roebling reached Fredericksburg at a very late hour. Finding everything closed and dark, he put up for the night in an old, abandoned jail. As he started groping around, feeling his way here and there in the dark, he stumbled over something lying on the floor and his hand suddenly came into contact with a stone-cold face. Upon making this gruesome discovery, the tired young officer simply moved to the other side of the room and went to sleep on the floor. When he awoke in the morning, he found that the supposed corpse was a statue of George Washington's mother; preparations for the erection of the statue had been interrupted by the war, and meanwhile it had been stored in the abandoned jail.

5

In February, 1863, Lieutenant Roebling was settled at General Headquarters, Army of the Potomac. Two months later, on April 21, he wrote on the eve of battle—just before the crossing of the Rappahannock for the clash of the two armies: "This is probably the last letter you will receive for some little time. Tomorrow or the day after, or some time subsequent, if we aren't licked we will be over the river at last. . . . It will be a hard fight, much more severe than the previous one here, as everything depends on it. . . . The storming of that line of hills will cost men, as every point is more or less entrenched. If we are successful in the first battle we will be before Richmond in a week. About the being *in* it, I am not so certain, although I have a bet of ten dollars that we will be in it by the 15th. . . . The cavalry has also had bad luck and is not up to time owing to the rain. . . . It is absolutely necessary to move now before the time of so many men expires."

Hooker's army, outflanked by Lee at Chancellorsville, was driven back across the Rappahannock, and the young engineer-officer lost his bet.

After the battle of Chancellorsville it became Lieutenant Roebling's duty to ascend every morning in a captive balloon to reconnoiter the enemy positions and to watch, from the swaying basket, the movements of the Confederate Army. On one of these flights he was the first to discover that General Lee was moving northward toward Pennsylvania, and he promptly reported his observations to General Meade.

These balloon ascensions commanded wide interest, as they supplied the most advanced information of large movements and coming developments. One of those who followed the reports with keen interest was J. P. Morgan; sitting in his office in New York, he was informed by telegraph of Lieutenant Roebling's daily sessions in the air. This story is told in Winkler's book, *Morgan the Magnificent*.

The young engineer's discovery of the new direction of the enemy's movement was of tremendous significance. It started feverish activity at army headquarters and launched the large-scale movement of troops which culminated in the most memorable battle of the Civil War.

Ten days before the Battle of Gettysburg it was discovered that general headquarters possessed no adequate detailed map of Pennsylvania. Lieutenant Roebling remembered that his father had a topographic chart of the area, and he was immediately ordered to go to Trenton and bring it back. During his absence the movements of both armies had been so rapid that, on his return, he was unable to locate either of them. Taking the precious map, he mounted his horse and rode in search of the Union Army. He was in such danger of capture by Confederate cavalry that he and his horse slept one night in a cave. It was not until daybreak of the first day's battle that young Roebling was able to reach General Meade with the map.

John Roebling, working at Cincinnati at the time, learned of his son's flying visit to Trenton. He wrote anxiously to Swan: "Mrs. R. wrote to me that Washington had been home on a visit. I knew nothing of this, no body has informed me of it. I am anxious to hear from Washington; have you any news from him?"

6

On the second day of the Battle of Gettysburg the young staff officer was on Little Round Top with General Warren, when the latter discovered the beginning of Hood's furious attack, which threatened to outflank and defeat the whole Union Army. Lieutenant Roebling helped with his own hands to drag up the first cannons, which did so much to save the critical situation. In fact, history regards this achievement of General Warren and his staff and men as the turning point of the war. In Jesse Bowman Young's book, the *Battle of Gettysburg*, the story is told in the chapter on "The Safeguarding of Little Round Top": "Then General Warren and his staff officer [Lt. Roebling] with some stragglers near by helped to pull the guns into position, no room being found for the horses just at that point."

The holding of Little Round Top on that critical day brought distinction to General Warren. A writer in *Captains of the Civil War*, published by the Yale University Press in 1921, affirms that "Warren was not only a good commander of engineers, but a good all round general, as he showed by seizing on his own initiative 'Little Round Top,' without which the left flank could never have been held." General Warren, in turn, has recorded this tribute to the staff officer who helped him in that memorable battle: "Roebling was on my staff and I think performed *more able and brave service* than any one I knew."

On the third day of the Battle of Gettysburg Lieutenant Roebling was studying a battle map on a table at General Meade's headquarters when a cannon ball carried away two legs of the table without injuring him.

Lieutenant Roebling was present with General Warren's division on the occasion when they saved Sickles' corps from annihilation, and he was close by when General Sickles lost his leg. In the ensuing controversy over the matter, which was carried on for years, Colonel Roebling never hesitated to express his opinion that "Sickles had no business to be where he was. When General Sickles went to Meade's headquarters, he was ordered to go to his own men, and it was in getting back that he lost his leg."

On another occasion a detachment of Northern soldiers was unsuccessfully trying to send a cannon ball into a barn under cover of which a number of Southern sharpshooters were sniping at the Northerners. Lieutenant Roebling came along and saw the shots from the old cannon going wild. By raising his hand he ascertained the force and direction of the wind; he aimed the cannon accordingly and directed the gunners to try another shot, which they did—and the barn was instantly demolished. The sharpshooters, deprived of their shelter, were all killed, while not a man on the Union side lost his life.

In August the engineer officer was with the headquarters of the Army of the Potomac at Germantown, impatient at the lack of action: "We are hard at our old work again, sitting still with all our might and main, and doing nothing worth talking about. No doubt Gen. Halleck is conceiving some mighty plan by which to crush the Rebellion at one blow. This standing still is the worst thing that could happen for us by all odds. No amount of future reinforcements will compensate us for it, because when an enemy is once on a retreat, don't stop until he absolutely compels you to."

7

During the pause in the campaign following the Battle of Gettysburg, the young bridgebuilder was not idle. He was making studies and computations for converting his father's unfinished Kentucky River bridge into a military crossing. The tall masonry piers and the anchorages had been built by John Roebling for the proposed railroad suspension bridge before the project was discontinued in the panic year of 1857 for lack of finances. The idea of erecting an army span at this site appealed to the son. In his studies he sketched and calculated a military bridge of 1,350-foot span with a twelve-foot roadway, wide enough to pass a wagon and horsemen alongside of it. It would be the longest span in the world! He estimated the cost at $150,000, and he pointed out that the Government could sell it to the railroad company after the war, since the light temporary structure could be used in the erection of the proposed railroad bridge. When he presented his plan to his superior officers, he found there were almost insuperable obstacles in the way of governmental red tape; General Burnside could not order the bridge to be built without the sanction of the Secretary of War and then of the Council of War in Washington authorizing the expense. The bridge would not be available before the next spring campaign, which might be too late. Lieutenant Roebling had to abandon the plan.

In mid-August he wrote: "Gen. Warren secured his commis-

THE OHIO RIVER BRIDGE AT CINCINNATI

JOHN ROEBLING'S DREAM—ORIGINAL DESIGN OF THE BROOKLYN BRIDGE

sion as Maj. Gen. the other day, dating back to Chancellorsville; he will temporarily take command of the 2nd Corps; I shall go with him." And at the end of the month: "We are getting along fine, making the most of the dull times. . . . Conscripts are coming in gradually; and the extensive shooting is stopping desertion."

From August, 1863, until March, 1864, Captain Roebling was attached to the general staff of the Second Army Corps, under the command of Major General Gouverneur K. Warren. Mutual respect and loyalty cemented this friendship. Washington met Emily Warren, his future wife, while she was visiting her brother the General in camp.

In his book, *Meade's Headquarters*,* Colonel Theodore Lyman, of General Meade's staff, recorded an incident under date of November 30, 1863:

"Captain Roebling from General Warren's staff galloped up. He is the most immovable of men, but had at that moment rather a troubled air. He handed a scrap of paper. General Meade opened it and his face changed. 'My God!' he said, 'General Warren has half my army at his disposition.' Roebling shrugged his shoulders. The note was to the effect that General Warren had made a careful examination of the enemy's works, had altered his opinion of last evening and considered an assault hopeless."

Roebling accompanied General Warren when the latter was transferred to the command of the Fifth Army Corps; and in this assignment he served through the bloody campaign around Richmond—from the Battle of the Wilderness, through Spotsylvania, the Crater fight at Petersburg, and nine other battles to that of Hatcher's Run.

About this time John Roebling wrote home from Covington, unwillingly revealing his anxiety about Washington's safety and his suspicion that something had happened at the front and was being withheld from him: "Why doesn't anybody tell me?"

After the Battle of the Wilderness young Roebling assisted in transporting the wounded to the army hospitals at Washington: "We are now on the Potomac and the steamer *Daniel*

---

* Reprinted by permission of Little, Brown & Company and Atlantic Monthly Press.

Webster, stormbound; it being so stormy and dark that we cannot go on to Washington, so we are now at anchor. We have about 400 wounded men on board. Dr. Corson is the only doctor on board. Nine of our party are with him. We carried every man on board ourselves and it was the hardest work that I have done for many years. . . . The guerillas captured 300 wounded soldiers the night before as they were on the road to Belle Plain from Fredericksburg, killing all the wounded and stealing all the ambulance horses. . . . We had no single convenience, no stretchers, stores, medicines, food or bedding. I got some bandages, whiskey, etc., from a surgeon, and Dr. Corson and myself have been busy all the evening dressing and bathing the wounds of the poor fellows on board. Some of them have had nothing to eat for four days, nor have they had their wounds dressed since they were dressed on the battlefield. We managed to make a few bucketsful of gruel and gave our own crackers in to pass around among the wounded. They devoured it like wolves and seemed to be contented after that to lie in the filth and waste straw that we have them lying on. The stench of their wounds is horrible, and the whole thing disgusting in the extreme. . . . There was nobody hurt on General Warren's staff except the Asst. Adj. General, who was wounded in the arm, but since then, or rather today, they have had a terrible battle. We have heard the shooting all day. . . . I feel pretty tired and have to go around among the men before turning in, so I'll stop. I will come home as soon as we dispose of our men in Washington."

8

On May 26, 1864, Washington Roebling was promoted to the rank of major and aide-de-camp on General Warren's staff. A fortnight later, thanking Swan for a newspaper item from Trenton, John Roebling disclosed his pride in his soldier-son and wistfully hinted that he would like to hear from him: "I am surprised that Washington does not drop us a line. If you meet with any more such notices in the *Herald* about the 5th Corps, please send it to me."

Colonel Theodore Lyman has left us a graphic description of Major Roebling's appearance and manner:

"General Meade rode along the whole front of the new line and carefully examined it, accompanied by his staff and the taciturn Roebling.

"Roebling is a character, a major and aide-de-camp, and factotum to General Warren. He is a son of the German engineer Roebling, who built the celebrated suspension bridge over the Niagara River. He is a light-haired, blue-eyed man, with a countenance as if all the world were an empty show. He stoops a good deal, when riding has the stirrups so long that the tips of his toes can just touch them; and, as he wears no boots, the bottoms of his pantaloons are always torn and ragged. He goes poking about in the most dangerous places, looking for the position of the enemy, and always with an air of entire indifference. His conversation is curt and not garnished with polite turnings. 'What's that redoubt doing there?' cries General Meade. 'Don't know; didn't put it there,' replies the laconic one."

This pen picture gives us Washington Roebling during the war years—perhaps arousing smiles by the state of his apparel, as careless of his appearance as he was of danger, but with an unforgettable far-away look in his blue eyes "as if all the world were an empty show."

Just about this time Washington Roebling saw President Lincoln at close range, as he came to review the troops on the field of battle; and fifty-seven years later Roebling recorded the picture from his still vivid impressions:

"I was in the Civil War for four years and saw Lincoln on two occasions—the first time in May, 1861, when he spoke a few words of welcome from the rear portico of the White House to the newly arriving soldiers, one of whom I was, and secondly about April 1, 1864, when he came down to Culpeper County to review the army previous to the Battle of the Wilderness. I was at that time major and aide-de-camp to General Warren, commanding the 5th Corps, and joined in the cavalcade.

"The President was mounted on a hard-mouthed, fractious horse, and was evidently not a skilled horseman.

"Soon after the march began his stove-pipe hat fell off; next his pantaloons, which were not fastened on the bottom, slipped up to the knees, showing his white home-made drawers, secured below with some strings of white tape, which presently unravelled and slipped up also, revealing a long hairy leg.

"While we were inclined to smile, we were at the same time very much chagrined to see our poor President compelled to endure such unmerited and humiliating torture. After repairs were made the review continued, but was shortened on his account. I never saw him again and was in Covington, Kentucky, when I heard of his assassination."

9

In a letter early in September Washington proudly referred to his fiancée, who by this time had met the family at Trenton; and he mentioned the expected fight with General Jubal A. Early, the Confederate raider in the Shenandoah Valley: "In a few days now we expect Early back from the Valley, and then look out for a fight. We are building two large forts that will give him a warm reception should he attack us."

Five days after a bloody clash at Poplar Spring Church, Virginia, Major Roebling wrote home: "Not having seen my name in the papers you have doubtless come to the sensible conclusion that I came out all right in the last scrimmage. Otis Fisher [of Trenton, son of a prominent manufacturer of anvils] was hit in the forehead by a musket ball, and died subsequently in the hospital. He was out of his mind most of the time after he was hit. When troops behave badly and break, mounted officers are very much exposed in attempting to rally them, and that thing happens to the 9th Corps in almost every fight. Our own corps fought splendidly; a fort and line of entrenchments were carried by assault with very small loss, not more than 200 men. The 9th Corps then went on our left and when the enemy made their attack in the afternoon, the 9th broke and ran. Two brigades of our first division were sufficient then to drive the 'rebs' back and lick them handsomely. The enemy attacked the right of our line the next morning and were repulsed with ease. I think that will end the fighting for the present, and we are now building forts on the ground gained."

For his brave conduct during the campaign before Richmond, Major Roebling was breveted Lieutenant Colonel, U. S. Volunteers.

In December, 1864, in icy weather, he took part in a raid into enemy territory to destroy the Weldon Railroad, which he effectively accomplished. This was Washington Roebling's last duty as a soldier. On that front the fighting was over.

At this time he received word that his father needed him at Cincinnati. On January 1, 1865, the young officer obtained his honorable discharge from the army, and two months later he was given the full brevet rank of Colonel, U. S. Volunteers, "for gallant and meritorious service during the war."

Before leaving for Cincinnati, Colonel Roebling married Emily Warren, of Cold Spring, New York, the sister of his former commander, Major General Warren.

A prevision of fate seemingly guided the soldier in his recognition of the life mate he needed. Mrs. Washington Roebling was a woman of strong character, rare intellect, and loyal devotion—a staff to lean upon in crisis and adversity.

10

Through four years of war the soldier-engineer had risen in rank from private to colonel. He had ridden captive balloons above Chancellorsville, and had built millitary suspension bridges in record time at Fredericksburg and Harpers Ferry. He had personally dragged the first cannon up Little Round Top to halt Hood's famous charge at Gettysburg. He had received three brevets for courageous conduct. And he had been pronounced "ablest and bravest" by his commanding general.

In the tense years of the Civil War Washington Roebling had matured. At the campfire and on the battlefield, the man had found himself, and had added years of understanding and fortitude to his stature. He had been tested in the fires of war and not found wanting. Toughened in fiber and strengthened in soul, he was now ready for the next great ordeal that awaited him.

A generation later, when the war with Spain broke out and the Colonel was in his sixties, he expressed the wish to serve his country once more. He actually made an application to the authorities, but his services were declined on account of his age and his physical disabilities. In referring to the matter the Colonel remarked that he was sorry, as he would have liked to fight for his country in every war in which it was engaged during his lifetime.

Despite his reckless disregard of danger in carrying out his duties and assignments, Colonel Roebling came through the Civil War unwounded.

He was always reticent about his personal experiences in the war. Of his own part in actual battle, the letters he wrote from the front said little or nothing.

One day in 1920 Major Carl Lentz, chairman of the Essex County Republican Committee at Newark, was visiting Colonel Roebling at Trenton. In the course of the conversation, which was naturally one-sided, Major Lentz started reminiscing about his Civil War experiences. He related the story of a serious injury, which resulted in the loss of his arm, during a cavalry engagement near the end of the war.

"I have always wondered," he said, "how I got off the battlefield."

Whereupon Colonel Roebling quietly replied, "Why, Major, I carried you off."

He had never told anyone of this incident, in which he had saved the life of a wounded fellow officer under gunfire.

CHAPTER XV

THE OHIO IS SPANNED

SINCE December, 1858, work on the Ohio bridge had been halted. Lack of funds, the financial depression, the suspense of the impending national crisis, and finally the urgencies of the Civil War prevented the resumption of construction.

The chief campaign of 1861 had ended in the Northern defeat at the First Battle of Bull Run. But with the turn of the year and with a change in the command of the Union forces, there came a change in the tide of battle. The spring of 1862 saw the Confederates losing three of their great cities—Nashville, Memphis, and New Orleans—and a large part of the western territory.

A bold and desperate rally was needed to save the Confederate position in the West. Early in July Colonel John Morgan made a sudden dash from Tennessee into Kentucky. At Tompkinsville he surprised a small body of Pennsylvania cavalry, and then he advanced, with mounted troops and infantry, to Glasgow. The following day he issued a proclamation to the people of Kentucky, calling on them to join the Confederacy.

The same afternoon, Saturday, July 12, Adjutant General Finnell wired an account of Morgan's proceedings to Covington. Captain Shinkle at once called a meeting of the citizens at the armory that evening, in order to mobilize the home-guard companies and put Covington in a state of defense. When the meeting opened, Shinkle presented the latest additional information wired by Finnell. Volunteers were needed to go to Lexington at once, for Lexington was Morgan's objective. Who was ready to go? About 180 men stepped forward and gave their names. Half an hour after midnight the volunteers, headed by Captain Shinkle, left for Lexington on a special train. An hour and a half later, at two o'clock Sunday morning, a detachment of the Fifty-second Ohio Infantry, 240 strong, followed on another train, and at four o'clock in the morning 280 more left for Lexington.

The news of the raid threw Covington into great excitement. When Monday's newspapers appeared, bold-face type and a full column of news and rumors told of the raiders' northward progress. Morgan, with 1,500 men, was headed for Lexington, threatening the entire Bluegrass country.

Tuesday afternoon Morgan was at Midway, only fourteen miles from Lexington and fifteen miles from Frankfort. He found the telegraph cut, the bridges burned, and the railroad torn up. Thanks to the speed and energy with which troops had been collected for the defense of Lexington, there was no danger of his attacking that city.

Morgan changed his route and on Wednesday he captured Georgetown. Thursday evening, in his march toward the north, he reached Cynthiana, the last town of any size before Covington, and here a battle began. Covington remained in constant communication with Cynthiana. Then suddenly the wires went dead. Morgan's men had cut the line, and Covington was thrown into a panic. The battle lasted only about thirty minutes, and the raiders took Cynthiana and sacked it.

The following day, after the news of the fall of Cynthiana, Paris was evacuated by the small force of Union troops; and the same evening Covington was put under martial law.

The worst danger, however, was already over. With Lexington to the south still in Union hands, Morgan dared not move any farther north. Without Lexington his position was hopeless; and Lexington was too securely held. Morgan turned south, and within a few days a furious counterattack was on his trail. Before the end of the month the raid was over—Morgan escaped into Tennessee.

Morgan's first raid had failed. His success had been due to boldness and surprise; it ended as soon as the forces of resistance had become organized to fight.

But the military needs of the South were still acute. During August, 1862, another expedition was organized by the Confederate leaders, this time on a much greater scale. Braxton Bragg and Kirby Smith were given an army with which to invade Kentucky, reach the Ohio, seize Cincinnati, and cut the Baltimore and Ohio Railroad. With ten times as many men as

Morgan had led, Kirby Smith moved into Kentucky, and on September 1 captured Lexington.

Morgan's raid was child's play compared to this. Here was a real danger which required every effort of the community to avert impending destruction. Under General Lew Wallace Cincinnati was placed in a state of defense, with the slogan, "citizens for defense, soldiers for battle." All business was suspended, martial law was declared, and rich and poor alike helped and toiled desperately to fortify the Kentucky hills, assemble supplies, and prepare for a state of siege.

Meanwhile, at Governor Tod's call, soldiers from all over Ohio were being rushed to Cincinnati to share in the defense of the city. There was no time to lose; every hour counted. Now more than ever before, on both sides of the river, men realized the necessity for a better means of crossing from shore to shore. *If only they had a bridge!* Boats could not begin to handle the burden placed on them. Men by the thousands had to be taken to Kentucky, and without delay.

To meet the need, a pontoon bridge was quickly constructed from Cincinnati to Covington, the first bridge here across the Ohio. Across this unprotected, bobbing roadway of planks the picturesque "Zouaves" and "squirrel hunters" walked—careful to keep out of step, lest the bridge be destroyed—over to the Kentucky shore, to intercept the advancing enemy forces. Covington could not be allowed to fall to Kirby Smith; and there was no other way of moving enough soldiers and supplies across the river to protect it.

The threatened siege of Cincinnati was averted. Kirby Smith and his army did not approach the city. But the seriousness of the danger which had been so narrowly escaped at last woke the community to the impelling need for a bridge.

2

The intensity of the crisis in which the emergency pontoon construction had played so conspicuous a role made the population newly bridge-minded and resulted in a turn in the fortunes of the dormant project. After the September "siege" of Cincinnati events began to move faster than ever before. During

November subscriptions to the new preferred stock came flowing in to the office of the bridge company.

The change that occurred and the spirit that actuated it were feelingly noted by John Roebling: "When the whole power of the nation was absorbed in its struggle with that gigantic Southern rebellion, fresh endeavors were made by the friends of the work, in conjunction with some prominent capitalists on the Cincinnati side, to resuscitate their sleeping enterprise. The great exigencies of the war, by the movement of troops and materials across the river, made the want of a permanent bridge all the more felt. It is a fact, worthy of historical notice, that in the midst of a general national gloom and despondency, men could be found, with unshaken moral courage and implicit trust in the future political integrity of the nation, willing to risk their capital in the prosecution of an enterprise which usually will only meet support in times of profound peace and general prosperity."

From the very beginning of January, 1863, there was great activity around the office of the bridge company. Two boatloads of stone were ordered for delivery at the Covington tower; and bids were asked for supplying stone, timber, iron, brick, and cement.

The one change which the master builder now considered vital to the complete success of the project was the reduction of the required height of the span, in order to shorten the bridge approaches. He insisted that one hundred feet above low water was the maximum to which the company should agree. The desired amendment was promptly passed by Kentucky; and later, after spirited opposition, it was also passed by the Ohio legislature.

Roebling proceeded with a careful examination of the towers, on which no work had been done since the suspension of operations in 1858. Through the intervening four years of changing tides and seasons the masonry had been subjected to overtopping floods, and the upper courses had been tested by the impact of drift and of steamers, barges, and heavily loaded boats. The excellence of the masonry of both towers was proved by its unimpaired condition after all the exposure and buffeting it had received in its uncompleted state. Not a joint had washed

THE OHIO IS SPANNED

out; and the top course presented a smooth, even floor, in perfect condition to receive the new courses.

Next, the engines and machinery had to be examined, to see whether they were still fit for use after the lapse of years. Roebling found the old steam engines and hoisting machinery deteriorated and inadequate, and they had to be replaced or rebuilt. Shops had to be erected and stocked with new sets of tools. Finally, new contracts had to be made for the delivery of stone and cement.

The directors found that no satisfactory arrangement could be made for carrying on the actual construction by contract, as had been done in 1856. It was therefore decided to lay the masonry by hiring labor by the day, under the direct supervision of Roebling and his assistants.

The new plan for building the towers was a great improvement over the primitive system the contractors had used in 1856. Roebling himself had worked it out, and he later described it in his published report on the completion of the work. Parallel to the masonry of each pier, a hoisting tower was erected, a strong vertical framework supporting the hoisting sheave, tracks, and trucks. Each block of stone was rapidly raised to the top of the tower, placed on a wheeled truck, and then run either right or left to one of the two setting derricks.

With preparations for construction going on in the spring, a boom in building lots developed on the Kentucky side. "Never within the memory of the oldest inhabitant has there been such great activity in real estate in Covington as exists at the present time," the *Commercial* reported at the end of May. "As there is no longer any doubt about the early completion of the Covington and Cincinnati Bridge, we anticipate a still greater demand for building lots."

3

On the Cincinnati tower the laying of the masonry began about July 1. Within a week the derricks broke down; and just as the repairs were completed and everything was ready to start again, a new interruption caused a shut-down. Word came of another approaching raid by Morgan and his men. Roebling wrote: "I have today got through with the reconstruction &

erection of my derricks & would commence laying masonry, if martial law had not been declared on account of Morgan's raid into Indiana; he is about 50 miles from here with 5000 men."

During the early part of July Cincinnati had been hearing the usual rumors of Morgan's forces, but they were dismissed as mere fancies. But suddenly, a year to the day after his first appearance, the bold raider broke into the local headlines again. His men had seized a boat, the *Alice Dean*, at Brandenburg, Kentucky, and five thousand of them had crossed over into Indiana. There they had quickly taken Corydon, Salem, Vernon, and Versailles, and now they were marching rapidly toward Cincinnati.

Sunday evening all navigation was stopped on the Ohio; and on Monday morning, July 13, General Burnside proclaimed martial law. Once again alarm swept through the river cities. Never before had the enemy come so close. All business was suspended. Work on the bridge was brought to a complete halt while the military resources of both cities were mobilized.

Once again Morgan, though he had more men than before, had fewer than he needed. Cincinnati was the prize he wanted to take, but it was too big a prize. He dared not even attack it. On Tuesday and Wednesday Morgan's forces divided. They pillaged farms, stole horses, and tore up what they could of the Little Miami Railroad. Then they realized that their problem was how to escape.

Martial law came to an end in the river cities Thursday morning; and ten days later Morgan was finally captured in eastern Ohio, after a series of hairbreadth escapes.

Work on the bridge was quickly resumed, only to encounter new difficulties. No sooner was Morgan out of the way than labor troubles began for the company, and Roebling was faced with a strike. This too was eventually straightened out, and construction then proceeded more smoothly. The bridgebuilder told the story in a letter to his Trenton superintendent:

"My engagements here are such, that I *cannot* absent myself for some time to come. I shall be ready to lay down the anchorage on the Cin. side in about three weeks, and then do the same thing on the Cov. side & this will consume the whole of next fall. I do not care about travelling & spreading much time under ways. My present idea is to remain here until next

November when we quit work for the season, and then to go home to spend the winter. . . . I have some trouble here with my laborers who struck this morning for a raise. I have paid them so far 1.25 and do not feel inclined to raise their wages. If I do every other builder in the City will be compelled to follow suit. I want to get rid of the Cincinnati wharf rats at any rate, and engage Germans in their places. I think after the draft at N. Y. has been enforced at the point of the bayonet, there will be no more trouble. The Germans about here are mostly loyal, the Irish alone are disloyal. We are informed this morning that the last remnant of Morgan's horse-thieves with Morgan himself have been caught. Things in the Southwest & West are quite promising; the backbone of the rebellion has been effectually broken. You are doing well by keeping your own counsel with our men about public affairs. No democrat can be trusted; they are all disloyal & treacherous more or less."

Despite the long series of interruptions, the Cincinnati tower was doubled in height by the end of the year, when work was finally suspended because of freezing weather. On December 8, Roebling wrote home: "I shall not be able to leave here before the middle of this month. The weather is still good and we are still laying masonry. . . . I have not heard from Wash since the last campaign."

The bridgebuilder arrived home in time for the Christmas holidays. The new year was celebrated with a gala dinner, to which the faithful superintendent, Charles Swan, was invited. "Mrs. Roebling and myself request the pleasure of your company at dinner at my house on New Year's Day, the 1st Jany., at 2½ P.M." Washington was away, fighting at the front, but all the rest of the family were gathered at Trenton for a happy reunion—the last family gathering Mrs. Roebling was to share in.

4

In the spring of 1864 the bridge company was faced with a new financial problem. All prices were rising steadily. Paper currency was rapidly depreciating.

Roebling was now in the market for wire for the bridge cables. It was necessary to turn to England, because no manufacturer in America could supply the large tonnage and high quality required in the specified time. A contract was drawn up with Richard Johnson & Brothers, of Manchester. Every pound of wire in the bridge cables had to be brought three thousand miles and more, and had to be paid for in gold at a premium.

Prolonged high water in the spring delayed the resumption of work. During the summer the stone setting was continued on the main towers, the anchorages, and the abutments.

The anchorages were built with limestone bases, and finished with light-colored freestone brought from the quarries near Portsmouth. Embedded in these masses of masonry were the enormous cast-iron anchor plates, weighing over eleven tons each, to which the ends of the cables were secured through giant chains of wrought-iron links. This anchorage construction followed Roebling's patented design.

In a popular description of the work, E. F. Farrington, master carpenter of the bridge, graphically stressed the abundant stability of the anchorage design as an answer to dubious laymen: "Timid individuals have sometimes asked the question: 'How are the wires fastened?' 'When a heavy load goes on the bridge, won't the wires pull out and let the whole thing down?' We recommend such people to make a little calculation as to the resistance offered by a pile of rock and cement, of the size of these anchor piers, and the lifting power required to move them. We think it will give them confidence in the strength of the structure."

The wrought-iron eyebars and pins for the anchorages were forged in the bridge company's blacksmith shops, under Roebling's supervision.

Both towers had now reached a height of more than ninety feet, at which point the solid construction changed to provide an arch through which the bridge roadway would pass. Enclosed between two of the high timber hoisting frames were a series of wooden ladders by which workmen ascended to the top of the towers. In the evening, when there were few observers, adventurous visitors climbed up these ladders to test their nerves.

The scarcity of mechanics, riggers, stonecutters, masons, and stone setters seriously retarded the progress of the work. The needs of the army had taken away a large proportion of skilled workmen. Roebling found himself severely handicapped by the lack of an efficient force.

In April Roebling became much alarmed over unfavorable reports of his wife's health. He wrote to Swan: "Your remarks about Mrs. Roebling's condition and her own letter to me are not satisfactory, and leave me to infer, that the worst is not over. . . . My wishes and advice will be powerless & of no avail to her, because she never has paid the least attention to anything I have said or done on this subject of health. Who is her physician? Nobody has informed me. . . . In my own personal case during the last month I have again had ample proof of the efficacy of judicious water-treatment. I have been very sick & reduced, but have always been on my legs and about the river, taking new & fresh colds, as it were, every day. At home, where I could do as I please, it would have been a small matter —but not so here, exposed to the cold damp chilly river air from morning to night. I have had a violent cough for 4 weeks. Homeopathic medicines were without effect, I *had to return* to my own treatment, and have at last succeeded. . . . Give my love to all & read this to Mrs. Roebling."

In November, in the midst of the strain of the job and of the war, the bridgebuilder received the news of his wife's death. He came home for the funeral. The recognition and homage he had failed to pay in life he now paid in death. In the family Bible he inscribed this tribute:

"Of those angels in human form, who are blessing the Earth by their unselfish love and devotion, this dear departed wife was one—she never thought of herself, she only thought of her children. . . . And what a treasure of love she was towards her own children! No faults were ever discovered—she only knew forbearance, patience and kindness. My only regret is that such pure unselfishness was not sufficiently appreciated by myself."

5

In the early spring of 1865 Colonel Washington Roebling joined his father at Cincinnati, to assist him in carrying the Ohio bridge to completion. With the title of Assistant Chief

Engineer, he remained on the work for two years, adding his skill and energy to his father's, until the work was done.

The younger Roebling wrote few letters during these two years of strenuous activity. In his first letter he gave his first impressions of the great bridge: "The size and magnitude of this work far surpasses any expectations I had formed of it. It is the highest thing in this country; the towers are so high that a person's neck aches looking up at them. It will take me a week to get used to the dimensions of everything around here."

Writing from Cincinnati John Roebling recorded his relief at the approaching end of the national conflict: "The army news are glorious; the rebels are now in the last ditch." But joy over the ending of the war was quickly turned into grief. The news of Lincoln's death struck the cheers of victory from the lips of the people. On April 17 the master builder wrote: "The sad bereavement of the Nation by the foul assassination of Pres. Lincoln has produced a strong and deep feeling of bitterness against the Rebels about here. The Rebels have lost their best friend, and the North has to mourn over the fall of a great and *good* man."

For the next two years the father and son had their hands full, speeding the completion of the Covington-Cincinnati bridge.

The financial problem of increased costs was met by the company with an issue of $250,000 additional preferred stock. Within a week the entire amount of the new stock had been subscribed. The construction went steadily forward without accidents or hindrances, except the spring floods. Recurring freshets submerged the machinery and drove the men around the tower from their work.

By the end of September, 1865, the two massive towers were completed. Two hundred and thirty feet they rose above the river. From the top of either tower a panoramic view could be had of Cincinnati, Covington, Newport, and the surrounding country.

Shoreward from each tower stood the huge anchor piers, solid blocks of stone and concrete. In them lay embedded the anchor plates and chains. From the abutments at the anchor piers the floor of the bridge would spring to the passage through the towers.

There the great stone shafts stood. From center to center of towers was a distance of 1,057 feet, the span of the bridge. But the gap was yet unbridged. The two towers were still isolated, like strangers staring at each other.

6

With the end of the war laborers became more plentiful, and wages that had doubled with the scarcity of workers and with the devaluation of the paper currency now began to come back to normal. In May, 1865, Roebling wrote: "My opinion is that material and wages are bound to come down very rapidly. . . . I have commenced reducing wages of laboring men from $2.00 to $1.75 & after a while shall make a further reduction. Mechanics & laborers are very plenty & applying for work."

In September the first wire cables to span the gap—the footbridge ropes—arrived from the Roebling factory in Trenton. The 2½-inch wire rope came from the factory on large reels. With the free end of the rope passed over the Cincinnati tower, the rest was coiled on a flatboat and towed across the river to the Covington side, the rope being paid out and sunk to the bottom of the river. A powerful tackle, operated by a steam engine, was attached to the rope, and, at a time when there was little river traffic, the double rope was drawn, glistening and dripping, out of the water and up to its suspended position. The ends were then made fast to the anchorages.

From these two footbridge ropes oak crossbeams, three feet long, were hung at short intervals by means of smaller wire-rope suspenders. An oak flooring was then laid lengthwise across the beams. Thus the first bridge was built, a miniature suspension bridge, complete in almost every detail.

The footbridge was first crossed an hour before noon, Wednesday, October 4—the first time any man had walked on a span above the river from Cincinnati to Covington. It was put up for the convenience of the workmen in carrying on the construction, but soon hundreds of persons were begging for permission to cross. All were refused, as their crossing would interfere with the operations of the workmen. In still weather this light, swaying structure could be crossed in comparative comfort and safety; but on windy days steady nerves, good balance, quick hands and feet, and some knowledge of navigation were necessary.

The time had now come when thousands of persons, curious and fascinated, regularly came to watch the progress of the work. It was no longer simply a matter of hoisting stones and setting them on top of a tower, but of activity which stretched through space more than a thousand feet across the water. One old gentleman, looking at the twenty-seven-inch footbridge swinging high over the river, remarked, "After all the talk about a fine bridge, it is a very flimsy affair."

Some amusing stories of the footbridge were later told by the versatile master carpenter. Adventurers ambitious to cross the suspended footwalk offered many a bribe and excuse to pass the watchman in charge. Some favored ones were permitted to make the attempt; a few succeeded in crossing, while others soon retreated with their knees shaking. Farrington told of a "fair and fat Methodist preacher" who essayed the feat on his hands and knees; but after he had gone a short distance, the flesh proved weaker than the spirit, and he crawled back, trembling. Another brave fellow from Cincinnati eluded the vigilance of the watchmen by the aid of a heavy fog one morning. He clambered up the wooden ladders then standing at the towers, and started across. But he, too, quickly came to his knees in prayerful retreat.

Several ladies, at various times, ventured to climb the ladders to the top of the towers, and a few were courageous enough to attempt crossing the span. Farrington admitted that he had been coaxed into gallanting two young ladies over the footbridge and up to the top of the Covington tower one starry night. The damsels behaved bravely, he reported, and made no demonstrations of nervousness beyond a little hysterical hugging and squealing at the giddiest places.

Three times the temporary footbridge, in its exposed position, was badly broken by high winds. Once the suspended planking was ripped apart for a distance of twelve feet near the Cincinnati tower. The last time the whole stationary portion through the Covington tower was lifted by the wind and thrown into the river. But the wire ropes from which the footbridge was hung remained firm and undamaged.

As soon as the footbridge was completed, the work of stringing the permanent cables was begun. With physical connection established between the two towers high above the river, neither high water nor cold weather could stop the work. With the goal in sight every energy was directed toward pushing the construction in speedy and uninterrupted progress.

7

John Roebling, approaching his sixtieth year, was beginning to feel the strain. For two decades he had been drawing on his capital of strength and vitality in a superhuman effort to prove his ideas by heroic accomplishment. And for two decades he had carried the tension of this Ohio bridge project, suffering the alternations of hope and disappointment, setbacks and advances, heartbreaking interruptions and feverish toil. The outpouring of physical and nervous energy could not continue indefinitely. The man's grimly determined physical resistance was beginning to succumb to the rigors and exposures of the job. And the last few years had added their burden and anguish: the emotional tension of the Civil War, the anxiety for his son on the field of battle, and finally the shock of losing his wife.

His daughter Elvira had done her best to make their home at Trenton cheerful and inviting; but, much as he needed a rest, Roebling could not leave the work at this time to spend a few days with his children. In May, 1865, he wrote to Charles Swan: "Miss Elvira informs me that our grounds look beautiful, and that the new dining room will be the most cheerful & pleasant room in the house. I shall not be able to enjoy it much before this Bridge is completed. After that I expect I shall take all the comfort I can get at home & leave Bridge building to younger folks."

The pathos of a human life is packed into a few lines. Yes, the master builder was weary! The man of iron had spent himself in ambition and striving, in toil and strain. He was now ready to pass his few remaining years in the restful peace and comfort of home, and to "leave Bridge building to younger folks."

8

During the early part of January, 1866, the weather was mild and favorable to the prosecution of the work. Three-fourths of the wire for the cables had been received from England, and most of it had been prepared for erection—oiled, passed through straightening machines, and wound onto large wooden drums. Each drum, mounted on a shaft, held one ton of wire.

The next step was the "running out" of the wire across the span for the actual stringing of the cables. Much ingenious equipment was required, all devised and developed by John Roebling. The driving machinery was mounted on the Cincinnati anchorage.

After the first wire was taken across and adjusted to a precalculated position and tension, greater progress was made. The average number of wires taken across daily was about eighty. They were carried in loops on spiderlike wheels, which ran constantly back and forth, suspended from an endless traveling rope driven by an engine at one end.

At mid-span, forty feet above the footbridge, a wooden crossbridge, called a "cradle," rested on two wire ropes which were stretched over the towers. The engineers and bridgemen knew by personal experience that there were winds in those days which made this "cradle" rock like the nursery fiction in the treetop. A man was stationed at each end of the platform to aid by flag signals in regulating the successive wires as they were run over.

The cable builders kept steadily at work through heat and cold, stopping only at night or when strong winds sweeping up the river made it impossible to continue. Hundreds of fascinated spectators lined the river banks, watching the operation.

Toward the end of May a severe windstorm struck the cities of Covington and Cincinnati. The footbridge was broken near the Covington tower, but fortunately it took only a few hours to repair the damage.

By the beginning of June the sixth strand in both cables was completed and lowered to its place, and the stringing started on the seventh and last. The light spinning wheels carrying the wires across the river moved ceaselessly to and fro. They had crossed a thousand times, two thousand times, eight thousand times.

At 11:10 on the morning of June 23 the last wire was run over. The wheels had crossed the river 10,360 times, carrying that number of wires across the river from Cincinnati to Covington.

The wrapping of the cables was begun immediately. The whole mass of 5,180 wires in each cable was compressed into a true circle by means of powerful clamps, and then covered with a continuous spiral wrapping of galvanized iron wire. The ingeniously contrived wrapping machine, devised by Roebling, was operated by men on suspended traveling scaffolds. A continuous, circling thread of soft wire was spun from the mechanical organism as it progressed spirally along the cable. A small wire rope suspended above each cable was used as a hand rope by the engineers and the workmen as they walked to and fro on the curving cables.

Thus the powerful wire cables were completed—each 12½ inches in diameter—larger than any the world had previously seen.

As soon as the cables were finished, a large number of men went to work fastening the suspenders. Flat wrought-iron bands, cut and forged in the company's shop in Cincinnati, were heated in a portable forge on the footbridge and passed up, at a dark red heat, to the men on the staging, who fitted them and shrunk them around the cable. The suspenders were then attached to these iron bands and allowed to hang down in readiness to receive the floor beams. By the end of August, with all of these suspenders hanging in place, the structure began to look "very much like a bridge indeed."

The work of erecting the wrought-iron beams closely followed the hanging of the suspenders. Soon the three hundred suspended floorbeams were ready for the floor.

Six hundred thousand feet of kiln-dried oak and pine lumber were used in making the floor. Two tracks for streetcars were laid, and it was hoped "that at no distant day, these vehicles will run between the three cities of Covington, Newport, and Cincinnati." These streetcar rails were widened to fourteen inches by the addition of flat plates, so that the wheels of other vehicles could also run on iron. "Anyone who has seen a heavy loaded team toil up the hills from the ferry landing will appreciate the benefit to horseflesh from this arrangement," remarked the designer.

Most of the nuts on the bolts that held down the first course of oak flooring were put on by men lying at full length, face downward, and reaching by hand under the beam flanges; and in this awkward position it was not always possible for them to keep hold of their tools. Could the bed of the Ohio under the bridge be laid bare, the heart of many a dealer in old iron and junk would rejoice. Sledges, hammers, wrenches, punches, bars, pincers, chisels, saws, and tools of every description used in construction might there be found. One floorbeam was lost on the Covington side, but was recovered by divers. But a length of wire rope, which one of the workmen let slip and which caused the death of another by hitting him as it went down, still lies in the mud near the Cincinnati tower.

9

Following the procedure he had inaugurated and used in all of his earlier suspension bridges, and with such conspicuous and noteworthy success in the great Niagara span, Roebling provided a generous system of inclined stays in the construction of the Ohio bridge. Nearly one-half of the total weight of the roadway and the live load was carried by diagonal stays of wire rope, running straight from the tops of the towers to successive points along the floor; so that the main cables, themselves stiffened by this arrangement, really had to carry only about one-half of the total weight of the roadway and load. There were a total of one hundred of these "overfloor" stays, made of 2¼-inch wire rope. They were fastened to all the suspenders, where they crossed, by annealed wire lashings. The stays served effectively to strengthen the floor and to prevent or check vibration during the passage of heavy loads and in high winds. Other inclined wire-rope stays ran from the towers below the floor, and were attached to the bridge underneath, in a sufficient number and in a manner to prevent or check all lifting and lateral motion. All of the stays were attached by stirrups and nuts, so that they could be tightened for future adjustment.

Besides the support and stiffness derived from the cables and stays, two heavy wrought-iron trusses were provided, extending the whole length of the bridge, separating the carriageway from the sidewalks. These trusses were made of such depth and strength as to be capable of supporting a great weight, independently of the cables and stays.

The overfloor stays carried loads and forces acting on the floor of the bridge to the tops of the massive masonry towers. The three longest stays at each side of the span passed over the towers directly to the anchorages, to which they were fastened.

And then—something unique and amazingly significant—another set of diagonal stays was installed, running in the *opposite* direction—from the stonework of the towers to points of the *cables* some distance out from the towers—in both main span and side spans.

The full significance of this comprehensive system of diagonal stays was missed by the rest of the profession at the time. In his penetrating insight into the problem of stiffening a suspension bridge, not only against normal loads but also against destructive undulations producible by wind, Roebling's genius was manifested. He not only realized the critical necessity of such a stiffening, but he also devised an efficient and effective means for supplying the requisite bracing. And in all of this he was three-quarters of a century ahead of the profession; it was seventy-five years before modern bridgebuilders grasped the idea and recognized its supreme importance.

Someone has said: "There are not wanting, in any age, men with an insight sufficiently keen to pierce the fog that envelops their fellows." Far ahead of contemporary thought and practice, Roebling's structures were stable, while numerous spans built by others had been blown down, some of them being rebuilt and failing two or more times in succession. Even after his time other suspension spans failed. At length bridgebuilders learned part of the lesson Roebling had taught them: the importance of adequate stiffening trusses. But they failed to grasp the other part: the significance of diagonal stays. And in time a later generation of engineers, thinking they knew better, began to forget Roebling's first lesson, and started reducing their stiffening trusses to smaller and finally negligible proportions. Then, around 1940, came the rude awakening. The proud span across the Tacoma Narrows—the third longest span in the world—was destroyed by wind, and other new bridges of slender proportions developed disturbing and dangerous aerodynamic oscillations. Emergency measures were applied—*diagonal stays!* After seventy-five years the farsighted genius of the master builder received startling demonstration.

With full knowledge of the weakness of other bridges, but with a remarkable record of building stable suspension spans, from his first small aqueduct structures to his record-breaking Niagara railroad bridge, John Roebling had come to Cincinnati to build the greatest suspension span in the world. He was certain that he could build firm and enduring suspension bridges; and he was convinced that his deviations from previously established design were of fundamental importance in securing safety and stability. He recorded these convictions, and the reasoning that guided him, in his report upon the completion of the bridge. In simple language for laymen he wrote:

"The object of these remarks is to explain the difference between the French system and mine. The principal feature of the French plan is to suspend the floor to a number of small cables. By my plan only two large cables are suspended, one on each side. The Cincinnati Bridge is supported by two large cables, $12\frac{1}{3}$ in. diameter. The cables in my bridges are always assisted by stays, and this feature is entirely wanting in the French system. . . . In the divided system one cable may break after the other, because they cannot be expected to act in perfect unison. The correctness of this view was fully demonstrated by the catastrophe which overtook the Wheeling Bridge on the 17th of May, 1854. . . . The floor was torn by the gale into three sections. . . . All the cables but two broke loose from the anchorage, *one after another*. . . . For want of stays and of stiffness the floor was subjected to very considerable oscillations, and thus was broken up and destroyed *by its own momentum.*

"I do not expect professional men to credit the statement, without seeing with their own eyes, that the center of the Ohio Bridge [Cincinnati] is not deflected by heavy gales, neither horizontally nor vertically. This remarkable horizontal stability is in great measure due to the powerful bracing of the cables."

And on his innovation of diagonal stays, a lesson missed by succeeding generations of bridge engineers until very recent years, Roebling showed his intuitive grasp and straight thinking when he wrote: "Thus a network is formed that occupies the same inclined plane which coincides with that of the cables. . . . The office of these stays is twofold. They not only assist the cables powerfully in the support of the bridge but they also supply the most economical and most efficient means for stiffening the floor. Every stay constitutes the hypotenuse of a rectangular triangle, whose short sides are formed by the tower and by the floor. . . . This looks like a very simple proposition indeed, and is readily comprehended by sailors, who are accustomed to stays on board ships. . . . I have always insisted that a suspension bridge built without stays is planned without any regard to stiffness, and consequently is defective in a most important point."

Realizing and emphasizing the importance of the new inclined check stays or *cable stays* which he introduced, for the first time, in the Cincinnati bridge, the master builder wrote: "To further guard against injurious oscillations which might be produced by the blow of a hurricane, eight heavy counter or check stays are anchored to the towers below the level of the lower cornice (just below the level of the roadway) and run up to the cables. Their sole office is to counteract any undulations which might be imparted to the cables by a severe gale."

The bridgebuilder's sublime and prophetic faith in the suspension type of construction, in which he pioneered and achieved when others doubted, stands recorded in his final report on the Ohio bridge: "It is in large spans, where the principle of suspension asserts its full claims, and where it will forever stand without a rival, no matter what novelties may turn up in the future. Within the whole range of engineering, no system of construction presents such favorable conditions as is offered by the suspension system. Fully aware that opposite opinions are taught in textbooks on engineering, I know that I am exposing myself to criticism, but I also know that the future history of engineering will do me full justice."

10

On the last Saturday in November Amos Shinkle, now president of the company, announced that the bridge would be opened one week later, on December 1, 1866. Tickets were put on sale, and during the following week the newspapers were filled with accounts of the great bridge, *the largest span in the world.*

The last few days in November were cold and unpleasant, with light drizzling rains and temperatures near freezing. On November 30 the rain stopped, but the sky was still cloudy.

Saturday morning, December 1, was clear and mild. At sunrise a gun squad from the Newport Barracks fired a salute of a hundred guns with two brass twelve-pounders, to signal the opening of the bridge. From the first hour on there was a mighty rush of people—cheering, shouting crowds. Up the unfinished approaches they swarmed—men, women, and children—to avail themselves of the first opportunity to cross the great span.

By noon the mercury registered forty degrees. It seemed as though the entire population of Cincinnati, Covington, and Newport had turned out for this long-awaited over-the-water promenade. Newspapers advised all promenaders "to go very warmly clad, as it is very cold out there, over the water, with a keenly cutting wintry wind whistling along the river." By sunset the number who had crossed was 46,000.

The following day, Sunday, was pleasanter. The sky was clear, and another rise brought the noonday temperature to fifty. Eight or ten men were busy all day long on the Covington side, selling tickets and taking them up, while on the Cincinnati side fourteen men were similarly employed. Yet at three o'clock the crowd was so great at the Cincinnati end that people had to wait in long lines before they could buy tickets.

The residents of Covington were amazed at the hordes that descended upon them. "The beautiful weather and the attraction of the great bridge brought out thousands upon thousands, the most of them from Cincinnati, and on some of the principal thoroughfares and sidewalks it was so densely crowded that locomotion was difficult." By the evening of that second day an additional 120,000 persons had crossed the bridge. From morning until night the span was literally black with pedestrians. "It is the great attractive feature of the city."

The bridge sidewalks were as yet uncompleted, so the foot passengers had to use one of the roadways intended for vehicles. The gaslight fixtures were not yet installed; the heavy iron trusses separating the carriageways from the footways were unfinished; and the turrets covering the saddles at the tops of the towers had not yet been built to their full height.

During the rest of December further work was speeded to complete the bridge. Work was done, too, on the approaches to the structure.

Now real winter descended on the city. On December 30 the ice in the river was so heavy that the ferryboats could run only with great difficulty. The *Cincinnati Belle* and the *Newport Belle* were carried down the river a mile or two by the ice early in the morning, and did not get back to their docks until three o'clock in the afternoon. In previous years this absence of the ferryboats would have created hardship for the citizens; but now all who wanted to cross the river went by the new bridge.

The last day of the year was cold and foggy. A raw southwest wind threatened snow, and the river was still full of heavy ice. As the ferryboats could not run, it became necessary to open the bridge for vehicles sooner than was originally intended.

On Tuesday morning, New Year's Day, 1867, the bridge was opened to traffic. The newspapers carried the announcement:

The Covington and Cincinnati Suspension Bridge will be opened this morning for vehicles of all kinds. The arrangement, it must be understood, is only temporary, as the bridge is not yet completed, and the crossing of vehicles will interfere with the workmen very materially, if it does not put a stop entirely to their operation. The necessity for the opening, however, is apparent. The present condition of the river shows the importance of the bridge; and the managers are doing their best to accommodate the public, even to the detriment of their own interests.

The formal opening of the great structure spanning the Ohio took place at eleven o'clock in the morning. The ceremonial procession of carriages and horsemen was formed at the office of the bridge company on Greenup Street in Covington. Between lines of cheering spectators the procession came down through the city streets to the bridgehead, and then, against the freezing wind, it moved across the bridge toward Cincinnati. At such ceremonies the engineer is too often forgotten, but it was not so on this occasion. First came the carriage containing the president of the company, Amos Shinkle, and *the engineer, John A. Roebling;* then the carriage containing the vice-president of the company and the *assistant engineer, Colonel Washington A. Roebling;* then followed the carriages containing the directors, and those containing the members of the City Council of Covington. Behind these came the body of the procession, led by two horsemen, and consisting of carriages bearing the proud citizens of Covington. When this parade reached Cincinnati, it was joined by a much larger procession, including a long string of vehicles filled with the local officials and employees of the five leading express companies—the Adams, American, Harnden, United States, and Merchants' Union expresses—and a large number of omnibuses carrying railroad men and other commercial and industrial groups. To the music of Menter's and Heidel's bands, the processions from the two cities, thus symbolically united, marched back over the bridge, amid the rousing cheers of the thousands who lined both shores of the river. Upon reaching the Covington side, the parade marched through the principal streets, and then returned to Cincinnati for further celebrations, many of them lasting into the night. A carnival spirit prevailed.

In the afternoon a large number of carriages and other vehicles had to stand in line for an opportunity to cross the crowded span. Over fifty thousand persons, in carriages and on foot, had the proud honor of crossing the structure on the first day of its official opening.

Thus, more than ten years after the work was actually commenced—ten years of hardships and tribulations that would have broken the spirits of weaker men—the Ohio bridge was completed and opened as the greatest bridge in the world.

11

The entire work had been so carefully planned and supervised that, despite the employment of men inexperienced in bridge erection, a remarkable record of accident prevention was established during the long period of construction. "One noticeable feature," wrote Farrington, "has been the scarcity of accidents among the workmen. But two workmen have lost their lives during the past two years; a remarkable fact, when the nature of the work is considered and the large number of men employed. It is not often that workmen are so exposed to cold and heat, to wind and storms, as they have been in the construction of this bridge; and it is due them to say that, as a body, they have done their duty faithfully, and they may ever look on this work as a monument of their skill and endurance."

The report of President Shinkle and the directors recognized the greatness of the accomplishment: "It is scarcely deemed necessary to say to you that the bridge is a complete success. Perhaps no work of its magnitude has ever been more so, the wonder and admiration of all who cross it; and in the future there will doubtless few visit the city of Cincinnati or its vicinity who will not cross the magnificent structure."

The same sentiment was recorded, in an unusual introductory tribute, in the otherwise dry and dusty official report of the Chief of Engineers of the United States Army covering the history and description of the bridge: "It is a magnificent structure, and one of which the citizens of the two cities are justly proud."

An editorial in *Engineering*, the leading technical journal published in England, also recognized the outstanding character of the achievement:

> The largest existing span of a suspension bridge is that of 1057 ft. over the Ohio River at Cincinnati, lately completed from the plans of Mr. John A. Roebling, and under the superintendence of his son, Colonel Washington A. Roebling. This bridge is undoubtedly the finest work, structurally and architecturally, of its class.

The roadways had been opened before all of the construction details were finished. There was still much to be done—installing the ornamental work on top of the towers, finishing the sidewalks and railings, painting the cables and the ironwork, and building the necessary offices for the company's use.

In the spring of 1867, as the finishing touches on the great bridge were slowly progressing, John Roebling decided that he could safely leave the rest of the work to his assistants, as he was needed in Trenton and in New York. On April 21, Washington wrote home: "Our work here is dragging along slowly; it would take the whole year to finish everything. I shall leave here the first week in June whether we are done or not. Father goes tomorrow, a week."

By July the bridge was finished at last.

12

The bridge traffic gradually increased, reaching a high level in five years. Pedestrians always formed a large part of the traffic; as many as four and a half million crossed in a year. Other sources of revenue were rents from the telegraph companies and, after 1882, from the telephone company, for carrying their wires across the bridge. Gradually the loans which the company had made were paid off, and the various bond issues were retired.

In a spirit of public service the company made contributions to civic enterprises and to local charities. In the flood of 1883, when the river reached a record height of sixty-six feet, the company assisted in the relief work and made generous contributions to the flood sufferers.

Only a year later, in February, 1884, a still higher flood record was established, the river rising to a height of seventy-one feet. It was impossible for the ferryboats to land on either shore, even if they could have made their way across the rushing water. There was no way of crossing except by the bridge. It had to be kept open, so that fire-fighting apparatus and medical and relief service could be sent to the cities of northern Kentucky. Human lives depended on the bridge.

Long before the crest of the flood was reached, the icy water was sweeping around the entrance at the Cincinnati end. Colonel Shinkle gathered a force of men and quickly built a fleet of flatboats, with benches for passengers. With iron spiked poles managed by two or three men in each boat, the flatboats were poled from the bridge, against the current, up to Walnut Street, and then up Walnut Street until they came to rest where the cobbled pavement sloped out of the water. A saloon there, just outside of the reach of the flood, took its future name

—the Landing Saloon—from the landing of the boats at its door. Throughout the week of the flood constant communication was maintained in this way between the two sides of the river.

The most outstanding service rendered by the bridge in its three-quarters of a century was its performance during the great flood in 1937, the most terrible catastrophe in the history of the Ohio Valley. It was no ordinary flood. Thousands of acres were under water in both Cincinnati and the Kentucky cities. Tens of thousands of persons were homeless. The downpour of rain gave no sign of stopping.

Other bridges along the Ohio were already closed. The Cincinnati end of the suspension bridge was high and safe, but the Covington approach was threatened by the rising water. Sandbags, cement sacks, sand, and gravel were piled all day long into a heavy dike. The lives of hundreds of thousands depended on the maintenance of this one vital link.

All day Saturday the river held its level, a little over seventy-two feet. Then once more it began its terrifying march upward, undermining buildings, driving more and more people from their homes. Already more than twice its normal width, the river spread destruction over huge new areas for every foot that it rose.

Then came a "Black Sunday," which Cincinnatians will never forget. Heavy rains were beating down on a six-inch snowfall, and the river went wild. Mill Creek Valley was already a vast lake, with buildings crumbling and disappearing one after another in the water; and tremendous fires were raging at the Crosley Radio and Standard Oil plants. Telephone service was interrupted, water supply cut off, streetcars stopped, and the electric power failed.

Faster and faster the river roared upward. The temporary dam at the Covington end of the bridge could not keep the water back. Soon the mounting river passed the seventy-six-foot stage.

At four o'clock on Sunday afternoon the superintendent decided to build a causeway from the bridge to the highest land adjoining. With streets under water for blocks around, workmen began to lay bags of sand and gravel to build a solid roadway to Greenup Street, several hundred feet away. Day and night, through the icy, soaking rain, three hundred men toiled at the emergency task. Forty trucks hauled materials across the bridge, and five pumps forced water over the embankment. As the river kept rising, the causeway had to be built up to ever increasing height. Finally the race was won. Over a thousand tons of sand, gravel, and dirt had been used. A corduroy top was laid on the sandbags. From the bridge to dry land in Covington a solid road had been built fifteen feet high.

From Steubenville to Cairo the Cincinnati bridge was the only crossing open on the Ohio River. Fire engines from Michigan and Illinois crossed here to go to the aid of cities in western Kentucky. Electric generating units and emergency supplies for the hospitals of stricken Louisville, Red Cross workers speeding to Louisville, Paducah, and Memphis, truckload after truckload of medicine and hospital equipment, all were rushed across the bridge to the prostrate communities along the Ohio.

And most pressing of all was the cry for food. At midnight came a telephone call from a city in the Bluegrass. "Is the bridge open? Can trucks come through from Cincinnati?" It was the mayor of Lexington calling; there was no food in his city. Trucks were loaded in Cincinnati with bread, meat, and vegetables—food for the stricken city—and the next morning it was there. Many thousands of people thanked God for the Ohio bridge.

And there the great bridge still stands, serving generation after generation, in normal times and in critical emergencies, in peace and in war—a monument to human faith and vision, and to the fighting spirit and genius of its master builder.

WASHINGTON A. ROEBLING

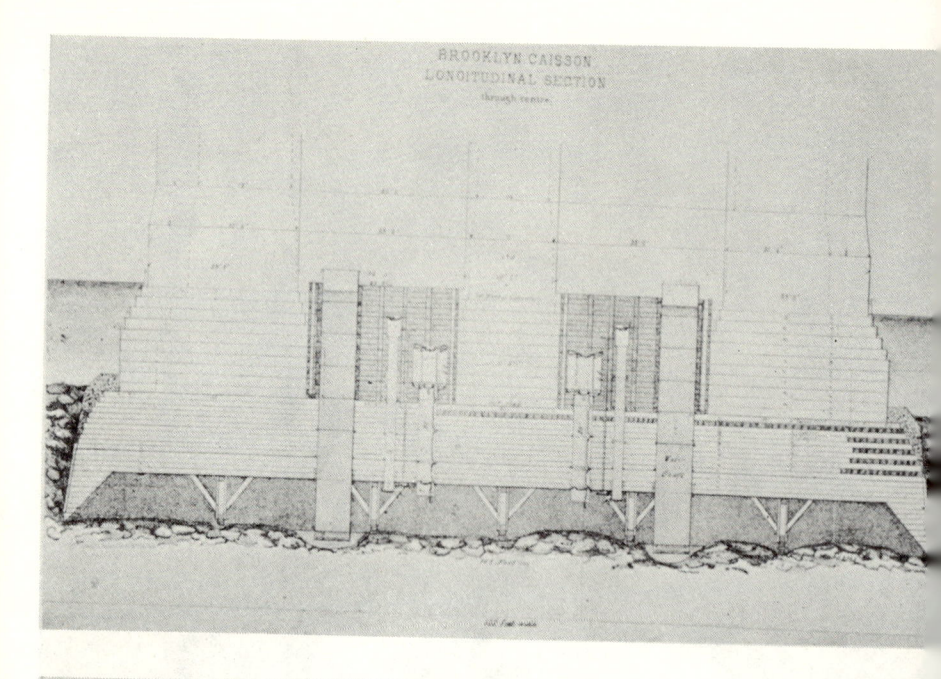

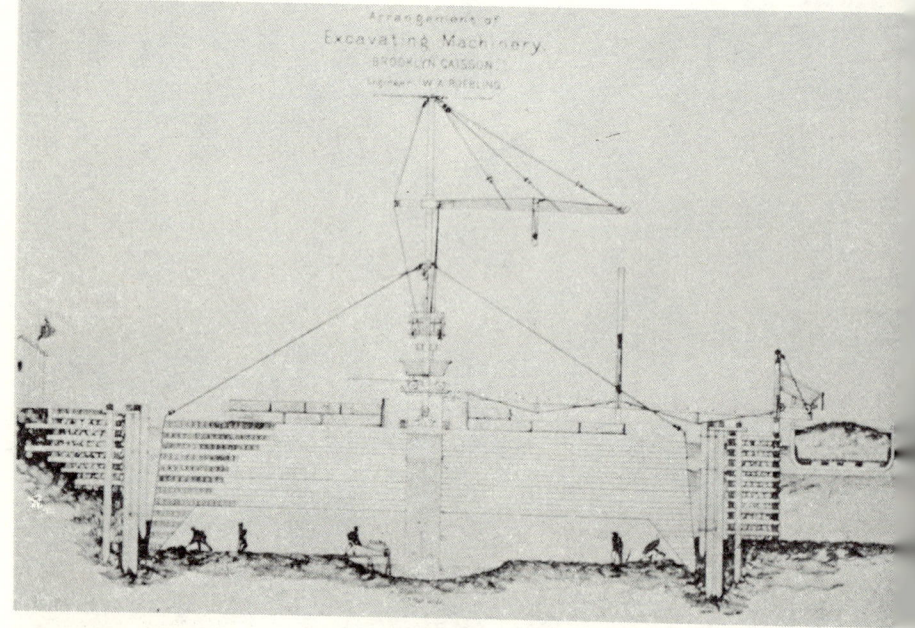

THE BROOKLYN CAISSON

Drawings by Washington Roebling showing, at top, the construction of the caisson and the masonry tower built upon it; and, at bottom, the method of excavation.

PART THREE
THE STRUCTURE

CHAPTER XVI

A CITY'S DREAM

Youth dreams of future glory.
A city, too, in youthful vigor and growing strength, has its visions of future greatness. And with the dawning awareness of potentiality, there is born an impatience of barriers that hamper and confine.

The first settlement within the present limits of Brooklyn was made in 1636, when some Dutch farmers settled along the shore of Gowanus Bay. Six years later the first ferry service to the island of Manhattan was inaugurated. Operated by Cornelius Dircksen, who kept an inn and farmed a small piece of land at Peck Slip, this early ferry was a rowboat. Service was irregular and intermittent, several days sometimes elapsing between crossings.

The two-oared skiff of 1642 in time gave way to other forms of primitive propulsion by manpower; and later an ingenious treadmill and paddle-wheel arrangement, set in the center of a boat, utilized teams of eight horses for motive power. To the citizens of that day this "New Ferry," which ran to Catherine Street, constituted a fascinating marvel of mechanical invention applied to transportation progress.

After a century and a half of ferries propelled by muscular effort of man or beast came Robert Fulton's invention of the steamboat. With the startling trip of the *Clermont* up the Hudson, a new era in transportation was born. Fulton himself made the substitution of steam for horsepower on the ferry service between Brooklyn and Manhattan. The line was known as the "Fulton Ferry," and the ferry line, in turn, gave its name to the two streets it connected. The first steam ferryboat in this service was the historic *Nassau*. On the very day that the *Nassau* began operating, in May, 1814, the chief engineer, a Mr. Lewis, was caught in the machinery and was instantly killed. The pilot, Peter Coffee, had better luck; he served the line for fifty years.

The *Nassau*, judged by today's standards, was an odd craft. It was a double boat with two hulls, side by side, the paddle wheel between them. The machinery was located on the deck in a small engine house; and one of the hulls contained a cabin for passengers, the other being used for vehicles and teams. Aside from the routine use of the ferry the voyage across the river was widely regarded as a pleasure trip and the ferry was much patronized for this purpose. There were frequent moonlight sails to the accompaniment of a band. The young maidens and swains of "little old New York" danced and romanced to the strains of music and the chugging of the paddle wheel. On those pleasant excursions of a summer evening the trip across the river was all too brief.

As the population grew, the demand for transportation increased. In 1835 the South Ferry was opened; in 1849 the Broadway Ferry; and then, in rapid succession, came the opening of the Wall Street Ferry, the "Roosevelt Ferry" running from Peck Slip, and the Twenty-third Street Ferry, followed by other lines.

2

Despite the fact that communication with New York—and through New York with the mainland—was restricted solely to the ferries and subject to all the inconveniences, delays, and hardships imposed by fog, ice, rain, and storm, the population of Brooklyn grew by leaps and bounds. By 1860 the census of the erstwhile village stood at 266,000, and by 1870 at 396,000, an increase of 50 per cent in a single decade. Brooklyn had become the fastest-growing city in the country.

As the village became a city and the city continued its amazing growth, its residents began to feel the need of some more substantial and dependable connection with New York than the ferries. The people of Brooklyn began to dream of a bridge! And in this dream many farseeing citizens of New York also joined.

As long as a century and a half ago the idea of connecting the city of New York with the opposite shore of Long Island began to engage public attention. General Jeremiah Johnson, writing in his notebook before 1800, recorded his conviction of the feasibility of spanning the East River. Three generations ahead of his time, he wrote: "It has been suggested that a bridge be constructed across the East River to New York. This idea has been treated as chimerical from the very magnitude of the design, but whosoever takes it into their serious consideration will find more weight in the practicability of the scheme than at first sight he imagined."

The idea of bridging the East River received new impetus with the birth of steam navigation—for the spirit of unbounded scientific progress was in the air and nothing seemed impossible.

In 1811 Thomas Pope, a carpenter-shipbuilder who had become engrossed in ingenious ideas for wooden bridges, published a book of bold imagination under the title, *A Treatise on Bridge Architecture*. Many of the most illustrious names in the New York of that day were listed among the subscribers. Pope had made an impressive scale model of his conception of a "Flying Lever Bridge" that was to consist of a single far-flung span from Fulton Street, New York, to Fulton Street, Brooklyn, and under which the largest vessels might pass unhindered. In his unfettered imagination he pictured a span of the unprecedented length of 1,800 feet, soaring 223 feet above the river. This he proposed to construct with light timber ribs built out in successive tiers from timber anchor cribs on both shores until the two tapering cantilever arms met over the middle of the stream to form a slender and graceful span of arch form. In a prose description he explained the design, and in lyric poetry he rhapsodized over his "rainbow bridge."

His proposal, however ingenious and poetic, was not taken seriously. When Pope solicited funds for his project, he only provoked considerable mirth. The story is told that on a certain showery day he was the guest of Robert Fulton on one of the latter's newly invented steamboats. As the vessel rounded the Battery, a rainbow spanned the East River. "See, there's your bridge, Pope," exclaimed Fulton. "Heaven favors you with a good omen." Whether Fulton was really impressed, or whether he thought Pope's idea as much of a chimera as the rainbow itself, is not known.

Although the ambitious carpenter's hopes were doomed to disappointment, the idea of spanning the East River persisted in the minds of more than one bold pioneer—and in the thoughts of the more enlightened and farsighted of the Brooklyn and New York public as well.

By someone's prophetic imagination, or as an expression of the popular hope, one of the Brooklyn streets running down to the river was given the name of Bridge Street, a name it still retains. By the eternal fitness of things that street, half a century later, was actually to receive the Brooklyn abutment of the great bridge.

3

Popular interest in the dream of spanning the water barrier between the two cities continued to grow. About the middle of the 1840's Prime, the New York historian, observed that "the erection of a bridge between New York and Brooklyn had become the great topic of conversation."

As the years passed, the limitations and hardships imposed on the two growing communities by ferry transportation became more and more apparent. This was particularly true in winter months of intense cold when the river was frozen from shore to shore, and in the subsequent thaws when it was an almost impenetrable mass of huge floes and jagged blocks of ice. At such times the ferryboat had to work its way through by main force, splitting and grinding the enormous masses of ice. The crossing then became a hazardous and uncertain adventure. A trip ordinarily accomplished in ten minutes took an hour, two hours, and sometimes from four to six hours. And there were occasions when the ferry, unable to break through to the distant shore, had to turn around and make its way back, as best it could, to its starting point, or to some other landing, and unload its shivering passengers.

One such occasion was in the unusually severe winter of 1851–52, when it was possible to walk across the river on the ice—something that had not happened for nineteen years. Crowds trooped in a long procession across the bridge of ice, a precursor of the more enduring bridge that was to come. When ferry service was resumed, the boats had to force their way through the floating ice field. By a coincidence an eyewitness account of that experience was recorded in a letter written by a nephew of John Roebling—August Meissner, Jr.—who had recently arrived in New York from Mühlhausen. Coming over from Brooklyn one morning, the ferryboat struggled aimlessly for an hour and a half without forcing a passage. At last it came near enough to a city dock to land its passengers by means of a plank; "but many carters, with their carts, wagons and chaises, had to remain on the boat and, on account of the freezing cold, were compelled to unharness their horses and bring them into the ladies' and gentlemen's cabins and give them their fodder there."

The ferry companies had to face these difficulties in a brave effort to maintain transportation across the river, for any sustained interruption of communication between the two cities would have brought untold confusion and hardship.

On the Manhattan side of the river the idea of a bridge was generally regarded with apathy. From their eminence and sophistication as residents of the metropolis, New Yorkers were inclined to look on Brooklyn somewhat deprecatingly and with a certain amused toleration. In those days not one New Yorker out of a hundred had made the tortuous ferry voyage to Brooklyn; and that community of 250,000 people across the river was both an unexplored region and an enigma to the vast majority of those who made their residence on Manhattan.

With Brooklynites it was different. The "sidewalks of New York" were as familiar to Brooklyn men and many Brooklyn women as the gas-lighted streets and unpaved highways of their own community; for a vast number of them worked in New York's marts of trade and commerce and had to make the trip across the river twice daily in pursuit of a livelihood.

4

For half a century the idea of a span across the East River had been but a nebulous dream. Those who had vaguely advanced the proposal were not engineers; and the art of bridge design had not yet progressed far enough to bring such an ambitious crossing within the realm of feasibility. For more than fifty years, men of imagination and citizens of public spirit had talked and dreamed of a bridge, but only vague and impractical proposals had been made. So long a span seemed to be beyond the range of human possibility. Men shook their heads and said, "It cannot be done."

But finally, after a half-century of longing and anticipation, there appeared a man who said, "It can be done!" John A. Roebling knew whereof he spoke—for he was ready and able to build the great span. There was no other bridgebuilder in all the world as well equipped as he for this task. He had developed the art of constructing suspension bridges beyond all previous conception, and he had courageously and successfully completed an epoch-making structure that men had said could not be built, the world-famous suspension span *carrying railroad trains* over Niagara.

In June, 1857, Roebling addressed a historic letter to Abram S. Hewitt, suggesting the feasibility of a bridge joining Brooklyn and New York, to be so constructed as to preserve the freedom of navigation on the river unimpaired. Mr. Hewitt, a leader in New York civic affairs, was associated with Peter Cooper in the iron business at Trenton. Knowing the bridgebuilder's record of achievements, the future mayor of New York recognized the significance of the communication he had received, and he promptly had it printed in the New York *Journal of Commerce*, where it attracted great attention because it came from an engineer who had already demonstrated his ability to conquer the impossible in the building of great spans. Twenty-six years later Hewitt, then a member of Congress, was one of the principal speakers at the opening ceremonies of the Brooklyn Bridge, and in his address on that occasion he told of the prophetic letter he had received in 1857 and of his own instrumentality in bringing it to public notice at that time. The speaker significantly observed that "Mr. Roebling was the first *engineer* to suggest the feasibility of a bridge between the two cities." He added, "This letter was the first step towards the construction of the work."

The bridgebuilder's thoughts were directed to the project as early as 1852, probably as a result of a personal experience that winter when he himself had been compelled to spend several uncomfortable hours marooned on a drifting ferryboat in the ice-choked waters. During the next few years, while he was absorbed in the task of spanning Niagara, the idea of an East River bridge was germinating in his mind. But not until he had a crystallized plan, with design calculations and estimates of cost to support it, did he make his proposal public.

The popular enthusiasm aroused by the publication of Roebling's letter was of brief duration. A few civic leaders were impressed, but as the ripple widened, it was quickly dissipated. A bridge was a beautiful dream, but people could not cross the river on a dream. Many were incredulous, others were ignorant or prejudiced. And the years rolled on. The Brooklynites continued to make their way across the river in times of storm and tide and ice, as the Lord gave them strength, and the sacred ferryboats still paid dividends.

5

For another decade the project slumbered. The Civil War intervened, absorbing the energies of the people. In 1865 Roebling, with characteristic persistence and with sublime faith in the project, prepared a set of plans. For the modest structure he contemplated at the time, the designer put his estimate at $4,000,000. He laid the plans before some Brooklyn civic leaders.

The initiative toward making the bridge a reality did not originate in the larger city of New York. Nor did it come from any public officials. It had its birth in Brooklyn, among a small group of public-spirited citizens of vision. In every such project there must be a leading spirit, personally fired with enthusiasm and capable of inspiring others with it. In this case the leader was William C. Kingsley, a prominent contractor of that day and one of Brooklyn's leading businessmen. He had become inspired with the dream of a great bridge to span the gap between the two cities; and now, after seeing Roebling's plans, he was convinced that the undertaking was feasible.

On December 21, 1866, a historic meeting took place, and the enterprise was launched. A group of three civic leaders met that evening in a house situated in what is now Owl's Head Park in Brooklyn. A tablet commemorating that significant occasion reads:

Owl's Head—At a conference held in this room on the 21st day of December, A.D. 1866, between Henry C. Murphy, William C. Kingsley, and Alexander McCue, an agreement was reached which resulted in the passage of an Act by the Legislature of the State of New York on April 16, 1867, providing for the construction of the New York and Brooklyn Bridge.

Kingsley had previously converted his friend, Alexander McCue, to the necessity and the feasibility of the bridge. On that extremely bitter December night he surprised McCue by calling on him about dusk. "Will you drive with me to Mr. Murphy's this evening?" he asked. "I want to get him to consent to prepare a bill for the bridge." McCue knew that Senator Murphy, a former mayor of Brooklyn, was then far from persuaded of the practicability of the enterprise; at the same time he did not believe that any man could withstand the onset of William Kingsley, with his facts and figures and his masterly way of using them. Curious to witness the effect of that strong personality on the cautious and subtle mind of the Senator, McCue joined his friend in the waiting carriage, and they drove down to Bay Ridge. Senator Murphy received them hospitably. The night was so cold that, at the host's suggestion, the horses were unhitched and put in the stable and the coachman was made comfortable in the servants' quarters.

At first the Senator showed his visitors the collection of rare books and historical manuscripts which was his pride. After a while Kingsley brought up the subject of the bridge. The Senator listened with polite attention, but not at first with more interest than, as a practiced listener, he was accustomed to give to any speaker in an interview. Soon, however, Kingsley riveted his mind as it had never been riveted before; the Senator listened like a man under a spell. Then, as though resenting the dominion of another over him, he began to interrogate, to criticize, to doubt. To everything he advanced, Kingsley gave the most respectful consideration. No sooner would the Senator stop, however, than Kingsley would meet him with new arguments, illustrations, and rejoinders, which were persistent, comprehensive and unanswerable. The two men were enjoying the intellectual battle. They went over the whole matter again, from beginning to end. Never in all his life had William Kingsley been in better condition for mental combat than on that winter night. He had resolved to capture Senator Murphy's assent, and he did capture it. Murphy avowed himself a convert to the feasibility of the project. That was a great point gained. Finally, he promised to draw the enabling bill. The time was far toward morning when the two callers left. But on that night, and in that talk, the Bridge, as a fact, was born.

And from that night, Murphy and Kingsley became "the Fathers of the Brooklyn Bridge."

6

Senator Murphy kept his promise. He drafted a bill, which he promptly introduced in the state legislature, to charter a company of private citizens to build a toll bridge across the East River. Before the bill was passed, and on faith and conviction that it would be passed, Kingsley, on his own responsibility, made contracts amounting to many thousands of dollars for construction materials.

While the bill was pending at Albany, a strong impetus was given to the enterprise by the arctic conditions prevailing in the East River during that memorable winter of 1866–67. It was the coldest, bitterest, longest winter the two cities had ever known. The river was so choked with ice that the ferries were helpless. The hardships of plowing through the ice gorges in the vicious cold at last wrung forth a cry for relief. Brooklynites, freezing on the ferry, cast mental eyes on the great bridge just opened at Cincinnati, and saw no reason why they should not have one as good or better. The demand for a bridge rose to a clamor. The newspapers that had previously scoffed now took up the popular appeal and added their support to the pending legislation. It was pointed out that passengers by rail from Albany completed their 150-mile journey to New York in less time than it now took the unhappy voyagers from Brooklyn to reach their destination in Manhattan. This graphic picture was grand ammunition for the advocates of the bridge, and was used with telling effect to impress the legislature.

On April 16, 1867, Senator Murphy's bill was passed by the legislature: "An Act to incorporate the New York Bridge Company, for the purpose of constructing and maintaining a bridge over the East River, between the cities of New York and Brooklyn." In this charter, thirty-nine incorporators were named, to constitute the first board of directors. A time limit—June 1, 1870—was set for the completion and opening of the bridge.

Exactly a month later, on May 16, a meeting of the incorporators was held to organize the company. Senator Murphy was elected president, and three committees were appointed—a committee on by-laws, a temporary finance committee, and a committee on plans and surveys. The list of incorporators formed a roster of the leading citizens of New York and Brooklyn at that time.

7

On the day of its appointment the Committee on Plans and Surveys convened for the most important assignment of all, the selection of an engineer to plan and build the bridge. At first someone suggested Colonel Julius W. Adams, who two years earlier had brought a plan to Kingsley. And then, by acclamation, the name of John A. Roebling was placed before the meeting—a name that was forever after to be inseparably linked with the great achievement. With no dissenting voice the committee promptly accepted it. At that time the master builder was just completing the Cincinnati bridge, which was then the greatest span ever built by man. His fame was at its zenith. He had established himself, by actual accomplishment, as the world's foremost exponent of the design and construction of suspension bridges.

The committee reported that it was guided in its choice of Mr. Roebling by the consideration that "the construction of a suspension bridge of a magnitude unprecedented in the previous history of such works demanded the greatest experience and developed ability that could be obtained." On the basis of prior inquiries and conferences with others who knew the record, the committee was satisfied that the distinguished constructor of the Niagara Railway Suspension Bridge and of the Ohio River bridge at Cincinnati was the one man who measured up to the task. The committee concluded its report by stating that "confidence on the part of the public and of those whose money was to be invested in the undertaking would best be insured by employing the Engineer who had achieved the most successful results in that type of project" and who was thus most pre-

A CITY'S DREAM 305

eminently qualified to accomplish the heroic enterprise upon which they were now embarked.

The incorporators immediately approved the committee's report and appointed John A. Roebling chief engineer of the proposed bridge, at a salary of $8,000 per annum, his appointment to take effect a week later, on May 23, 1867. There was no hesitation, for they knew that in selecting the master builder they were making no mistake. Destiny had provided the man, and wisdom guided the choice.

Only twenty-three years before, John Roebling had pleaded for a chance to build his first suspension structure, with spans of only 162 feet. Now he had become the world's foremost authority on suspension bridges and was entrusted with the responsibility of planning and building the world's greatest bridge, with an epoch-making span length close to 1,600 feet.

The stage was set for the greatest drama in the history of bridgebuilding, the drama of the building of the Brooklyn Bridge.

CHAPTER XVII

THE GOAL IN SIGHT

UPON being notified of his appointment, the bridgebuilder left Cincinnati and came to New York to begin his surveys and plans for the great span. A few weeks later, his son sailed abroad to see and study all that could be learned of the newly developed pneumatic method of sinking caisson foundations—knowledge necessary in undertaking the difficult task of sinking the foundations of the East River bridge. He remained for a year and made a thorough study of the subject, visiting England, France, and Germany, and consulting the foremost engineers in those countries. Besides inspecting all the important engineering works then being carried on in Europe, he made a special study of the new structural material, steel, visiting the works of Krupp at Essen, as well as the most important ones in England.

During his travels Washington Roebling and his wife visited Mühlhausen, his father's birthplace. And there, in an inn across the street from the family's ancestral home, Colonel Roebling's son and only child was born. In honor of his distinguished grandfather the child was named John A. Roebling, II.

The fame of the Ohio bridge had traveled across the ocean. Upon its heels came the exciting news of the still greater span which John A. Roebling had been commissioned to build. In July, 1867, an editorial in *Engineering*, the distinguished technical periodical published in London, featured the completion of the record-breaking span at Cincinnati and heralded the commencement of the larger and more startling project at New York. To give their readers a conception of the staggering proportions of the new undertaking, the editors described it in terms of known spans:

Having completed this work, Mr. Roebling is about to undertake one of far greater magnitude, that for connecting the city of New York with Brooklyn, across the East River. It is necessary, here, that a headway of one hundred and thirty feet be maintained in mid-channel, and the great suspension span will be one thousand six hundred feet in the clear, or much more than twice that at Clifton, nearly three times that of the Menai suspension bridge, and nearly six times that of either span of the Lambeth suspension bridge. This great opening will be approached, on either side, by a succession of arches, the whole length being between one and a half and two miles, and the total estimated expense is six million dollars, or, at the present value of American currency, nearly one million sterling.

The editorial then proceeded to describe the proposed design: the number, size, and arrangement of the cables; the stiffening trusses and the roadways; the design and dimensions of the towers; and the proposed cable-operated railways for carrying passengers across the span. In all of these features the description coincided remarkably with the structure as finally completed, showing that the designer already had the essentials of his plan clearly formulated at the time of his appointment or within a month or two thereafter.

The details now given [the editors explained] have been supplied to us by Colonel Washington A. Roebling, who is now in London, and who, we trust, before his return in January next, will read a paper at the Institution of Civil Engineers upon the great suspension bridges executed and in hand by his father in America.

There followed a technical description of the Brooklyn Bridge plans, after which the conservative British journal concluded with an unusual tribute of glowing and prophetic appraisal:

As a whole, no work of the same character, yet executed, approaches this in magnitude. It will undoubtedly be the crowning work of Roebling's life, and we may, not unreasonably, conclude that it will hand his name down to posterity as one of the greatest engineers of modern times.

2

John Roebling, who all his life had been working at high pressure, now actually surpassed his accustomed fevered pace, as though he were increasingly conscious of a personal race with death. In the incredibly short time of three months after his appointment he had completed the preliminary location surveys, the design studies, and the estimates of cost for all parts of the monumental structure; and on September 1 he had his "Report" completed, with drawings showing the location, the plan, the anchorages, and the approaches of "The East River Bridge." On the following Saturday night, he read this classic document, the last important document of his life, at a meeting of the New York Bridge Company.

The "Report" began by stating that its object was to establish the practicability of bridging the East River without interfering with navigation, to point out the best location for the crossing, to explain the design and the different features of the construction, to estimate the cost, and to ascertain whether the undertaking would pay as an investment.

With sound judgment of the future development of the two cities, Roebling fixed the location of the span. Of the several routes he had surveyed and studied he selected the line to City Hall Park because it afforded the best grades and the best accommodation for traffic. "For the next fifty years to come," he wrote, "the City Hall Park will remain the great focus of travel, from which speedy communications will ramify in all directions." Balancing comparative usefulness and prospective revenues against first cost, he recommended the adoption of the line "as the true route which will most favor the interests of the community, as well as of the Bridge Company."

Seeing into the future, Roebling predicted that if a first bridge were built across the East River, others would follow sooner or later. The second bridge, he suggested, would connect to Williamsburg in Long Island; and a proper route for another would be across Blackwell's Island. The master builder's predictions were subsequently fulfilled, even sooner, perhaps, than he anticipated.

Continuing his presentation, Roebling vigorously defended the feasibility of his proposed span against the argument that it would exceed all precedent: "Any span inside of 3000 feet is practicable," he declared. "A span of 1600 feet or more can be made just as safe, and as strong in proportion, as a span of 100 feet. The larger span is not a question of practicability, but simply a question of cost."

He observed that the exact location of the Brooklyn terminus "should not be determined before final surveys have been made and the line has been actually traced on the ground." Pending those surveys—which were to be of such tragic consequence—Roebling tentatively fixed the governing dimensions, almost identical with those of the structure as finally built.

He described and discussed the clear height of the river span, the resulting approach grades, the design of the land approaches, the crossings over the streets, the fireproof and waterproof construction of the floor to serve as a roof for the warehouses underneath, the cross section of the bridge, and the arrangement and dimensions of the roadways, railways, and promenade.

To illustrate the capacity of the bridge, he pictured a hypothetical invasion by a foreign enemy, landing on Long Island and making it necessary to transport large masses of troops and trains of artillery from New York to Brooklyn in as short a time as possible. He showed by figures that the total number of men that could be passed in twenty-four hours over the railways, roadways, and central walk, without crowding, would reach nearly half a million, in addition to artillery and baggage trains.

Perhaps the best known feature of the Brooklyn Bridge is its wide, elevated promenade. To millions who have enjoyed it through the years, it has been associated with memories of pleasure, romance, and thrill. Roebling realized the future usefulness and drawing power of this great walkway high above the river and the flanking streets. "This elevated promenade," he said, "will allow people of leisure and old and young to stroll over the bridge on fine days. I need not state that in a crowded commercial city such a promenade will be of incalculable value."

In this initial report, the master builder made the revolutionary proposal that steel wire be used instead of iron wire for the cables! This represented true pioneering courage, for never before had steel been so used. Today, when it is almost the only structural metal employed by engineers, we find it difficult to realize the initial resistance to the adoption of steel for bridges and the courage that was required to choose that material in its early days. As late as 1877 in England, for supposed reasons of safety, Board of Trade regulations actually prohibited the use of steel in any structures. Yet ten years earlier, at a time when no steel bridge had ever been built, John Roebling had the daring and the vision to propose this untried material for the cables of his Brooklyn Bridge. If anything had gone wrong in consequence of his decision, his strenuously achieved reputation would have been wrecked, together with the great span that represented the culmination of his lifework.

In his plans for his bridge Roebling utilized the cumulative wisdom of all his prior contributions to the art, so as to secure rigidity and stability in a measure never before attained. Profiting by the sad experiences of others whose spans had been destroyed by vibrations or had been blown down by the wind, the master builder developed a design that would defy the elements and stand immovable through years to come. The features by which this enduring resistance was secured, he described in his classic report: "To guard against vertical and horizontal oscillations, and to insure that degree of stiffness essential to meet the effect of violent gales in such exposed situation, I have provided six iron trusses, which run the whole length of the suspended floor from anchor wall to anchor wall. . . . A most effective framing is thus obtained, which will be found to possess ample stiffness even in the greatest emergency. I am not disposed to underrate the great force of a severe gale, and I am also fully aware that herein lies the great danger to all suspension bridges of ordinary design. But my system of construction differs radically from that formerly practiced, and I have planned the East River Bridge with a special view to fully meet these destructive forces. It is for the same reason that, in my calculation of the requisite supporting strength, so large a proportion has been assigned to stays in place of cables."

Perhaps the most distinctive feature of the Brooklyn Bridge is the system of inclined stays radiating downward from the tops of the towers to the floor of the span. The magic of the resulting weblike tracery of lines has been the delight of artists and poets, and of all who love beauty. But these radiating lines were not gratuitously added for artistic effect; the genius of Roebling introduced them primarily for the critical function of adding rigidity to the span, and then took advantage of the additional load-carrying capacity which they incidentally supplied. This contribution to the strength of the bridge was explained in simple terms by the designer: "The floor, in connection with the stays, will support itself without the assistance of the cables. If the structure is viewed in this light, it will make a different impression upon those who are in the habit of associating with the idea of a suspension bridge the picture of a free-swinging pendulum. . . . The supporting power of the stays alone will be 15,000 tons, ample to hold up the floor. If the cables were removed, the bridge would sink in the center, but would not fall."

Half a century or more before other bridge engineers learned the lesson, Roebling's clear thinking had crystallized the basic principle governing the economy and rigidity of suspension designs—namely, that any desired degree of stiffness can be secured if one is willing to pay for it and, conversely, excessive economy in first cost may mean dangerous reduction in stability. As he lucidly expressed it: "If a rigid economy was to be exercised for want of capital, less strength and stiffness might answer. By a reduction of weight and of strength, a great deal of expensive material and work might be saved. But in the same proportion would the stiffness of the work be also decreased, and its safety in the case of a great hurricane less insured."

Roebling realized, as he said in his report, that the dominant feature of the finished bridge would be the two towers. He gave their dimensions and described their design and construction, revealing the careful thought he had given to various means for keeping the cost down while securing the maximum impressiveness of appearance. Among these means were the hollow openings in the granite mass, and the avoidance of expensive stone-cutting. "The shafts," he added, "are provided with projecting buttresses, for the purpose of saving masonry, to gain strength, and to improve their architectural appearance. In a work of such magnitude, and located as it is between two great cities, good architectural proportions should be observed. . . . The impression of the whole will be that of massiveness and strength."

Turning next to a discussion of the tower foundations, the master planner pointed out that the foundations for the support of those large masses of masonry must be unyielding. He stated that solid rock bottom was, of course, preferable to any other; but if this could not be found at any reasonable depth, then other means must be used to make the base secure. He reported on the borings then in progress to determine the character and depth of the tower foundations on the two sides of the river.

He then proceeded to describe the proposed construction of the huge anchorages to resist the pull of the cables at the two ends of the bridge. This supplies us with a terse description in his own words of the unique and effective design of anchorage which he had invented and perfected, and which he had used on a smaller scale in his earlier bridges. In the case of the Brooklyn Bridge, a problem of new magnitude was presented. "No rock will be found available," Roebling explained, "for anchoring the suspension cables on either side of the river. Each of the four cables enters the anchor-walls at an elevation of nearly 80 feet above tide, and . . . a connection is formed with the anchor chains. . . . Each chain is attached to a massive anchor plate, which supports the superincumbent masonry, whose weight must be great enough to balance the greatest tension to which the chain can ever be exposed. By the curvature of the chain a portion of its tension is conveyed into pressure, directed upon the wall."

On the basis of his preliminary design Roebling estimated the cost of construction at about $7,000,000, exclusive of the cost of the land for the approaches. He put the question, "Will the future intercourse between the two cities justify such an outlay?" And he supplied the answer when he affirmed prophetically that "the future growth of Brooklyn will keep even steps—no faster, no slower—with the additional facilities of travel which will be created. If the proposed bridge shall possess capacity of 40,000,000 annually, these 40,000,000 will be there as sure as the bridge is built." This prediction may have sounded fantastic at the time, but the reasoning was sound and the incredible conclusion was later generously vindicated. In subsequent years the bridge actually carried more than 150,000,000 passengers and pedestrians annually.

With invincible logic the bridge planner affirmed his conviction that New York must share in the natural and legitimate increase of population of the whole country, in order to carry on the commerce of the world. "This general increase of population," he continued, "is sure to take place, *bridge or no bridge*. It remains for the people of Brooklyn to decide whether they will attract their legitimate portion of this additional increase or not."

Finally, Roebling showed how the bridge would quickly pay for itself in railway passenger fares, without considering tolls from the vehicular roadways and the promenade. He further showed, by estimates that later were amply fulfilled, how the resulting increase in valuation of property in Brooklyn would pay for the whole cost of the work in three years. The public benefit would far transcend the private profit.

That the master builder realized the monumental character and enduring significance of the crowning work he had planned is clear from the closing passage of his report. With justifiable pride, he ended: "The contemplated work, when constructed in accordance with my designs, will not only be the greatest bridge in existence, but it will be the great engineering work of this Continent and of the age. Its most conspicuous feature—the great towers—will serve as landmarks to the adjoining cities, and they will be entitled to be ranked as national monuments. As a great work of art, and as a successful specimen of advanced bridge engineering, this structure will forever testify to the energy, enterprise, and wealth of that community which shall secure its erection."

No false modesty here, and no exaggerated appraisal. The man knew what his brain had wrought, and in this—his last testament—he was presenting his masterwork to his fellow men and to posterity.

3

The directors were impressed and reassured by Roebling's report. It supplied them with a clear picture of the technical features of the project and a keen analysis of its broader aspects. It was thorough, sound, and convincing.

But the general public still remained incredulous. Even prominent men like Horace Greeley of the New York *Tribune* and Mayor Kalbfleisch of Brooklyn voiced doubts and misgivings. People were staggered by the great length of the span, far greater than any previously attempted. It was "a leap in the dark."

Roebling had presented his plans for flinging a mighty span across the East River, and now he had to fight his way, inch by inch, against incredulity, ignorance, and prejudice, before the right to try was given him. There arose about him a chorus of protest, voiced by engineers more eminent than he, who denounced his plan as visionary and impracticable. Officials and citizens, frightened by these voices, joined in denunciation and discouragement. But Roebling remained steadfast and confident, and the men who had selected him stood by him.

At one of the many hearings that were held on the bridge question, a famous engineer who came out in favor of the wire type was asked what reason he had for believing it would do the work. "I believe in it," he answered, "because John Roebling says so."

4

The surveys, the foundation borings, and the preparation of contract drawings proceeded, with the bridgebuilder giving every detail of the work his energetic, personal attention.

The Committee on Plans and Surveys approved Roebling's report, including his recommendation for a New York terminus at City Hall Park. His structural plan of the bridge was accepted, subject to such modifications as further study or other circumstances might prove to be advisable. The committee recommended that the work be commenced immediately and that the necessary steps be taken to raise the required capital from the general public, from banking houses, and from the cities of New York and Brooklyn.

Under the enabling charter of the company the two cities had been given legal authority to subscribe to its capital stock. Toward the end of the year 1868 both municipalities availed themselves of the privilege—with a little persuasion. Brooklyn subscribed for $3,000,000 worth of stock, and New York for half as much. The smaller city took the larger subscription because it naturally expected to derive the principal benefit from the undertaking.

In addition, individuals with faith in the enterprise took stock in the amount of $500,000, bringing the total capital issue to $5,000,000. The stock was to be paid for in installments. Both the public and the private subscriptions were secured largely through the indefatigable efforts of William C. Kingsley, who was also the largest individual subscriber—in the amount of $160,000.

Among the other early subscribers to the stock, at $100 a share, were Senator Henry C. Murphy, 100 shares; Isaac van Anden, 200; Seymour L. Husted, 200; Alexander McCue, 100; William M. ("Boss") Tweed, 560; Peter B. Sweeney, 560; Hugh Smith, 560; Henry W. Slocum, 500; J. S. T. Stranahan, 100; and Mayor Martin Kalbfleisch, who had previously opposed the project, now personally subscribed for 200 shares.

The government of New York City at that time was in the control of a corrupt political ring headed by the notorious "Boss" Tweed. This band of politicians grew rich and powerful by looting the city's treasury through extravagantly administered contracts. From the construction of the notorious "Tweed Court House" in City Hall Park in 1868 and 1869 the "Ring" siphoned some $9,000,000 into their own pockets. The privately selected contractors simply inflated their more or less legitimate claims and split with Tweed afterward. The hope of similar graft in the construction of the Brooklyn Bridge may have been an inducement that prompted the "Boss" to approve the city's financial participation in the enterprise. Fortunately, however, the corrupt gang was soon turned out of office and Tweed was convicted and put in prison, where he died some years later. He had played his part in making the bridge project a reality by directing the city's stock subscription in 1868, but any hope he may have entertained of political control and illegitimate profit was thwarted. Perhaps, during his dying years in prison, he derived some satisfaction from the memory of the one good deed of his political career—the municipal underwriting of the Brooklyn Bridge, without which the great work would not have been built.

Tweed's early connection with the enterprise left in the public mind seeds of suspicion that lingered and grew long after that politician's elimination from the municipal scene. In time this unfortunate feeling of distrust developed into open charges and demands for an investigation. Although the builders of the bridge were ultimately proved to have honestly and conscientiously administered their trust, the burden of popular suspicion and accusation added to the other strains of a heartbreaking task.

5

In the meantime, back in the years 1868 and 1869, the master builder was continuing his lone battle against opposition and prejudice. On all sides he had to meet attacks challenging both the feasibility of the unprecedented span and its economic value. The span was 50 per cent longer than the longest yet built—Roebling's Cincinnati bridge. But of greater significance than the length was the tremendous load the bridge was designed to bear—a weight of 18,700 tons! Millions of dollars of public and private funds were to be risked on this "wild, untried scheme"—"a bridge that could not be built"—"a structure that would collapse under the first loading"—"a project that could never pay for its enormous cost."

In order to set the growing doubts and apprehensions of the public at rest and to melt the frozen opposition of the incredulous, Roebling asked that a board of the most eminent members of the engineering profession be invited to review his plans and estimates. He considered it proper that his design should be subjected to careful scrutiny by experienced engineers of established standing, in order that, if the plans should receive their approval, the enterprise would stand before the public sustained by collective professional judgment and not solely on the unsupported affirmation of one engineer. In compliance with this sound and courageous suggestion a commission of engineers to review the plans was authorized by the company in January, 1869. Seven eminent engineers were invited by the designer to sit in judgment on his work. This jury of experts included competitors of Roebling's and engineers who had previously criticized the design as unsound. The Board of Consulting Engineers consisted of Horatio Allen, Chairman, Benjamin H. Latrobe, William J. McAlpine, John J. Serrell, James P. Kirkwood, J. Sutton Steele, and Julius W. Adams—the latter having submitted his own design in 1865 and having subsequently indulged in attacks on his competitor's plans. The professional eminence of this group of seven experts is indicated by the fact that four of them—Kirkwood, McAlpine, Allen, and Adams, in the order named—served successively as presidents of the American Society of Civil Engineers between 1868 and 1875.

Beginning in March, 1869, the consultants held many meetings, at which Roebling ably defended the thoroughness of his design and its practicability. They discussed the project from every angle and studied the plans for the foundations, towers, anchorages, cables, superstructures, and approaches, and even the novel plan for cable propulsion of the rapid-transit trains on the bridge.

At the second meeting of the consulting board the designer eloquently defended the feasibility of the proposed span length of 1,600 feet. He appealed for an open-minded consideration of his arguments and figures on this point, since it was the unprecedented length of the contemplated span that staggered most of those who doubted the success of the venture.

An inspection trip was made to Cincinnati and Niagara Falls to view the suspension bridges Roebling had built. The members of the board were thus enabled to see these large-scale examples of the proposed type of design and to observe the actual functioning of the significant improvements the master builder had made in the art of suspension-bridge construction.

Finally, after two months of exhaustive study and discussion, the Board of Consulting Engineers unanimously confirmed the feasibility, strength, and durability of the proposed structure as planned and designed by John A. Roebling. In their report, issued in May, 1869, they publicly announced their conclusion "that it is beyond doubt entirely practicable to erect a steel wire suspension bridge of 1600 feet span across the East River in accordance with the plans of Mr. Roebling, and that such structure will have all the strength and durability that should attend the permanent connection by a bridge of the cities of New York and Brooklyn." The fact that a public work of such difficulty and magnitude was thus unanimously approved by an assembly of recognized experts, in the face of probable professional jealousy and a natural desire to display a superior critical faculty, showed that Roebling had been able to convince all of the judges of the thoroughness, safety, economy, and practicability of his design.

At the same time a commission of three army engineers, appointed by the Secretary of War, was studying the plans from the point of view of possible obstruction to navigation as well as for general feasibility. On May 22 this commission submitted its report to the Chief of Engineers of the United States Army; they recommended an increase of five feet, from 130 feet to 135 feet, in the clear height of the bridge; and they stated that there was no doubt of the entire practicability of the structure or of its strength and stability when completed. On June 21, 1869, Brigadier General A. A. Humphries, Chief of Engineers, officially informed the company that the Secretary of War had approved the plan and location of the bridge at the increased clear height.

With the War Department endorsement, a Federal bill authorizing the construction of the bridge was passed by Congress and approved by President Ulysses S. Grant. The last obstacle to the actual beginning of the work was removed.

6

All doubts had now been set at rest. Professional opposition had been overcome, official approval had been secured, and public confidence had been won. The plans were accepted and approved, funds had been subscribed, and construction was authorized to proceed.

Here at last the aspiration of a man's lifetime was about to be fulfilled. The bridgebuilder stood on the threshold of achievement.

It was to be his dream bridge—the consummation of his life's ambition. From his first sight of a small suspension span in his student days all of his longing, all of his preparation, all of his striving, were pointed toward this goal. By iron determination, by relentless concentration, by unsparing energy, by achievement after achievement, he had won this opportunity to create the world's greatest span. And now, from his brain and his soul, combining the genius of the mathematician, the builder, and the artist, he had crystallized the vision—the lines and the form, the power and the grace, the beauty and the magic—of this masterwork. He had battled and overcome the forces of doubt and prejudice, and he had finally won the right to go ahead with the building of the Bridge.

In his heart and in his mind he saw the great bridge of his dreams—the magnificent pylons, enduring as the pyramids, founded on solid rock deep below the rushing tides; the powerful cables of steel sweeping downward and upward in the arc of Nature's law; and the arching roadway carrying the multitudes over a span greater and more beautiful than the world had ever seen.

He saw it clearly as it was to be, "a great work of art," "the great engineering work of this Continent and of the age." Time was to justify his prophecy, and posterity was to behold the magic and the power of the thing he had created. But he, who had dreamed and planned all this, was not to see the time-defying towers, or the vast sweep of the mighty cables, or the rushing thousands crossing on the far-flung span. He was not to see his bridge take form. He was not even to see the first stone laid.

From ancient times the tradition has been passed down from generation to generation: "The Bridge demands a life!" In this case the Bridge demanded the life of its creator. At the end of the race, with the goal in sight, fate exacted its toll—a toll doubly cruel by its timing.

7

In the early summer of 1869 a final detailed survey of the entire line of the bridge was under way.

On July 6 the accident occurred. The Chief Engineer was taking observations to determine the exact location of the Brooklyn tower. His assistant, Colonel William H. Paine, was stationed with a transit on the distant New York shore to help by giving the line across the river. Roebling was standing on a cluster of piles at the rack of the Fulton Ferry slip, receiving signals from Colonel Paine. Engrossed in the work, the bridge-builder was oblivious to a ferryboat just entering the slip. The boat, laden with passengers, crashed heavily into the fender. The fender rack was suddenly forced against the piles, and Roebling's foot was caught and crushed between the timbers.

The injured Chief Engineer was taken to his son's home at 137 Hicks Street, Brooklyn. Immediate amputation of the toes of the right foot became imperative. Roebling gritted his teeth, told the surgeon to go ahead, and went through the painful operation without an anesthetic.

But the amputation did not end the trouble. Lockjaw set in, and then followed two weeks of extreme suffering. This time the power of the mind was unable to vanquish physical disease and pain.

One of Roebling's visitors during this crucial illness was the family minister, the Reverend John C. Brown, of St. Paul's Episcopal Church, Trenton. Referring to the injury, which Roebling's friends then thought would confine him to his sickroom but a short time, the rector remarked, "It seems a strange accident, that at the beginning of so great a work you should be laid up for two or three months." "There is no such thing as chance," the engineer replied. "All is wisely ordered."

Clinging to his lifelong prejudice against conventional medicine, Roebling obstinately refused to let regular physicians attend him and endeavored to prescribe his own treatment.

And then came an instance of the man's wonderful endurance, illustrating as well an amazing retention of inventive faculty to the very end. On the last day of his life, as he lay helpless and suffering, he devised and drew an apparatus to be used in lifting and moving himself in bed. This design he fully explained to his attendant and nephew, Edmund Riedel, and directed that the apparatus be constructed forthwith.

But early the next morning, that of July 22, 1869, the master builder was no more.

8

John Roebling was twenty-five years old when he came to America, and sixty-three when he died. He had come a poor immigrant youth, but he left the imprint of his genius to enrich the nation's heritage.

In the last few years of his life the bridgebuilder felt physically spent. He had exhausted his vitality in his unstinted outpouring of energy into his work. And in the last few months of his life, even before his fatal accident, his mind turned to thoughts of the immortality of the spirit of man. In a conversation with the rector of his church Roebling stated that he then felt stronger in spirit, although his body had passed its prime. "Yes," he replied to a question, "this body may be torn limb from limb, but the soul cannot be injured; it shall live forever."

On Sunday, July 25, the funeral service was held. Thousands, both rich and poor, came to the Roebling home in Trenton to look for the last time upon one whom all respected and loved. Bands of workmen came, not alone, but with their wives and children—for they had known him as the friend of their families. Tears coursed down the cheeks of men who seldom wept.

9

At the unveiling of the Roebling monument in Trenton in 1908 the speaker of the day said:

"I am expected, by those having these ceremonies in charge, to translate into words what the sculptor has so admirably expressed in bronze, namely the type and quality, the idiosyncrasy of your famous fellow townsman. But words are too plastic for such a task. As if his nature had been subdued to what it worked in, the Iron Master of Trenton was a man of iron. Iron was in his blood, and sometimes entered his very soul.... John A. Roebling as he was, as you knew him, head bared to the blows of fortune or the storms of heaven, eyes fixed unwaveringly on whatever object he had in hand, poised, confident, unyielding, imperious and proud—John A. Roebling is there—seated forever on yonder pedestal."

The cost of the monument was subscribed by the men, women, and children of Trenton. The sculptor, William Couper of New York, studied the life and personality of the man he undertook to portray, and in the features of the statue he succeeded in expressing the bridgebuilder's dominant will. "I set out," said the artist, "to bring out the characteristics of a man of action, a man of dogged determination. If I have done this much I have only brought out the character of John A. Roebling."

But neither sculptor nor orator gave the whole picture of the man. Beneath the obvious dominant characteristics of pride of conviction and unconquerable will there were other, more human, facets of personality, little known to the world. Under an armor of hardness and reserve there were softer emotions and a strange loneliness.

Posterity will remember John Roebling for his technical achievements and will picture him as a man engrossed in his computations, his inventions, his industrial developments, and his bridgebuilding—a man in a grim race to achieve, a "man with iron in his soul." But it is good to know that the man also had his warmly sensitive human side, that we can also remember him as a man who left his homeland and came to America because he was inspired by a passion for human liberty; who would not establish his colony in the South because his freedom-loving heart revolted against slavery; who loved music and found time from his great labors to play the flute and the piano; who found mental refreshment through studies in the things of the mind and of the spirit; who saw beauty and poetry in sea and sky, in mountain streams and in sunsets, and who found a thrill in being able to say "the Bridge will be beautiful"; and, above all, whose kindliness and sympathy won the love of his fellow men, so that even the wives and children of his workers wept at his passing.

CHAPTER XVIII

DOWN IN THE CAISSONS

WITH its inspiration gone, the Brooklyn Bridge seemed impossible to build, for John Roebling had been more than the designer. His had been the vision, the conception, the knowledge of the whole and of every part, the anticipation of every problem of execution. A new guiding spirit was needed to carry the great work forward.

By destiny or prevision the master builder had prepared a deputy who was indeed a part of himself—his own son, Colonel Washington A. Roebling. This young but able engineer had been trained and prepared for this task. He had been given the best engineering education of the time; he had received his practical training under the greatest bridgebuilder of his day—his own father; and his own judgment, initiative, and courage had been tried and tested on the battlefield.

For years he had shared his father's professional confidence and labors. They had worked together on the construction of the Pittsburgh-Allegheny bridge and the great Cincinnati span, and the disciple had become thoroughly schooled in the master's methods. And in the Civil War he had devised and built his own suspension bridges under conditions testing both resourcefulness and fearlessness.

When the elder Roebling began planning the Brooklyn Bridge, the son took an immediate interest in the great adventure. He helped his father in the preliminary studies; and then he journeyed to Europe, visiting all the important engineering and metallurgical works and studying caisson work and the fabrication of the new structural material, *steel*. With the mass of data and information he had collected, he returned to America to serve as his father's aide for twelve months before the latter's death. He was thoroughly familiar with the plans for the bridge, having helped to create them.

And, most important of all, Washington Roebling had the

endowments of character and personality that the grave responsibility demanded. He had inherited not only his father's true engineering instincts but also his concentrated devotion and unconquerable tenacity of purpose.

To this son the great engineer on his deathbed turned over the completion of the task. The following month—August, 1869—at a meeting of the directors Colonel Washington A. Roebling was officially appointed to succeed his father as Chief Engineer of the work.

Courageously and loyally the son assumed the gigantic responsibility of carrying forward the construction of this, his father's greatest dream. To this consecrated task the young engineer gave all he had—his health, his strength, his career.

He resigned as president of the wire company at Trenton to devote all his time and energy to the bridge, and his two brothers, Ferdinand and Charles, took charge of the wire-rope plant. From that day until the job was ended, Washington Roebling lived only for the Brooklyn Bridge.

But many were the problems, the difficulties and the struggles which fell to the young chief engineer during the building of the great span. Tragedy, stark and grim, marked both the commencement of the project and its consummation. In the approximately thirteen years from the actual beginning of the work, when laborers started to clear away the site for the Brooklyn tower on January 2, 1870, to the opening of the structure to the public on May 24, 1883, deaths and casualties were many. Into the construction of the monumental span went blood, sweat, and anguish in equal measure with cement, stone, steel, and treasure.

2

The first problem was the construction of the foundations for the two massive piers in the river. In order to carry these granite masses to their towering height, a firm foundation was indispensable. For this work pneumatic caissons were adopted. This was then a new method of building deep foundations under water, just being introduced into this country and involving many unknown elements of difficulty and hazard.

Even Colonel Roebling was not sure how it would work out in practice. He was just as much a pioneer as Kit Carson or

DOWN IN THE CAISSON

Contemporary engravings showing, left, feeding the bucket of the water shaft; and, right, the mouth of the supply shaft.

THE NEW YORK ANCHORAGE

The diagram at top left shows the method of anchoring the cables; at top right is one of the anchor plates.

Daniel Boone. He had to sink his caisson far beneath the river mud and fill it with concrete, all while building tens of thousands of tons of masonry upon it.

A pneumatic caisson for sinking subaqueous foundations is merely a diving bell on a vast scale. It is essentially a huge airtight box, having a roof and sides but no bottom, in which men can work under the protection of the compressed air which keeps the water from entering and filling the chamber. The "sand-hogs," as the men working within the confined space are called, handle the excavating of the material below them so as to permit a regulated sinking of the caisson under the superimposed weight of the masonry which is simultaneously being built up in successive courses upon its roof.

At the Brooklyn tower site a trial boring, made by the elder Roebling in 1867, showed solid rock at a depth of over ninety feet below high water. To carry the foundation down to that depth would have added enormously to the cost. Fortunately this was not necessary. At a depth of about forty feet the material was so hard and compact as to be entirely satisfactory as a foundation.

It was necessary, however, to provide a uniform distribution of the load over the entire area, whatever the depth might be, so as to insure an even settling from the inevitable, though slight, compression of the underlying material. The magnitude of this aspect of the problem became evident when it was considered that the base had to support a weight of eighty thousand tons, in the form of a masonry tower extending to a height of more than three hundred feet above the foundation.

Special consideration had to be given to the design of the roof of the caisson, for adequate strength and rigidity, since it would have to carry the superincumbent mass of masonry during the sinking and would furthermore have to serve as a permanent foundation platform to carry the full load of the tower after completion.

Colonel Roebling decided that a solid timber platform of sufficient thickness to act as a beam for carrying and distributing the superimposed load would meet the requirements of the case. He knew that timber permanently immersed in water is imperishable, and if sunk beneath the river bed would be safe from the ravages of sea worms. This massive timber platform would constitute the roof of the caisson and would later remain as a permanent part of the foundation under the tower. The buoyancy of the timber would moreover facilitate the lowering of the caisson without the use of powerful and expensive mechanical rigging.

Work was begun on the caisson for the Brooklyn tower. The roof was made a solid mass of timber, initially five feet thick, to be subsequently increased to fifteen feet to support the weight above. The walls, also sturdily fashioned of timber, were made nine feet thick at the top and tapered to form a cutting edge at the bottom. This cutting edge, or shoe, to penetrate the sand and clay, was shod with a heavy rounded iron casting and was armored with boiler plate.

To make the caisson airtight, the seams were all thoroughly calked with oakum inside and out. In addition, to protect the timber from the teredo and other marine borers, an unbroken sheet of tin extended over the whole caisson between the fourth and fifth courses of the roof and down the four sides to the shoe.

The caisson was constructed close to the Brooklyn shore at Greenpoint, on seven launching ways, sloping down into the water. It was built like a huge inverted flatboat, and with the broad side toward the water. The two end walls and the five intermediate partitions framed inside the caisson provided direct bearing on the seven supporting ways. Full complements of wheelbarrows, crabs, winches and other tools were placed within the caisson for future use. The total launching weight was three thousand tons.

On March 19, 1870, the launching took place. It was considered a noteworthy feat of engineering and a success in every respect. As soon as the last block was split out and the control cams were released, the caisson began to move. The impetus it acquired in the first part of its course proved sufficient to overcome the immense resistance offered when it struck the water. Exactly as planned the air caught inside of the air chamber assisted materially in buoying up the mass. The deck was not submerged, nor was a large wave formed in front.

The launching of the caisson, which had been under construction at the Greenpoint shipyard for months, was regarded by the public as the real commencement of the structure and the first visible and tangible indication of what was to follow. The ceremony was attended by several thousand. The caisson and the speakers' stand were gaily decorated with flags and bunting. A band played and the politicians supplied the oratory. After the launching the contractors furnished refreshments.

An air pump and boiler which had been set up on deck before the launching were promptly put in operation, and in a few hours the water was all displaced from the air chamber, the surplus air blowing out at one corner. This provided proof that a satisfactory state of airtightness had been secured. Afterward, when the air was all allowed to escape, the top of the caisson settled down to within seventeen inches of the surface of the water, which agreed with previous calculations of its buoyancy.

3

In the meantime the permanent site was being made ready to receive the caisson. This was the old ferry slip where John Roebling had received his fatal injury. The preparation of the site consisted in establishing a rectangular basin, open on the side toward the river, the remaining three sides enclosed by a wall of timber sheet piling, and the bottom leveled by dredging to a uniform depth of eighteen feet below high water, a depth sufficient to float the caisson at all stages of the tide.

The dismantling of the old Fulton Ferry slip—pulling out piles, tearing out fender sheathing, removing heavy cribwork filled with stone, and dredging off the loose material on top— required about one month's labor.

All the old timber and piles taken out were found to be infested with thousands of sea worms; their ravages, however, were confined to the space between low water and the mud line. A pile which was sixteen inches in diameter below the mud line, and perfectly sound and free from worms, would be eaten away to a thin stem of three inches just above, all timber being affected alike. This experience showed the necessity of going below the river bed with the timber foundation, and also proved its entire safety in that position.

After about ten thousand yards of old fill and surface mud had been removed, a line of soundings was taken that showed three thousand cubic yards yet to be removed before the level of eighteen feet would be reached. The character of this material was next to solid rock; none of the dredges could make the slightest impression upon it. In fact, all the old harbor charts showed this point as a reef of rock. Under these circumstances recourse was had to powder. Holes were first made in the bottom, by driving down a pointed pile and then withdrawing it. Three such piles were used, shod with iron at point and head. Into each hole a canister of powder was inserted by a diver, and the charge was then exploded by electricity.

After about one week's work the canisters were made heavier, using thick shells of cast iron. These possessed the advantage of dropping to the bottom of the hole by their own weight. After a thorough blasting the dredge could work to advantage for a time. Boulders too large for the dredge to handle were slung with wire rope under water by divers, and either raised to the surface or floated under water beyond the enclosure.

The casual observer above, to whom the surface of the water appeared the same, day after day, would think that nothing was being done. But the divers who slung boulder upon boulder, night after night, had a different story to tell.

The driving of the iron-shod piles afforded a thorough knowledge of the entire ground. On the easterly side a few blows would force the pile through soft clay. In the center, however, there was a broad ridge of hardpan of varying thickness. It was a compact mass of sand, clay, and boulders cemented together so hard that frequently a hundred blows of a fifteen-hundred-pound hammer were required to drive the pile three feet into the material.

While the dredging progressed, the enclosure proceeded. An outer row of piles was first driven, and anchored back to resist a high bank of fill. Within this line a row of timber sheet piling was driven, space being allowed to tow in the caisson.

On May 1, the site having been sufficiently leveled, the strange upside-down craft was towed down the river from Greenpoint, a distance of five miles, by six tugboats, under the command of Captain Maginn. During the trip the air chamber had to be kept fully inflated, as for part of the journey there was but a foot of space between the river bottom and the lower edge of the caisson. On the following day the caisson was warped into position and carefully centered with the aid of surveying instruments. By June 20 the ten additional courses of roof timber, to complete the fifteen, were laid.

As the roof of the caisson was built up, additional sections of the large wrought-iron shafts were put in. These included two "water shafts" for taking out excavated material, two "man shafts" for entering and leaving the caisson, and two "supply shafts" for sending down supplies during the sinking and for delivering the material to fill the caisson when it was sunk to its final level. There were also a number of smaller pipes passing through the roof, for supplying gas and water and for blowing out sand by air pressure.

All of the doors to the air locks and shafts fitted closely, and were mounted to swing downward into the space or chamber having the greater air pressure.

On May 10 the air chamber was first entered by the engineers and workmen, and explored. Removing the temporary wooden bulkhead used in the launching, clearing away boulders and all loose material from under the edges, and cutting doorways through the main division frames occupied several weeks. During this period the caisson was rising and falling with every tide, and not until three courses of masonry had been laid was it weighted sufficiently to rest firmly on the bottom and resist the action of the tides, its top still visible above the water. Now the operation of sinking the caisson could proceed. By this time the force of men had been increased to over one hundred, and the work could be carried on continuously. The men worked in three shifts of eight hours each.

The work of excavation was carried on from the air chamber. Any obstructions encountered under the shoes and under the partition frames were removed to permit the sinking to progress. At the same time the masonry was being laid on top, with the aid of boom derricks and hoisting engines.

4

Crossing the East River by ferry, passengers noticed an area of strange activity near the Brooklyn shore—puffs of steam engines, the clanking of chains, and the groans of huge derricks. Squads of workmen were hustling about their tasks on a mounting pier of masonry. Immense blocks of stone were being landed from a schooner, hoisted, and set in position. But the most fascinating part of the work was invisible, the execution of a great engineering feat deep below the bed of the river.

The novelty of the subaqueous chamber attracted newspaper men and others whose curiosity outweighed their fears. A breezy account of the unique experience, "Down in the Great Caisson," was published in July, 1870, by the editor of the *American Artisan:*

Strolling one day by the Fulton Ferry, we thought it well to visit the huge caisson upon which the Brooklyn Pier of the great East River Bridge is destined to rest. After a brief parley at the gate, we entered the yard and found our way to the cylindrical shaft projecting from the top of the caisson and through which access is had to the interior. Noting, as we went, the workmen laying the first course of stone blocks upon the caisson, our attention was soon diverted by the appearance of a gentleman with a very red face slowly emerging from the shaft and forcibly asserting "Never again!" Anxious to learn to what the gentleman thus decidedly objected, we passed into the shaft. The attendant shut down its circular cover, and told us not to be alarmed if we felt strangely. We told him we wouldn't, and he turned the brass handle of a valve. The compressed air from below came in with a sound like the rush of waters, and our ear drums felt as if to split was our "manifest destiny." This lasted but a very little while. A plate was removed from the center of the floor on which we stood, and under us was revealed the deep round shaft, having within it an iron ladder, upon which we descended to the space below, where the work of excavating underneath the caisson was being carried on.

This work of excavation is very simple, albeit very hard for the laborers. No inconvenience is experienced from the density of the atmosphere, exceeding the pressure of the external air by about eight pounds to the square inch. The inauspicious sensations experienced on entering remain but a few minutes. The bed of the excavation is slush and mud, slowly carried by shoveling to the apparatus by which it is removed. Numerous large boulders are met with, and are drilled and broken to fragments before being taken out. Small calcium lights throw steaming luminous jets into the corners where the workmen are busy, and sperm candles stuck in sockets fixed on iron rods serve as torches for visitors threading the planks laid down for paths in the submarine cavern. The mud and silt under foot, the dank wooden walls at the sides, the chambers dim in the scattered lights, and the laborers moving like good natured gnomes in the shadows, complete the picture that we carry away in our memories. After ten minutes we came up the ladder to the air lock, the opening in the floor was closed, another brass valve handle is turned, and not without some of the uncomfortable feelings incident to entering, the air is reduced to normal pressure, and the cover of the shaft is thrown open. We come out with opinions not quite so decided as those of the gentleman we met on entering, for a sight of the manner in which so important an engineering structure is sunk to its bed below the restless waters of the East River is ample recompense for the tensioned eardrums, the climb on the iron ladder, and the half-dozen handfuls, more or less, of mud the editorial coat-tails unwittingly brought from a score of feet below the level of the river bottom.

5

Within the caisson the laborers, with pick, shovel and barrow, kept feeding the dredges with the mud and stones they dislodged. When the earth had been sufficiently removed, the workmen drove out alternate wedge blockings from under the partitions on which the mass rested, leaving its weight on the remaining blocks; these slowly settled under the load, and in the most uniform manner. The earth was leveled, and the wedges previously removed were again tightly driven in. Then the other wedges were driven out, and in the whole operation the caisson had gone down a fraction of an inch. The weight and strength of the caisson and the load it carried were so enormous that any difference in the density of the bottom material or in the driving of the wedge blockings was irresistibly equalized, and the mass sank, inch by inch as the process was repeated, plumb down. Of course care was perpetually taken that no boulder or other obstruction should be left under the descending shoe of the caisson's edges, and no pains were spared to remove any cause that might tend to deflect the perpendicularity of the descending mass.

But for all that, the work was anything but easy. Often the material to be excavated was so hard and tough that the picks and shovels could not loosen it, and steel bars specially prepared had to be driven in and small portions at a time picked off. Boulders kept turning up in the most unfortunate places, in all sizes up to ten cubic yards, most frequently of traprock and sometimes of quartz or gneiss. They represented a complete series of the rocks found for a hundred miles to the north and northeast of Brooklyn. These boulders had been transported and deposited there by the ancient glacier. Men in the working chamber constantly probed under the water for these hidden obstructions, in order to prevent the caisson from settling on them to its possible injury, and to prevent the certainly increased difficulty of their removal under its weight. Sometimes even powerful winches and three ten-ton hydraulic jacks were ineffectual to remove the large stones from under the edges and the partitions. Boulders had to be undermined and rolled into the caisson, and there split by drills and wedges.

A vivid picture of the sensations and impressions within the caisson has been left by E. F. Farrington, the master mechanic on the construction of the bridge, a veteran of Roebling's bridges at Niagara and Cincinnati: "Inside the caisson," he wrote, "everything wore an unreal, weird appearance. There was a confused sensation in the head, like 'the rush of many waters.' The pulse was at first accelerated, and then sometimes fell below the normal rate. The voice sounded faint and unnatural, and it became a great effort to speak. What with the flaming lights, the deep shadows, the confusing noise of hammers, drills, and chains, the half-naked forms flitting about, with here and there a Sisyphus rolling his stone, one might, if of a poetic temperament, get a realizing sense of Dante's inferno. One thing to me was noticeable—time passed quickly in the caisson."

The results of the first month's work were not very encouraging. The hardpan under the caisson was difficult to excavate, and the penetration attained was insufficient to seal the cutting edge securely. Frequent "blowouts" occurred—a large quantity of compressed air suddenly escaping under the edge. The body of compressed air was always trying to escape. Before the caisson became firmly fixed by side friction in the material penetrated, it was liable to lift up at one edge and thus set free a portion of the confined air. Sometimes the wave of a passing steamer would be sufficient to upset the equilibrium of the caisson and cause such a blowout. This sudden rush of escaping air would send up a large waterspout, thirty to sixty feet high, flooding a part of the work on top and stampeding the workmen laying the masonry. Inside the caisson the roaring was also heard, and with it came a powerful rush of air toward the place of escape. After each blowout the bottom of the working chamber would be flooded and the remaining air would be filled with a thick fog which persisted for some time.

The rate of descent under these handicaps—hardpan, boulders, and blowouts—averaged less than six inches a week. At this rate it would have taken more than a year and a half to sink the Brooklyn caisson, and about three years for the one on the New York side. And the wear and tear on tools was enormous.

But Colonel Roebling, who spent much time with the men in the caisson, was not disheartened. He realized that the initial rate of progress was disappointing. "On the other hand," he pointed out, "we are gaining daily in experience. The workmen are becoming more accustomed to the novel situation and more practised in the particular kind of work to be done."

6

There were new, increasing, and almost endless difficulties.

After the caisson reached the depth of twenty-five feet below the outside water level, the boulders encountered were so large and numerous that blasting was necessary. Although the idea of using powder had been entertained for some time, it had not yet been attempted as it appeared to involve too many hazards. There was no previous experience as to the possible consequences. What would be the effect of the explosions in the dense atmosphere? Would the concussions rupture the eardrums? Would the powder smoke in the confined space become suffocating? Would the blasts break the valves and doors in the air locks, letting the air escape and thus causing a flooding of the caisson? Would the pressure of the explosion cause a disastrous blowout of the water shafts? This last was the principal apprehension, as it would be fatal to both the men and the caisson.

Colonel Roebling courageously determined to settle these questions by actual personal experiment. He spent hours in different parts of the working chamber, firing off his revolver with successively heavier charges. Then he proceeded to fire small blasts of powder with a fuse, and finally he tried heavy charges of explosive. When the concussion in all cases was found to be harmless, blasting became an established procedure. The good effects were at once apparent in the expedited lowering of the caisson, twelve to eighteen inches a week instead of six inches. As many as twenty blasts were fired in one watch, the men merely stepping into an adjacent chamber to escape the flying fragments. One convenient way of disposing of boulders encountered under the shoe was to drill a hole through them, plant the explosive charge at the bottom, and shoot the rocks bodily into the caisson where they were broken up at leisure.

When blasting was used, great care had to be exercised against setting fire to the yellow-pine roof of the working chamber through the flash and the burning fuse. The gas pipe was broken several times, but the flame was extinguished before any real damage was done. In blasting under the shoe there was danger of injuring it, but nothing serious resulted. In fact, the shoe was already injured; armor plates were bent and crushed and partly torn off by jagged points of rock, the inner casting was cracked, and in many places the whole iron shoe had been forced in; yet no air escaped, because the clay was tight outside.

The powder smoke from the blasting was a decided nuisance. It would fill the chamber for half an hour or more with a thick cloud obscuring all the lights. A sprinkling jet of water to throw it down mechanically was of little avail. The use of fine rifle powder ameliorated the condition somewhat. The sulphur smell was not disagreeable, simply because the sense of smell is almost entirely lost in compressed air. "A wise provision of nature," observed Colonel Roebling, "because foul odors certainly have their home in a caisson."

The engineer was using the experiences here as a basis for a solution of the problems to be tackled in the more difficult foundation on the New York side. "The use of powder," he remarked, "has proved so efficient that no difficulty is expected in leveling off the irregular surface of the gneiss rock upon which the New York caisson will rest."

7

The work was carried on day and night by relays of men. The majority of the men took their meals along and remained down the full eight hours without any injury.

On deck there were double shifts of enginemen and firemen to run the excavating engines and the engines for the dirt cars, as well as two gangs to dump the cars. In addition there were the enginemen for the air compressors and the stone-hoisting engines, besides blacksmiths, machinists and gas men, one gang to remove the boulders brought up by the buckets, a carpenters' force of twenty-five men, and thirty men for setting the masonry. The total daily force at this tower foundation amounted in all to 360 men.

The work was so hard on the dredge buckets that the night shift was devoted to repairs, not only to the buckets but to the cars, engines, and other machinery as well. The men in the night gang also devoted themselves to getting out boulders from under the caisson in difficult places where only a few men could work, or they would run one dredging shaft alone, or be digging out the "well" under the other shaft.

The men were provided with dressing and eating rooms aboveground, and every means was used to insure their health and comfort and to push the work forward. Two shanties were put up, provided with numbered hooks and pegs for the men's clothes, most of which were left above on account of the warm temperature below. In front of the dressing shacks were sets of wash troughs, with hot and cold water.

As each air lock for entering the caisson held thirty men, two sets of lockings were required to let down a shift of a hundred and twenty men. In changing shifts, the old gang remained in the working chamber below until relieved by the new shift.

For the work in this caisson no trouble was experienced in getting all the labor required. When one man left, a dozen were ready to take his place. New York was the best labor market in the country. There was one small strike in the beginning, but it amounted to nothing. The wages paid at first were $2.00 per day for eight hours' work. After the caisson had reached the depth of twenty-eight feet, the rate was increased to $2.25 for eight hours' work, and remained at that figure to the end. The earlier stages of the work, on account of the constant fog in the chamber and the influx of water under foot, were in reality far more disagreeable than the later stages, when the work was dry.

Each of the two large water shafts came down through the roof of the caisson and extended about two feet below the level of the cutting edge into a "shaft well," or pool. By having the lower end of each shaft sealed in the well at the bottom, a balanced column of water was maintained in the shaft to serve as a seal against the escape of air from the working chamber. Into this well all excavated material in the caisson was dumped, and then removed to the top with a clamshell dredge or grab bucket operating through the shaft.

Gradually but surely the caisson sank toward its final resting place. At the end of five months twenty thousand cubic yards of earth and stone had been removed.

Colonel Roebling was disappointed with the rate of progress. "In regular harbor dredging," he observed, "one bucket alone will raise one thousand yards per day. Our buckets therefore should have removed the material quite comfortably in one month's time. In place of one month five were required, and these were five months of incessant toil and worry, everlasting breaking down and repairing, and constant study where to improve if possible. We had in fact a material which could not be dredged."

The bucket could easily lift out any stone it could grapple, even up to one or two yards in size, provided the stone was shoved into the right position. Sometimes a large boulder would get stuck under the water shaft where the dredge could not reach it. Then someone would dive under the water and move the boulder—this for $2.00 a day and because of devotion to Roebling. "When the lungs are filled with compressed air, a person can remain under water from three to four minutes," said the Colonel. He knew—he had tried it.

The buckets were provided with heavy steel teeth. Different patterns of teeth were required for grappling stone, for scooping up mud, and for other materials. Constant repairs were necessary; a supply of five buckets was required to keep two in working order.

When the two halves of the bucket closed tight, it would bring up water and mud. But the smallest stone coming between the jaws would permit all the sand or other fine material to escape before the bucket reached the top. "And yet, with all our drawbacks," the Colonel remarked, "our daily experience confirmed us in the assurance that we had selected the only instrument capable of disposing of all the material at hand, no matter how large or ill-shaped or badly packed the boulder; no matter how tenacious the clay and hardpan, nor how flowing the occasional veins of quicksand."

At the beginning the air pressure was governed entirely by the tides, and regulated itself according to their height. During the falling of the tide it was practically unnecessary to run the air pumps; but with a rising tide the pumps had to be run at full speed to build up a balancing pressure. A declining pressure in the working chamber was always attended by a thick fog, which lasted until the tide changed. These fogs could not be entirely overcome, although pumping in a large excess of air partially dispersed them. Every change in the air pressure produced either a rise or fall of the column of water in the water shaft, as in a giant barometer; and constant attention was required to prevent a blowout from the lowering of the water in the pool or from a washing away of the dam surrounding the pool.

When the caisson finally entered the watertight and airtight stratum of clay, the tides no longer had any effect upon the air pressure. The clay proved so tight that the pressure could easily be raised to four or five pounds higher than that called for by the outside head of water.

This satisfactory condition continued until fresh-water springs were encountered. Then there was an end to regularity of air pumping. Sometimes three pumps were sufficient, then again all six compressors running at their maximum speed could not maintain an adequate pressure. And as the pumps were overtaxed, they broke down, and the pressure fell still further. There was one critical period when the pressure fell to eleven pounds in place of twenty, where it should have stood.

A minimum of three pumps were needed at all times, in order to supply the necessary fresh air for 120 men and the numerous candles and gas lights, and to prevent a rise in temperature. The thermometer in the working chamber stood uniformly at seventy-eight degrees, day and night, winter and summer, whether the temperature outside was ninety or zero.

With all the discomforts and exposure of working and blasting so far under water and the ever-present dread of being crushed under the great weight overhead or of being trapped in a flooded caisson, another grave danger arose from living in the compressed air—the danger of caisson disease, commonly called the "bends." The symptoms of an attack were cramps and severe pains in the joints, followed by paralysis. The caisson was no place for men suffering from heart or lung disease, or enfeebled by age or intemperance. But a man of good health and physique, with sound head, heart, and lungs, temperate in all things, and observing a few simple rules, had no difficulty in working under a pressure of twenty or thirty pounds.

In the depth reached by the Brooklyn caisson, the highest pressure ever attained in the air chamber was only twenty-three pounds to the square inch, added to the normal atmospheric pressure. Such a pressure was not sufficient to produce any serious cases of caisson disease. Only six men suffered attacks amounting to temporary paralysis, and in each case this occurred on their first visit and after staying below for only a short time. None of the old hands were affected to any extent, even when staying down eight hours.

The only physical discomfort normally experienced in the work on the Brooklyn caisson was in the effect on the eardrums in passing through the air lock when entering or leaving the caisson. During the period of sharply changing pressure in an air lock a severe pain in the ear is usually felt until the pressure on the two sides of the eardrum is equalized. To some the pain is insupportable, while others are hardly inconvenienced. But this is only a transient discomfort, although it may sometimes result in deafness.

As the weather became colder, the men became subject to colds and congestion of the lungs by reason of the severe drop in the air temperature from eighty to forty degrees which attended a passage out of the caisson through the air lock. A simple and effective remedy was provided by equipping the air lock with a heating coil, through which steam was allowed to flow as soon as the outlet air cocks were opened. This overcame the drop in temperature and, at the same time, eliminated the formation of the disagreeable mist which otherwise attended decompression.

8

As the caisson slowly descended, the tower masonry was being built up above by means of three large derricks with horizontal booms, standing on the caisson itself and guyed from the land. They controlled all parts of the foundation. For the lower courses Kingston limestone was used. Above low water granite was used exclusively for face stone. A large stone yard was established three miles below the bridge. As the stones arrived from the quarries in sailing vessels or barges, they were unloaded and assorted in courses, and then reloaded on stone scows and sent to the tower.

The load on top and the pressure within kept slowly increasing as the sinking progressed, until one Sunday morning, about six A.M., the "Big Blowout" occurred. Every particle of compressed air suddenly blew out through the south water shaft, leaving the caisson in an instant. The immediate cause was the washing away of the dam surrounding the pool of water under the shaft, and was the result of a watchman's negligence. As Colonel Roebling later pithily remarked: "To say that this occurrence was an accident would certainly be wrong, because not one accident in a hundred deserves the name. In this case it was the legitimate result of carelessness."

Luckily it occurred when no one was working below. What effects were produced inside the caisson at the time is therefore not known. Eyewitnesses outside heard an explosive roar and suddenly saw a dense column of water, fog, mud, and stones thrown up five hundred feet into the air, accompanied by a shower of falling fragments and yellow mud covering the houses for blocks around. This column was seen a mile off. The accompanying roar was deafening. The effect was like a volcanic explosion. The noise was so terrifying that the whole neighborhood was stampeded and made a rush up Fulton Street, to get away from the waterfront. Even the toll collectors at the ferry abandoned their tills and ran. The stampeding crowd soon collided with another, running in the opposite direction to see what had happened.

There were three men on top of the caisson at the time, including the watchman. The current of air rushing toward the blowing water shaft was so strong as to knock him down; while down, he was hit on the back by a stone, and then he lost consciousness. One of the other men jumped into the river, and the third buried himself in a coal pile.

It was all over in an instant. Both doors of the air lock automatically fell open with the release of interior pressure, and the dry bottom was visible through the air and water shafts. For the first and only time daylight entered the caisson and the interior could be seen without artificial illumination. Not a particle of water had entered under the shoe! As soon as possible a stream of water was passed into the water shafts from above, the locks were closed, and in the course of an hour the air pressure was restored to fifteen pounds. It required some courage to re-enter the caisson.

The total settling that had taken place during the few moments of this blowout amounted to ten inches. Every block under the frames and posts was absolutely crushed, the ground being too compact to yield; none of the frames, however, was injured or even out of line. The brunt of the blow had been taken by the shoe and the sides. As the loaded caisson suddenly descended, one sharp boulder under the cutting edge had cut the armor plate, crushed through the shoe casting, and buried itself a foot deep in the heavy oak sill. At this point, and in a number of other places, the sides of the caisson had been forced in some six inches. The marvel was that the airtightness was not impaired in the least.

Subsequent examination showed that the roof of the air chamber had assumed a permanent depression or sag of four or five inches. This deflection never increased.

As the caisson proceeded on its downward course, the disproportion between the dead weight above and the supporting air pressure below became greater and greater; and in order to meet this overweight, a large number of additional shores were introduced; each prop rested upon a block and wedges and supported a cap spiked against the roof. These shores, requiring repeated readjustment, added considerably to the labor of lowering the caisson; they also diminished the available working space. They gave, however, a positive assurance against any crushing of the caisson from above and could be readily removed when a boulder was to be taken out.

The downward movement of the caisson was usually so impulsive that the blocks under the posts were allowed to crush and were subsequently dug out. "In fact," remarked the Colonel, "their crushing was the only indication we had that any portion of the caisson was bearing particularly hard. The noise made by splitting of blocks and posts was rather ominous, and inclined to make the reflecting mind nervous in view of the impending mass of 30,000 tons overhead."

When the caisson had arrived within three feet of its proposed resting place, the Colonel decided to erect for its support seventy-two brick piers, symmetrically located. Their capacity was just sufficient to support the whole weight above in case the air should blow out. These pillars, blocking up the roof of the working chamber, were completed in three weeks. The wise foresight of this precautionary provision was later demonstrated.

9

In those days there were no electric lights—they were not invented by Mr. Edison until nine years later. But the men in the interior of the caisson could not work in darkness. The problem of satisfactorily illuminating the working chamber filled with compressed air presented many serious difficulties. In the absence of reflecting surfaces, a powerful light was required to penetrate the thick mist usually present and to illuminate every foot of the uneven soil.

Colonel Roebling tackled this problem with scientific practicality. Candles were tried first, but they were expensive and inefficient, and, from their rapid and incomplete combustion, they emitted an amount of smoke that was intolerable in the confined space. The inhaling of so much floating carbon was injurious, as the lampblack would remain in the lungs for weeks and months. Nevertheless, candles had to be used for all special work requiring close illumination. The Colonel found that he could eliminate much of the nuisance of the candle smoke by redesigning the candles and by special chemical treatment of the tallow and the wick.

Oil lamps he promptly discarded since they smoked more than candles, and the oil was dangerous as a fire hazard.

The existence of an establishment in New York for the

production of compressed oxygen gas in large quantities and at moderate prices made the introduction of calcium lights quite feasible. A double system of pipes was put up in the air chamber, one for oxygen and the other for coal gas. Fourteen calcium lights were installed, two in each compartment. In addition there were sixty burners for common street gas, which was used whenever the supply of compressed oxygen failed. The total lighting cost was $5,000, of which the candles cost more than half.

The calcium lights had the advantage of producing less heat than gas or candles and practically no smoke, the product of combustion being principally water. The ordinary gas lights were far more economical than either calcium lights or candles, but they produced an intolerable amount of heat and vitiated the air even more than candles, although producing but little visible carbon.

To overcome the air pressure within, all gas had to be compressed before being fed through the pipes entering the caisson. Quite an apprenticeship was necessary to adapt the calcium lights to all the new conditions. The chief danger lay in leaky pipes and in carelessly leaving cocks open. One gas explosion took place below, sufficient to singe off whiskers and create some alarm. The sense of smell became so blunted that leaking coal gas was not easily detected.

Naturally, one of the gravest dangers attendant upon caisson work before the days of the electric light was fire. The menace from fire, accelerated by the compressed air, was ever present. In the compressed air the flame of a candle would return when blown out. Every precaution had to be taken and all inflammable materials banished. Several minor fires at the beginning showed the necessity of constant caution. One of these early fires was of sufficient magnitude to demand the flooding of the caisson; this was easily accomplished at that time because the water entered freely under the shoe as the air escaped above through emergency release valves provided for the purpose.

But when the caisson entered more impervious material and the fact was settled that the river water was permanently excluded from the interior, it became a matter of increased importance to provide every possible safeguard against fires. Two large hose connections were installed, throwing high-pressure fire streams; and steam pipes were introduced, for smothering a blaze. In addition all seams between the roof timbers were carefully pointed with cement, and iron shields were placed over the permanent lights. It was made the special duty of two men to watch continually over all lights.

10

Notwithstanding all these precautions, there was a seam, where the supporting frames joined the roof, that had not been pointed; an empty candle box was nailed under it, in which some man kept his dinner, and while getting it he probably held a candle too close to the timber roof. This proved to be the heel of Achilles. The oakum calking ignited, and the air pressure drove the fire into the interior of the roof timbering and out of sight. The fire was discovered on the evening of Friday, December 2, 1870. Being directly over the frame, it had remained undetected until the latter had been partly burned through and, judging from the headway attained, it had been burning for some hours. A fire under the waves of the East River! A fire in timbering precariously supporting many thousands of tons of masonry overhead! A slight panic ensued among the men, but no one left the caisson.

Colonel Roebling was summoned, and he took charge of the situation. All appliances for putting out the fire were brought to bear. While the hose was being got ready, two large cylinders of carbonic-acid gas under high pressure were discharged into the burning area without producing any effect whatever; as soon as the stream of carbon dioxide was stopped the timber would immediately reignite. The two high-pressure streams of water were then turned on and these soon extinguished all fire that could be seen. There was a violent draft of air through the burned aperture; this was stopped with cement. The streams from the two hose connections were kept playing for two hours. At the end of that time one of them was replaced by steam at high pressure and allowed to run for half an hour; but it could not be ascertained if this was of any benefit, and the water was again turned on.

Colonel Roebling remained in the subaqueous chamber for seven hours, from ten at night to five in the morning, until the fire appeared to be conquered. Then he collapsed and was brought out unconscious. But four hours later he was back in the air chamber.

The question of flooding the caisson had been seriously discussed by the Chief Engineer and his assistants. To extinguish the fire without having recourse to flooding was very desirable, since that operation involved the risk of permanent injury to the caisson; on the other hand, if the fire was not out, it was simply a question of time before the entire structure would be destroyed.

The only way to ascertain the presence of fire was to bore for it at random through the solid timber. Colonel Roebling had therefore given instructions, before he collapsed, to have a series of auger holes bored upward into the timber roof.

A number of holes were bored for a distance of two feet. They showed no fire. Others were then bored three feet, still showing no fire. This result was, of course, encouraging. Time was lost in lengthening out augers and also in the boring, because the draft carried the chips up into the holes. At 8 A.M. a hole four feet deep revealed the dreaded fact that the fourth course of timber was one mass of living coals!

The Colonel, still weak and exhausted, reached the spot. He at once decided to flood the caisson as the only way to save it. The city fire department was called out and all available fire engines were soon at work pouring water down the shafts. Additional fire-fighting forces were brought up. The *Fuller*, a harbor fire boat, supplied eight powerful streams; the *J. L. Tebo* three, and a Navy Yard tug two more. By ten that morning thirty-eight streams of water were pouring into the caisson, beside the water from the emergency pipes in the caisson itself. In five and a half hours the air chamber was filled. The total quantity of water required was over a million gallons.

The caisson was allowed to remain flooded for two and a half days. The air pressure was then again put on and the water forced out. It took six hours, and the air pressure required was twenty-two pounds. An inspection below showed but little apparent damage, beyond blocks that were crushed and some posts thrown over. The structure proved tighter than before the flooding, owing to the swelling of the timber. The building of the brick piers under the roof was resumed and completed in about two weeks, when the caisson was lowered down to them through the remaining distance of two feet.

No injury appeared in the masonry above from the changes in pressure to which the caisson had been subjected, but it took three months to make repairs below.

II

For several weeks after the fire, odors of turpentine and of other products of the combustion of yellow pine were very strong above the caisson, being forced out with the air bubbles; and a large quantity of frothy pyrolignic acid, a product of the destructive distillation of wood, made its appearance on top of the masonry. These signs gave rise to very unpleasant suspicions.

About two hundred exploratory borings were made in the roof to ascertain the extent of damage from the fire, both laterally and vertically. It was found that the damage was confined to the third and fourth courses of timber, but had spread out laterally in many directions, covering a much larger area than had been anticipated, the remotest points being some fifty feet apart. The fact that the air rushed out through every bore hole indicated that the immediate roof must be relied on to retain the compressed air. With this object in view it appeared advisable to inject cement grout into the burned cavities through the auger holes until the leakage was stopped. Accordingly, a cylinder was prepared with a piston and injecting pipe, and by this means six hundred cubic feet of cement were forced into the cavities. The escape of air then ceased.

A number of trial bore holes failed to disclose any space not filled with hard cement; but in order to make sure, one large hole was cut up into the roof through five timber courses, directly over the place where the fire originated. It was found that the cement grout had, indeed, filled all vacant spaces, but that the timber was covered with a layer of soft and brittle charcoal varying from one to three inches in thickness. There was no alternative now but to remove all the cement and carefully scrape the charcoal from every burnt stick of timber. As the cement was removed, the full extent of the fire was gradually revealed, and in place of one opening in the timber, five were needed in order to reach the remotest points. This task required eighteen carpenters, working day and night for two months, in addition to the attendance of common labor.

Above the first opening the fire had destroyed the third, fourth, and fifth courses, having burned through the tin that had been placed between the fourth and fifth layers, and had also damaged parts of the sixth course. The fourth course, under the tin, had been the principal sufferer. The various ramifications of the fire had evidently been caused by air leaks. In several places one stick was burned away while the adjoining ones remained sound. The "fattest" sticks had succumbed the soonest. The general combustion had been of the nature of a slow charring, progressing in all directions.

The timbers were cleaned and squared off. Whenever a stick was found only partially consumed, it was carefully scraped and the cavity rammed full of cement. The larger spaces were filled up with lengths of yellow pine forced in with screw jacks and wedges, and well bolted vertically and laterally.

It was a job of dentistry on a mammoth scale—cleaning and filling a cavity worm-eaten by the erratic, zigzag wandering of the fire over a space of fifty feet square. The workmen had to crawl and work in all sorts of cramped areas and uncomfortable positions, lighted by little bull's-eye lanterns, in an air full of coal and cement dust and smoke. After everything was filled up solid, a number of long bolts were driven up from below so as to unite both the old and new timber into a compact body, and forty iron straps were bolted against the roof from below for additional reinforcement.

Under Colonel Roebling's zealous supervision the repair work was scrupulously and conscientiously executed. To set at rest any lingering doubts concerning the stability of the structure after the repairs, he called attention to the vast reserve of strength he had provided from the start. "It must be remembered," he observed, "that there are still eleven courses of sound timber above the burnt region. These have abundant capacity to distribute any local deficiency in equal bearing. From the faithful manner in which the work has been done, it is certain that the burnt region has now been made fully as strong, if not stronger than the rest of the caisson."

12

Two weeks after the "Great Fire" the excavation under the caisson had been completed, with the cutting edge at a depth of 44½ feet below mean high tide, and with the timber roof bearing on the brick supports. The concrete filling of the working chamber was then started, commencing at the shoe and proceeding in layers six to eight inches thick. Each layer was allowed to harden for five hours before the next was put on.

When the concreting had been in process a fortnight, one of the two supply shafts suddenly blew out. A charge of broken stone and gravel had jammed in the shaft; the men dumped in another load without measuring the depth, and then gave the signal to the man below, without shutting the upper door. The second charge loosened the first, and the two together overcame the air pressure against the lower door when the release lugs were turned. As soon as this happened, the air rushed out of the caisson, carrying an eruption of stone and gravel with it. Panic-stricken, the men above ran away, leaving those below to their fate. Anyone above, with the least presence of mind, could have closed the upper door by simply pulling at the rope.

Colonel Roebling was one of those trapped in the working chamber in this terrifying situation, between the timber roof and the rapidly rising water. His coolness and presence of mind in the emergency saved the day. He closed the lower door of the supply shaft, which had been jammed open, and soon everything was all right. But perhaps his cool command of a situation showed more vividly in this adventure than it ever had on the battlefields of the Civil War. A vivid account of the experience is recorded in his own words:

"I happened to be in the caisson at the time. The noise was so deafening that no voice could be heard. The setting free of water vapor from the rarefying air produced a dark, impenetrable cloud of mist and extinguished all the lights. No man knew where he was going; all ran against pillars or posts or fell over each other in the darkness. The water rose to our knees and we supposed of course that the river had broken in. I was in a remote part of the caisson at the time; half a minute elapsed before I realized what was occurring and had groped my way to the supply shaft where the air was blowing out. Here I was joined by several men in scraping away the heaps of gravel and large stones lying under the shaft, which prevented the lower door from being closed. The size of this heap proved the fact of the double charge. From two to three minutes elapsed before we succeeded in closing the lower door. Of course everything was all over then, and the pressure, which had run down from seventeen to four pounds, was fully restored in the course of fifteen minutes. A clear and pure atmosphere accompanied it. The effect upon the human system, and upon the ears, was slight, no more than is experienced in passing out through the air lock."

It was fortunate that the interior brick piers had been provided and that the caisson had already been brought to a full bearing on them. Careful examination was made to see what effect the suddenly applied weight of thirty thousand tons had had upon these pillars and the other supports. Although the brick piers had been subjected to an enormous concentration of pressure, they showed no signs of yielding. In the neighborhood of the water shafts and supply shafts, where there were no pillars or other supports, a slight local depression had taken place in the roof. The occurrence demonstrated the fact that the brick props were strong enough to bear the whole load, and also proved the necessity for their erection. There was no increase of air leakage from the abnormal strains to which the caisson had been subjected.

The Colonel, in his account of this experience, raised the interesting question: "What would have been the result if the water had entered the caisson as rapidly as the air escaped?" And he drew this significant conclusion: "The experience here showed that the confusion, the darkness, and other obstacles were sufficient to prevent the majority of the men from making their escape by the air locks, no matter how ample the facilities." He pointed out, however, that there was an automatic safety provision: The supply shafts projected two feet below the roof into the air chamber; consequently, even if the caisson were flooded, one or two feet of air would be trapped in the top of the caisson so that "there would be enough air space left to save all hands who retained sufficient presence of mind."

13

The filling of the working chamber with concrete proceeded. A hundred cubic yards were laid per day, but as the space became smaller this rate of progress was reduced. A great saving in time as well as in concrete was effected by letting the edges of the caisson sink into the ground three feet deeper than the average level of the bottom. This diminished the volume to be filled about one-third.

During most of the time the weather outside was so cold that the concrete had to be mixed below. The gravel, being full of frost, frequently froze fast in the supply shaft; this was obviated by introducing a steam pipe and thawing out each charge. Later, when the weather became warmer, the concrete was mixed above and sent down directly for placing; much labor was thereby saved in the air chamber. The boulders which had been taken out of the caisson were broken up into square blocks, and again built in below with the concrete.

Next to the roof a shallow, sloping layer was rammed in with narrow, flat-faced iron rammers. As this was the most important point in all the filling and was apt to be slighted, it required the most careful watching.

The rapid influx from fresh-water springs prevented any material reduction of the air pressure. The water rising in the shafts was perfectly fresh, without a trace of salt, which showed the welcome fact that the timbers would ultimately be saturated with fresh water. Finally, when the working chamber had been filled in, the air locks were removed and the water shafts were filled with concrete.

When the sinking operation was started, the caisson had been loaded down with masonry, and additional courses were laid as the excavation in the working chamber proceeded. The building up of the tower masonry thus kept pace with the descent of the caisson, so that the work was always above water. When the caisson reached its final level, eleven courses of masonry had been laid. They were set in heavy beds of cement and all the spaces filled with either cement or concrete.

Colonel Roebling, scientifically trained, insisted on a careful planning and arrangement of engines, tracks, turntables, and the like for economical and efficient working. For these things he uniformly made a thorough study and worked out complete plans in advance. His engineering drawings, not only for the structure but also for the details of the construction equipment, were eagerly sought and published at the time in leading technical periodicals on both sides of the Atlantic. In his planning of the operations there was no fumbling, no cut and try, no guesswork. He handled the work like a problem of military supply. In consequence, the mechanics and laborers were never stopped for lack of materials, notwithstanding the difficulties of transportation and the limited space available for storage and handling.

He inherited his father's passion for safety. One rule which he followed throughout the building of the bridge was that no temporary work, such as derricks, scaffolding, and the like, should be used without a careful proportioning of the minutest detail to the strains involved. As a result it may truly be said that, throughout the thirteen years of construction, no man was injured by the breaking of any temporary work from insufficiency of design. Whatever accidents occurred resulted from invisible defects or from the apparently inevitable human equation.

With the boom derricks used for laying the masonry, numerous guys were required, with hundreds of connections, any one of which giving way might result in a serious accident. Colonel Roebling established a rigorous system of daily inspection of all the equipment. Despite all precautions, however, a serious accident occurred in October, 1871. A defective weld in one of the connections gave way, causing two derricks to fall—killing three men and injuring five others.

During the intense cold of December plentiful use of hot water and salt kept the cement mortar from freezing and permitted the work to continue. By May, 1872, the masonry of the Brooklyn tower had been built up to a height of seventy-five feet above the river, with only twenty feet more required to bring the stonework up to the level of the bridge roadway.

14

The sinking of the caisson on the Brooklyn side had involved many trying experiences, but the foundation for the New York tower presented even greater problems, difficulties, and hazards. Layers of quicksand had to be conquered, a depth nearly twice as great had to be penetrated, and new and unknown terrors of caisson disease had to be faced.

The turning point for the Bridge—as determining the feasibility of the enterprise—was reached down in the earth and under the bed of the East River. Everything hinged on the success of this submarine operation, and the engineering world awaited the outcome.

The first borings John Roebling had made for the tower foundation on the New York side gave the appalling depth of 106 feet below the water level—a depth never before attempted. Finally a slight change in location reduced this by thirty feet.

The problem to be tackled was no simple one. Given a structural mass weighing from thirty thousand to fifty thousand tons, of steadily increasing height and weight, and covering an area of seven city lots, it was required, by burrowing underneath through a variety of uncertain materials, to lower the entire structure, including the superimposed tower masonry, steadily, accurately, and safely, to a firm foundation nearly eighty feet below the surface of the river.

The foundation on the New York side was located in deep water, at a distance of four hundred feet from shore. It occupied the two slips of the Williamsburg Ferry. Nearly a year was consumed in negotiations for the removal of the ferry to an adjoining slip. Possession of the site was finally obtained in April, 1871, and the operation of dredging a level bottom for the reception of the caisson was immediately begun. The sloping river bed consisted of black dock mud above the sand, and was covered with sunken timber cribs filled with stone.

To explore the prospective foundation problems at this site, Colonel Roebling had made a series of borings down to rock, nine bore holes distributed around the area of seventeen thousand square feet. The conditions, as revealed by the borings, created a real foundation problem. Above the bedrock was a layer of cemented gravel of abundant resistance, interspersed with areas of flowing quicksand. And the rock itself was extremely sloping and irregular. To attempt to level it off by blasting under the caisson after that depth had been reached would be prohibitively costly in time, money, and hazard. It would add a year to the time of building the bridge and half a million dollars to the cost. To leave one corner of the foundation resting on the rock and the remaining area supported on more yielding material would invite disaster.

After studying the results of the test borings, the Colonel decided that the caisson must go nearly, if not quite, to rock. This determined the approximate depth to which the caisson would have to be sunk. The final question, whether to come to bearing on the rock or to stop a foot or more above it, would have to be settled when that point was reached. "The only course left open under the circumstances," he said, "is to proceed with the work and, when the caisson has arrived within a short distance of the rock, make a sufficient number of soundings, and then determine upon a course of action when we are face to face with the material."

As this caisson would ultimately be subjected to the load of a much larger mass of masonry than the one on the Brooklyn side, he made the roof twenty-two feet thick instead of fifteen, to carry the greater weight. The working chamber was made of the same height as before, nine feet six inches.

To guard against fire—profiting by the experience of the Brooklyn caisson, in which seven fires had occurred, the last almost proving disastrous—the interior of the new air chamber was lined throughout with thin boiler iron, riveted together and calked. As a second safeguard a system of pipes was distributed through the structure in such a way that a large stream of water could be readily turned on at any point.

In this caisson the Colonel sought to improve his design of the air locks. He provided two sets of double air locks, enabling a whole working shift of 120 men to enter the chamber at one locking. And he placed the air locks at the bottom of the shafts, with the thought of saving the men the additional exposure of making the climb in compressed air. Moreover, in view of the greater height, a spiral stairway was installed instead of a ladder.

15

The plans for the New York caisson were perfected in the summer of 1870, and the contract for its construction was made in October. A year earlier, at the time of letting the contract for the Brooklyn caisson, it was with difficulty that any bids were secured. To launch a mass of those vast dimensions upside-down and *broadside* was so contrary to the experience of shipbuilders, that they wanted large odds for the risk involved. But as a result of the increased confidence following the first successful experience, the second caisson, although larger and requiring much more work, was undertaken by the same builders for little more than half the cost of the first.

The caisson was built at the foot of Sixth Street, New York, the old yard across the river at Greenpoint having been abandoned for shipbuilding purposes. A severe winter, with delays in the delivery of the ironwork, prolonged the completion to May 8, 1871, on which day the launching took place. The caisson was then towed to the Atlantic Basin, where additional courses of timber and concrete were put on preparatory to its removal to its permanent site.

Owing to the vexatious delays in obtaining possession of the ferry slips, nothing of any importance could be done toward locating machinery and workshops at the tower site until August, when the ferries finally stopped running. During the next few weeks a pile dock was built, as well as a large square enclosure of sheet piling to receive the caisson. On the pile platform, between the shore and the caisson berth, were erected engine houses, hoisting engines, unloading derricks, railroad tracks and hauling engines, concrete mixers, engines for pumping illuminating gas, thirteen air compressors for supplying air to the caisson, buildings for machinery and offices, sheds for blacksmiths, carpenters, and machinists, and for cement and other stores, and wash rooms, dressing rooms, a hospital, and rest bunks for the caisson men.

All preparations for receiving the caisson were completed by September 11, and on that day it was towed from the Atlantic Basin to its final resting place. Four air pumps and boilers placed on the deck served to inflate the structure during the voyage. The total weight at this stage was seven thousand tons. Under the skillful guidance of Captain Murphy, and with the assistance of six tugs, the trip was safely accomplished in two hours and a half. A few days' work then sufficed to complete the pile enclosure and confine the caisson in its permanent location. The surrounding wall of sheet piling served to break the force of the tidal current, which otherwise would have produced a force of ninety tons against the floating timber structure at a time when it was most important to keep it stationary in its true position—just before touching bottom.

The surface covered by the iron lining inside the air chamber was pierced by six thousand bolts. Although great pains had been taken to make these tight, it was found that four compressors were required to overcome the leakage of air. Men were sent down inside the caisson through the air locks and, floating around on rafts, they succeeded in stopping the leaks so that one compressor running at half speed more than overcame the loss.

The final courses of roof timbers were then put on, and by November 1 the last of these was completed.

Washington Roebling was enthusiastic over the strength he had provided in the design of the caisson. "The great timber foundation is now complete! It contains twenty-two feet of solid timber above the roof of the air chamber, seven stories more than the Brooklyn caisson, and since the strength of such structures varies as the square of the depth, we may consider it to be nearly twice as strong as its Brooklyn brother." And in his final report, following the successful sinking of this foundation, he proudly called attention to the confirmation of his appraisal of the abundant strength of the caisson: "The result has proved this. At a depth of 78 feet, and a load on its back of 53,000 tons, not the slightest sign of weakness or crippling has been discovered! No deflection has been observed in the roof, even when the main frames and edges below were entirely dug out and not resting on the ground." Nowhere else in his report does he indulge in an exclamation point.

16

By the end of November sufficient weight of masonry had been placed on the caisson to prevent its rising at low tide when inflated. A gang of laborers then began work within the caisson for several hours every day, taking up the temporary floor of the air chamber and removing the principal obstructions in the shape of dock logs and stones under the edges and cross frames. As more weight was added above, the caisson became permanently grounded, after which three regular shifts were able to work full time.

After this the underlying mud was attacked. The character of the work at this particular time was more disagreeable than at any subsequent period. The location had for many years been the principal dumping ground for city garbage. The mud to be excavated abounded in decaying animal and vegetable matter. Although the offensive odor of these was checked while imbedded in the salt-water mud, it came forth in its full original strength when brought in contact with the caisson air. Even to the blunted sense of smell in the compressed air, the odors were overpowering. Moreover, the repulsive olfaction was only a warning of the real danger—the noxious effect of inhalation. During this work more men were overcome by the foul air than by the compressed air. The condition was relieved by keeping the material constantly covered with water, until it was gradually disposed of through the two water shafts.

The coarse river sand and beach gravel under the mud soon created a change for the better in the working chamber. The water was easily expelled by the compressed air, leaving it dry under foot. By this time also the gas lights were in operation in all the compartments, giving ample illumination in every part. Two coats of whitewash over the ironclad ceiling and walls aided in reflecting the light. The air chamber was now an agreeable spot compared with what it had been in the beginning.

Sixty double gas burners, divided among the six compartments, gave all the light needed. An interesting fact was observed: In the compressed air all of the gaslights became sensitive flames, leaping in answer to the stroke of a hammer on a piece of iron, or even to the tones of the voice.

The compressed air produced peculiar effects upon sound. Pitch was raised. A deep bass voice became a treble, and the prolonged heavy sound of a blast was so modified as to resemble the sharp report of a pistol. A curious fact, noticed under these circumstances, was the impossibility of whistling: the utmost effort of the respiratory muscles could not increase the density of the air in the cavity of the mouth sufficiently to produce a musical note on its escape. A similar difficulty, though in a less degree, was experienced in speaking; and for this reason protracted conversation was very fatiguing.

For communication between the upper and lower world the telephone would have been of great benefit; but it had not yet been invented. Instead, an ingenious mechanical signal system contrived by the engineers proved of great assistance. It consisted of a tube with index pointers attached at top and bottom; under each pointer was a plan of the caisson, showing the position of every pipe and shaft; by rotating the tube, attention could be called to any one of these points in a moment. In addition, an index attached to a rod, both above and below, traversed a panel of signal messages, such as "Stop," "Start," "Bucket is caught," etc., directing what was to be done.

Above the caisson the stones were laid by three boom derricks, similar to those employed on the Brooklyn foundation. They were guyed solely from the caisson itself, so that the settling of the latter did not disturb the guys. For every twenty feet of masonry the derricks had to be raised, an operation requiring a few days.

The granite came from Maine and the limestone from Kingston, Lake Champlain, and Canajoharie. Owing to the early winter much of the stone in transit for the New York tower was frozen in; but this had been anticipated, and at no time was there any stoppage for want of stone.

During the severest weather the work of laying masonry was suspended for several days at a time. Down below, within the caisson, the excavation and sinking continued uninterrupted. Day in and out 236 men in two shifts were engaged here in removing the mud and gravel.

The performance of the dredge buckets in the mud and in the coarse sand and gravel was quite satisfactory. They constantly maintained a hole about six feet deep under each water

CABLE-MAKING ON THE BRIDGE

THE FOOTBRIDGE USED DURING CONSTRUCTION

shaft, and removed from three hundred to four hundred cubic yards of excavated material a day. The men were equipped with rubber boots, supplied by the bridge company, since this work required them to stand in the water.

17

Some fifty-eight iron pipes to operate as "sand siphons" were located in the roof of the caisson, passing up through the timber and discharging above beyond the cofferdam. It was first intended to use water as the vehicle by which to eject the sand through these pipes; but, after a study of all the governing conditions, Colonel Roebling concluded that it would be more efficient to utilize the compressed air as the ejecting vehicle. He explained his reasoning to his staff and to the trustees: "Another strong reason in favor of the air is this—An air chamber with an iron skin can be made practically airtight, but a certain quantity of fresh air must be thrown in per minute to keep the atmosphere reasonably pure and fit to live in. This air would usually escape under the edges and do no work. Now, why not allow it to escape through pipes and at the same time carry out sand with it, and not be wasted? Any other mode of sending out the material would require extensive provision of additional machinery in the shape of pumps, boilers, and pipes, entailing an additional cost of at least $40,000, and would also involve difficulties of application for want of the required space around the foundation." In view of all these considerations he decided to give the air-siphon system a thorough trial. The result vindicated his judgment, although new problems were created demanding the exercise of further inventive resourcefulness.

In the final arrangement a straight pipe extended to within one foot of the bottom of the caisson. The sand or gravel was heaped around the lower end of the pipe until it was two or three feet deep, when the cock at the top was opened, letting the rush of escaping air carry the material out. The regular supply of air was sufficient to keep three sand siphons going at a time, and the work of throwing and feeding the sand to these pipes was enough to keep sixty or more men busy. The labor in itself was very fatiguing, making frequent resting spells necessary.

When two or more stones jammed in the pipe, they were loosened by dropping a bar attached to a rope from above. When the pressure reached twenty-eight pounds, it was found that too much air escaped, and the lower ends of the pipes were reduced to 2½ inches.

The material carried by the blast of air passed out of the vertical discharge pipe with tremendous velocity, stones and gravel often being projected at least four hundred feet high. As the sand was very sharp, the constant abrasion soon caused the pipes to wear through near the roof.

To deflect the sand at the top of the pipes at right angles, both wrought- and cast-iron elbows were at first used. The furious blast of sharp sand would generally cut through these in an hour or two, sometimes in a few minutes, although the iron was 1½ inches thick. The plan finally adopted was to use an elbow into which a back piece of hard, chilled iron, about two inches thick, was fitted and secured. This device was cheap, and solved the problem. Under the action of the sand blast it would last as long as two days, and then it required but a few minutes for a man to remove the holding key and insert a new back piece to receive the blast.

Great care was required to prevent accidents from the use of the sand ejectors, as, with a heavy pressure, small pebbles were discharged almost with the velocity of musket balls. One of the men, by incautious exposure, was struck in the arm by a large fragment, producing a wound more than five inches long. A boatman on the river had his finger shot off.

Toward the end of the work, when the excavated material from below was utilized above as back fill around the masonry, the horizontal elbows of the sand siphons were all removed, and heavy granite blocks were set above the mouths of the pipes. The sand was then discharged against these granite blocks and thus deflected into the cofferdam.

18

When the quicksand was fairly entered upon, it was found that the dredge buckets in the water shafts became useless. The quicksand, in combination with small stones and boulders, would compact to a mass as hard as rock. This mass could not be penetrated by the teeth of a grab bucket, and even the point of a crowbar could scarcely be driven into it. A trial remedy was the use of a hose under the shaft to loosen and stir up the material, but then the sand was so fine and quick-running as to escape through the slightest crevice in the buckets. The sand pipes became thenceforth the sole reliance, until the coarse gravel and stone became so plentiful as to choke the ends of the pipes, making it necessary to stop to remove the stones. On this account the work during the last ten feet was very slow and tedious. Before this the progress had at times averaged a foot a day, but toward the end this rate decreased to one or two feet a week.

At a depth of about sixty feet a bone of the domestic sheep was found under a large boulder, and still lower down some cannon balls and fragments of brick and pottery. For an enduring foundation it was necessary to go deeper, in order to reach strata that had not been disturbed in the geologic age of man.

At sixty-eight feet a number of boulders too large to be moved were encountered under one water shaft. To permit the sinking of the caisson to continue, the shaft was cut off near the roof of the air chamber, after blowing out the water and capping the top of the shaft. The same process soon had to be repeated with the other water shaft.

The downward movement of the caisson remained under perfect control throughout the whole of the sinking. It was very gradual, owing to the large bearing area presented by the wide cross frames and broad shoe which the Colonel had provided in his design for this caisson. While the caisson was passing through the mud, river sand, and gravel, the frames sank through the material without digging; but in the quicksand and harder material below, the whole areas of the frames had to be dug out underneath before settlement would take place. The caisson was also controlled to sink perpendicularly in its true place, any side or tilting movement being promptly corrected by digging under the high side. Toward the end of the excavation fully half of the caisson rested on quicksand and the other half on cemented gravel; but even under these extreme circumstances, the sand-hogs, under Roebling's direction, were able to put the hard side down lower, while the soft side was kept stationary.

When a depth of seventy feet was reached, soundings were begun in the air chamber for the location of bedrock, by means of a pointed rod ten feet long, driven in by sledges. These thorough probings were carried on daily for a month, until a clear idea of the form and depth of the bedrock was attained. The surface of the underlying rock was found to be very irregular, composed of alternate projections and depressions, the extreme difference in elevations being sixteen feet. Fortunately the top of the rock was found to be covered by a layer of very compact material—cemented gravel—so hard that it was impossible to drive in an iron rod without battering it to pieces. This was the first stratum reached that could sustain the weight of the tower and that bore evidences of not having been disturbed. It was good enough to found upon, as good as any concrete that could be put in place of it. Colonel Roebling therefore decided to stop excavation at a depth of seventy-eight feet and to rest the caisson on the hard material overlying the rock. The jagged points of rock were leveled off and all material was removed down to that depth.

The excavation and sinking were completed in May, 1872. Everything was kept under perfect control. The caisson was stopped exactly at the predetermined depth. The difference of level at the extreme corners of the cofferdam on top of the caisson was measured and found to be less than one inch.

The concrete filling was done as in the Brooklyn caisson, except that the greater supporting power of the thicker roof and the wider cross frames rendered the brick piers superfluous, and they were dispensed with. In filling the working chamber, the spaces under the frames were first made solid; then, wherever there was quicksand under the shoe, a trench was cut down to the hard gravel below and filled with concrete. All the remaining quicksand was dug out below, down to the hardpan, and worked into the concrete. The filling procedure was carried out in such a manner that the final exit could be made by means of the water shafts.

19

For seven months, through winter and spring, Colonel Roebling and his men had been working in the compressed air of the submarine chamber. As the pressure increased, the working hours below for each man had been gradually reduced from the initial eight hours until finally, when the pressure was thirty-five pounds, the men worked only two hours at a time, twice a day.

Men were now hard to get, wages were high, and the time of labor became so short that the progress of the work was slow and costly. To make any real headway required an immense amount of supervision. The young bridgebuilder fully appreciated the problem on his hands. "The labor question," he remarked, "becomes the most serious drawback. The forces of nature may be measured and brought under control provided they are properly understood, but human nature is not so amenable to laws." Early in his training he had learned that the management of men is an important part of an engineer's work.

Ascending the circular stairway above the air lock was found to be exceedingly fatiguing as the depth increased, and the stairway in one of the shafts was replaced by a steam elevator. This relieved the men of the exertion of climbing at a time when their circulation was embarrassed and their system unstrung by the sudden return to normal pressure.

The men in the caisson suffered serious discomfort from the large amount of unconsumed carbon which floated in the air in the form of smoke from the burning gas jets. The inhalation of this suspended carbon produced irritation of the air passages and gave rise to a characteristic black expectoration. Six months after the work in the caisson was finished, some of the men were still coughing up sputa streaked with black.

By increasing the number of compressors, the atmosphere in the caisson was brought to a degree of purity such that it contained only one-third of one per cent of carbon dioxide. To maintain this standard, 150,000 cubic feet of fresh air had to be pumped into the working chamber per hour.

The first observed physiological effects of the compressed air arose from the sudden unbalanced pressure on the eardrums. "Hold your noses," was the warning cry of the attendant before closing the door of the lock. The men were strenuously warned not to go into the lock unless they were able, when holding the nose and blowing forcibly, to feel air enter both ears through the Eustachian tubes. When this precaution was neglected, the individual was said to be "caught in the lock" or unable to "change his ears." Inability to equalize the pressure upon the two sides of the eardrum caused extreme pain, or acute inflammation, or sometimes a ruptured eardrum.

But the affliction to be dreaded was the painful paralysis, and sometimes death, that might follow upon a stay in the compressed air. In those days caisson disease was a new experience, and the medical profession was mystified. It was not until six years later that the nature of caisson disease was determined —by Paul Bert, a French physiologist. The paralysis was found to be caused by the liberation of nitrogen bubbles from solution in the blood and tissues of the body upon reduction of pressure. The danger lies, not in the compression within the air chamber, but in the decompression upon leaving the caisson. The rate of reduction of pressure is the critical factor; and in the case of an attack, the indicated remedy is prompt return to compressed air followed by slow decompression. In modern caisson or tunnel work "hospital locks" are provided at the site for this purpose.

In 1872, however, the cause and the treatment were still a mystery. A few facts were known: The onset usually came immediately upon emerging. Persons in a nervous and excited state, and those of a weak constitution, were most susceptible. While in compressed air, violent exertion such as climbing of ladders and other strenuous work must be minimized. At higher pressures the working period must be reduced.

When the higher air pressures were reached in the New York tower foundation, special precautions were taken against caisson disease; and medical attention was provided. Colonel Roebling engaged the services of Dr. Andrew H. Smith to attend to all caisson cases and to examine new candidates for work below.

Not until the pressure reached twenty-four pounds were any serious effects observed. Dr. Smith was in daily attendance. Every man on the job was examined, in order to exclude those suffering from weak hearts or lungs and those whose resistance was reduced by age or intemperance. All new men were required to secure a permit from the doctor before they were allowed to enter the caisson.

Strict "Rules of Health" were drawn up and were posted conspicuously for the guidance of the workmen in the caisson:

1. Never enter the caisson with an empty stomach.
2. Use, as far as possible, a meat diet and take warm coffee freely.
3. Always put on extra clothing on coming out, and avoid exposure to cold.
4. Exercise as little as may be during the first hour after coming out, and lie down if possible.
5. Use intoxicating liquors sparingly; better none at all. It is dangerous to enter the caisson after drinking intoxicating liquor.
6. Take at least eight hours of sleep every night.
7. See that the bowels are open every day.
8. Never enter the caisson if at all sick.
9. Report at once to the office all cases of illness, even if they occur after returning home.

The physical conditions to which the men were subjected were a strain on their constitutions and on their powers of adjustment. In passing through the lock going down, they endured a sudden rise in temperature of fifty or sixty degrees, coinciding with an increase of atmospheric pressure of from eighteen to thirty-six pounds, at the same time passing from ordinary air to air supersaturated with moisture. Within the caisson the breathing became deeper and faster. The work of the heart was increased and its action was therefore excited. The pulse rose to 120 immediately upon entering the caisson. After two to four hours of work in the compressed air the men had again to pass through the air locks, where this time they underwent a rapid reduction of pressure and temperature. The sudden chilling was alleviated by surrounding the interior of the lock with coils of pipe heated with steam, turned on only when locking out.

Hot coffee was always served to each man immediately he emerged from the caisson. It served as a warming stimulant to relieve the chill and the nervous prostration which marked the return to the open air.

A large room in the yard was fitted up with bunks to give the men an opportunity to rest while the changes of readjustment to normal were going on in the system.

20

Despite all the precautions taken, a large proportion of the workers suffered from the effects of the compressed air. From

January through May there were 110 cases of sufficient severity to require medical treatment; and in three of these cases, the victims died.

The sick were treated in the first instance at the yard, where a room was set apart as a temporary hospital. Remedies were kept on hand: ergot, morphine, and stimulants. Cases occurring during the doctor's absence were treated by the engineer on duty, according to a prescribed plan; or, if the case was severe, the doctor was summoned. Serious cases were sent to the public hospitals.

The ages of the sand-hogs ranged from eighteen to forty-five, and they were of almost all nationalities. The habits of many of the men were doubtless not favorable to health, but everything which admonition could do was done to restrain them from excesses. Many of them slept in crowded, vermin-infested lodginghouses, where the beds or bunks were arranged in tiers, one above the other, in rooms in which there was scarcely an attempt at ventilation. One of the men fell a victim to spotted fever, contracted from such surroundings; and his death was at once ascribed by his comrades to the effect of the compressed air.

The onset of the bends invariably developed after the air pressure was removed. The attacks were characterized by painful cramps in one or more of the extremities, and sometimes in the trunk, frequently accompanied by headache and vertigo, epigastric pain and vomiting. The access of the first sharp neuralgic pains was often very abrupt, as if the patient "had been struck by a bullet." The pain was excruciating—men of the strongest nerve were completely subdued by it. It felt "as if the flesh were being torn from the bones." The agonizing torture generally began in the knees, shifting to the legs or thighs, then creeping up the trunk to seize upon the shoulders and arms. The skin was of a leaden hue, and was covered with a profuse perspiration standing out in beads upon the surface. In 15 per cent of the cases, paralysis occurred. It most frequently affected the lower half of the body, but in some instances also included the trunk and one or both arms. Accompanying symptoms were headache, dizziness, double vision, incoherence of speech, and sometimes unconsciousness. The duration of the attack usually varied from a few hours to six or eight days, but in some cases it continued for weeks.

In the fatal cases a condition of profound coma was the usual forerunner of death. The occurrence of this symptom left but little hope of recovery.

The first fatal case occurred on April 22, when the caisson was at a depth of seventy-five feet. John Myers, about forty years old, a native of Germany, went to work in the caisson for the first time that day, the pressure being about thirty-four pounds. He was of a stout, heavy build, of "very full habit." When examined two days before, his lungs were sound. He worked through the morning shift of two and a half hours without inconvenience, and remained about the yard for half an hour after coming up. He then complained of not feeling well, and went to his boarding place, which was but a few rods distant. As he passed through the lower story of the house, on his way to his own room, which was on the second floor, he complained of pain in the abdomen. When nearly at the top of the stairs, he sank down insensible; and before he could be laid on his bed, he was dead.

Eight days later a second fatal case occurred. Patrick McKay, aged fifty, born in Ireland, had been working four months in the caisson and had not complained of ill health. On April 30 he remained in the caisson half an hour beyond the usual time, at the second watch, under a pressure of thirty-four pounds. The others who were with him in the air lock found, when about to emerge, that he was sitting with his back against the iron wall of the lock, insensible. He was at once carried up to the surface and removed to the Park Hospital. He was unconscious, with face pale and dusky, lips blue, pulse irregular and feeble. Under the administration of stimulants he recovered some degree of consciousness, and begged incessantly for water. Paroxysms of convulsions soon set in, and in one of these he died, nine hours after the attack. The autopsy showed that all of the organs were healthy, except the kidneys, which were the seat of Bright's disease and were very much altered in structure. "In this case," said the physician, "the effect of the compressed air was merely to hasten an event which, at best, could not have been long delayed."

The following month the third death occurred. William

Reardon, born in England, aged thirty-eight, corpulent and "of intempèrate habits," began work on the morning of May 17. He was advised to work only one watch—two and a half hours—on the first day; nevertheless, feeling perfectly well after the first shift, he went down again in the afternoon. The pressure at this time was thirty-five pounds. Immediately after coming up from the second watch—two hours—he was taken with severe pain in the stomach, followed by vomiting. In a few minutes the cramps seized his legs, which soon lost the power of motion though they continued to be the seat of extreme pain. He was removed to the Centre Street Hospital, where he gradually sank, and on the following morning he died.

These fatalities from the work in the pneumatic foundation aroused excited criticism in the public press. The type of operation was new, caisson disease was a mystery, and there was no yardstick of prior experience. The public jumped to the conclusion that criminal negligence was involved.

Colonel Roebling did his best to present a clear picture of the situation, in the light of the limited medical knowledge of the subject then available. "The few cases of death that have occurred can in but two instances be charged to the effects of the pressure," he pointed out. "It is true that scarcely any man escaped without being somewhat affected by intense pain in his limbs or bones, or by a temporary paralysis of arms and legs; but they all got over it, either by suffering for a few days outside, or by applying the heroic mode of returning into the caisson at once, as soon as the pains manifested themselves, and then coming out very slowly."

21

In his official report on the progress of the work at this time Washington Roebling paid generous acknowledgment to his assistants and to the workmen, adding these words pregnant with future significance: "*The labor below is always attended with a certain amount of risk to life and health, and those who face it daily are, therefore, deserving of more than ordinary credit.*"

During the days and nights that the work was going on under the bed of the East River, the young chief engineer was continually in the caisson, personally directing the efforts of his men and setting an example of devotion and courage. From him flowed the energy and impulse that animated the entire force. Mindful always that any slip, no matter how trivial, at this stage of the work, might prove disastrous, he actually spent more hours in the working chamber than anyone else. Night and day he worked with the men under the crushing air pressure, until it wore out his strength.

One afternoon in the early summer of 1872 Colonel Roebling had to be carried out of the caisson, nearly insensible, a victim of the dread caisson disease. All night his death was hourly expected.

In a few days he rallied and, mustering all his reserve of strength, he attempted to return to work. But he again collapsed and had to be brought home.

At the age of thirty-five his days of physical exertion were ended. He remained painfully paralyzed—doomed to lifelong suffering—with progressive blindness and deafness setting in, vocal cords affected, and every nerve and muscle tortured with pain.

Thenceforth this man, who had been full of life and hope and daring at the inception of the work, was a crippled invalid, confined to his home. Just as his brilliance as an engineer was at its height, and just as his masterwork was passing its crisis, he was struck down.

CHAPTER XIX

OF GRANITE AND STEEL

THE CHIEF ENGINEER was helpless, except for his fighting spirit and his active brain. From the hour of his breakdown in the caisson there was not a day when his injured body and nerves were not racked with pain.

But the work had to go forward, and, from his sickroom, the stricken man continued to direct it in every detail. Although his nervous system was wrecked, his mind was not affected; indeed, his intellect appeared to be quickened. Realizing how incomplete his plans and instructions for completing the bridge were, and fearful that he might not live to finish the work himself, he fought his pain and feverishly spent his time in writing and drawing. His plans, his ideas, his specialized knowledge, could not be allowed to die; they must be recorded and made available to his assistants. The papers and sketches he prepared contained minute and exact directions for finishing the towers, for building the anchorages, for stringing the cables, and for suspending the spans. Every difficult or specialized erection operation he anticipated and worked out, with calculations, diagrams, and specific instructions.

In this manner the work continued; and it can safely be said that no great project was ever conducted, before or since, by a man who had to work under so staggering a handicap. From an upper bay window of his home on Columbia Heights he anxiously followed the work through field glasses, while he prayed for strength and struggled for time.

Through the long winter he painfully wrote out memoranda and instructions covering future stages of the work—page after page of detailed notes. Writing so much in his weakened condition further impaired his eyesight. More and more he came to depend upon his wife to act as his amanuensis and to write out his notes and instructions.

His nerves were shattered. He partly recovered the use of his voice, but he was too weak to carry on a long conversation;

and some days he could not talk at all. He had become so morbidly sensitive that the mere sound of a strange human voice was unbearable. Not one of the assistant engineers could consult with him; and yet the most intricate technical problems, to which he alone held the key, had to be worked out for the erection operations.

His wife was as rare and strong as the man she tended. She met the emergency by becoming his capable aide. Instead of crying hysterically over her husband's misfortune and sinking down nerveless and helpless into the privileged softness of her sex, she set herself at once to the task of acquiring the knowledge necessary to become a real helpmeet to him in the critical situation. Under his guidance she studied higher mathematics, the calculation of catenary curves, strength of materials, stress analysis, bridge specifications, and the intricacies of cable construction; and she applied her new education to the problems of the bridge. She grasped her husband's ideas and she learned to speak the language of the engineers. She made daily visits to the bridge to inspect the work for the Colonel and to carry his instructions to the staff. She became his co-worker and his principal assistant—his inspector, messenger, ambassador, and spokesman—his sole contact with the outside world.

But in developing scientific and technical skills to meet the emergency, she did not sacrifice her womanly qualities. She remained the comforting companion and tender nurse. As Thomas Kinsella, one of the bridge trustees and editor of the Brooklyn *Eagle*, remarked at the time: "The most abstruse study has not interfered with Mrs. Roebling's ministrations at her husband's bedside. If he is restored to health, it will be largely through her patient and intelligent attendance upon him, and Colonel Roebling will be indebted to his noble wife even as the people are."

In the summer of 1873, on his physician's advice, the Colonel went abroad—to Wiesbaden—for several months in a vain quest to restore his shattered health, but on his return he suffered another attack of acute prostration.

During the following years the Chief Engineer continued to superintend construction from his sickroom. An artist of the day has left us a drawing, published in a popular magazine, depicting the paralyzed bridgebuilder seated at his window, gazing wistfully at the unfinished structure in the distance; his field glasses rest on the sill; and on a table near by is his beloved violin, its strings now mute—for never again will the invalid's fingers be able to produce the music he loved.

The Colonel's continued planning and direction of the work, despite constant pain and a nervous collapse which would have cut off an ordinary man from mental as well as physical activity, was an outstanding triumph of consecration and will power. His own self-sacrifice inspired the most devoted loyalty in the corps of engineers about him, and they directed all their energies to carrying out his wishes. Not the slightest advantage was ever taken of his illness to deviate from the letter or spirit of his plans; and Colonel Roebling, physically helpless, away from the work, always remained complete master of the enterprise. Like a wounded general who directs a battle from a hilltop, Washington Roebling issued his orders and superintended the completion of the bridge.

Through field glasses, he watched the building of the towers and later the stringing of the cables. Though he was absent in body, it was still his spirit that prevailed and his brain that directed the work. The bridge was his—by right of inheritance, by his own labors, and by his own sacrifice.

2

Physical hardships, dangers, and afflictions were not the only ordeals that had to be faced by the Chief Engineer in the execution of his task. He was also harassed and beset, through the long years of painful toil, by financial and political complications arising from conditions beyond his control. Political corruption, civic jealousies, lack of funds, injunction proceedings, and dishonest contractors were among the tribulations.

The revelations of municipal corruption attending the collapse of the "Ring" and the conviction of "Boss" Tweed in 1871 left an aftermath of popular suspicion, which extended to the bridge project. The unscrupulous Tweed Ring had spread its tentacles toward the prize and had thus cast it under a shadow. The Ring was smashed, its leader was in prison, where he subsequently died; but the cloud of public distrust persisted.

All of the evidence indicates that the officials of the bridge company were sincerely loyal to their trust. Their one transgression—to be judged in the light of conditions at the time—had consisted in yielding, in 1868, to the compulsion of a corrupt city administration as the price of making the bridge possible. It was not long before this one irregularity and the attempt to cover it started a train of embarrassing consequences. The engineer of the bridge was an innocent victim. Vice-President Kingsley, who had bravely put up his own money, silently bore the brunt of the subsequent repercussion.

The first major responsibility of the bridge directors had been to arrange for the construction of the tower foundations; and this also provided the first opportunity to reimburse Kingsley for his initial outlay. On account of the uncertainties and difficulties involved in the sinking of the caissons, contractors would not undertake the work except at an exorbitant price to cover the unknown gamble. The bridge company had therefore decided to purchase the materials, hire the men, and handle the construction operations with its own forces. Kingsley, who had had extensive experience as a contractor, was appointed general superintendent. He had been the moving spirit in starting the enterprise and, in addition, he had privately advanced the funds for the preliminary expenses and for securing the necessary legislation. In lieu of paying an extravagant profit to an outside contractor, the bridge company decided to pay a fee to Kingsley, ostensibly for his making his contracting experience available but—we may infer—principally to reimburse him for expenditures that could not be recorded. In July, 1870, the Executive Committee adopted a resolution to pay him 15 per cent on the amount of all expenditures for the construction of the tower foundations up to high-water mark, including the payment for the land, as compensation "for his services and advances to the company." By the end of the year he had received on this basis a total fee of $175,000. Protests arose from minority stockholders when they learned of this, and a committee was appointed to make an adjustment. In November, 1871, Kingsley's claim was liquidated in full at $125,000; he refunded the difference, and no further sum was ever paid him on that account.

This affair started a whispering campaign. The financial operations of the bridge company were thereafter constantly under the shadow of public suspicion. Of the $4,500,000 subscribed by the cities of Brooklyn and New York, it was rumored that large sums had disappeared into the pockets of public officials. Although the two municipalities had put up the bulk of the capital, the management of the bridge was entirely in the hands of the private incorporators and investors, who had subscribed only $500,000 and who had actually paid in not more than 40 per cent of their subscriptions.

As the cost of the work during the first two years greatly exceeded anticipations, new impetus was given to suspicions of extravagance and diversion of funds. Suspicions became rumors, and rumors became open charges. Where were the enormous sums of public money going? At this rate, how much more would be required to finish the bridge?

The officials of the company were under attack, and they needed a clear, unbiased, and authoritative presentation of the facts to disarm their critics. In June, 1872, shortly before he fell victim to caisson paralysis, Colonel Roebling was requested by the Executive Committee to examine and report "whether the prices paid for stone, lumber and other materials, and for labor, have been reasonable and just, and whether the same have been used in the construction of the bridge, or have been necessary therefor." Within four days the engineer responded with a full and satisfying reply to the questions asked:

"Notwithstanding an order from the President of the Company that the Chief Engineer should have nothing to do with the making of any contracts or the purchase of supplies, I am nevertheless perfectly cognizant of nearly every such transaction. I know that all contracts have been made in a judicious manner and have resulted to the best interests of the Company. . . . The stone for the East River Bridge is supplied at prices lower than any that have ruled for twelve years previous. . . . The same holds true in regard to purchases of lumber, cement, gravel, iron, etc. The contracts for iron work in the caissons have in two cases resulted in direct loss to the contractors. In a majority of instances prices have been so low as to result in constant trouble to the inspection department in keeping supplies up to the required standard of quality. . . . I can further say that every dollar's worth purchased for the Bridge has been expended in a legitimate manner, and for the proper purpose for which it was designed, and nothing whatever has, to my knowledge, been diverted into any outside channel. I am in daily attendance at the Bridge, give it my whole time and constant superintendence, and am therefore in a position to give an honest judgment on this question. . . . The prices paid for labor have been considerably below the prices paid outside for similar labor; whereas it might properly be inferred that they would be much higher, owing to the danger and exposure attending a large portion of the work. . . . No man has ever been employed, to my knowledge, by reason of any political influence, nor is there any man on the payrolls who does not honestly earn every cent he gets."

The Executive Committee passed another resolution asking the Chief Engineer to report "whether the expenditure thus far has exceeded the estimate, and whether the Bridge can probably be completed for the sum total originally estimated, and, if not, what was the cause of the excess."

In reply to the question of expenditures to date Colonel Roebling stated that the foundations had exceeded the estimate by $160,000. He explained: "There were three principal items that far exceeded the estimate and are more than sufficient to account for the excess of $160,000. These were the enormous cost of labor in the compressed air, the damage done by fire in the Brooklyn caisson and the necessity of making the New York caisson fireproof, and lastly, the unexampled hard digging on the Brooklyn side."

On the larger question of the total cost of the project the Colonel had his answer ready; for, two months earlier, he had revised his father's 1867 estimate from $7,000,000 to $9,500,000. He gave his answer most emphatically: "Finally, as regards the question whether the bridge can probably be completed for the sum total originally estimated for, my answer is *No!*" The design of the bridge had been widened to eighty-five feet, and two tracks, for horsecars, had been added. The clear height of the span had been increased to meet the War Department's requirements. A deep foundation had been adopted to carry the New York tower to rock, instead of the pile foundation which John Roebling had figured in his estimate. Allowance also had to be made for rising labor costs and for the great advance in the price of iron and steel.

No allowance had previously been made in the estimate for the cost of land and land damage. The executive committee now estimated the land condemnation at $3,500,000, bringing the total prospective cost to $13,000,000.

3

From the huge sealed caissons, over against either shore, rose the great stone towers, tall and grim, which were to carry the cables. In due time they stood complete, with their broad bases welded to their foundations by the bond of stone and concrete in the river's bed.

From the shoe of the caisson to the stone ridge of the roof, the New York tower reached a total height of 350 feet. Ninety thousand tons of limestone and granite went into its building. Above the surface of the water both towers rose nearly as high as the spire of Trinity Church, which was then the outstanding skyscraper in New York.

What is perhaps the most beautiful drawing of the great span is an early perspective rendering showing John Roebling's dream bridge as he envisioned it. At each end of the arching span the soaring masonry of the tower terminated in a lighter stone parapet, which provided just the right finishing touch for the inspired composition. Shortage of funds as the work approached completion resulted in the omission or postponement of these intended top courses. Few who view the bridge know or perceive that the tower masonry was never completed. Only by comparison with the master builder's original design is the deficiency revealed. Some bridge designers and artists who have seen the old drawing have expressed the hope that the city may some day find the funds to complete the towers as they were conceived by their designer.

4

The erection of the tall masonry towers was not completed without a generous quota of accidents, despite all the precautions taken. A system of daily inspection of the derricks was instituted by Colonel Roebling at the start and was faithfully kept up to the end of the work; this was entrusted to the foreman in charge of riggers, a man of large experience. But, as is sure to be the case, accidents occur only when least expected and at some weak point that has been overlooked.

In several instances the wrought-iron hooks of blocks under heavy load suddenly straightened, dropping the large stones being lifted. These failures of the hooks occurred only after long-continued use; prior repeated strain had weakened the metal, but visual inspection before the accident would not have revealed the weakness. One day a leading block on a derrick suddenly split, but the men below the crashing load fortunately escaped injury.

Several serious accidents, all the result of carelessness on the part of those injured, occurred to men guiding hoisting ropes on engine drums. A heart-stopping scream of agony would announce that another man had his hand caught between the wire rope and the winding drum, and that his arm was being wrenched from its socket—until a fellow worker could reverse the machinery and release the victim.

On two occasions injuries resulted from defective strap brakes on the hoisting drums, the wood blocks having become worn so that proper tightening was impossible. When the brakes failed to hold as a large stone swung in the air, the operator had some anxious moments before the inevitable crash.

The men working in the yard at the foot of the tower were in constant danger from falling objects—tools, tackle, timbers, derricks, and blocks of granite. Even a light hammer or an iron bolt dropping from such a height would crack a man's skull or break his back, to say nothing of the crushing effect of a half-ton block of masonry.

Careful design and vigilant inspection afforded such insurance as it was humanly possible to provide against failures of the equipment during normal use, but it seemed impossible to prevent accidents due to carelessness of the men in handling the rigging. Raising the derricks to new settings, as successive courses of stone were completed, was a precarious operation. One day, owing to the improper removal of the temporary blocking, a balance derrick careened, breaking the leg of a man caught underneath. When these derricks were in use, a sliding weight on the rear arm had to be adjusted to balance the weight being lifted. Two of the derricks, on separate occasions, turned over from over-balancing, with narrow escapes for the men working near them.

The limestone first received was provided with round lewises for lifting. These were found to be unsafe, but not until one or two serious accidents had resulted by stones so lewised falling when swinging over the work. In these instances the stones fell only a short distance, merely the height of the derrick arm above the course of masonry being laid, but it was enough to kill one man and to maim another for life. One day the accident was more dramatic: a heavy block of stone slipped from the fastening as it was swinging over the edge of the tower. "Look out below!" was the frantic shout as the limestone block went hurtling downward in its fall of two hundred feet. With three seconds of warning, the men below leaped to safety just in time to escape the crushing impact.

At such a height as that of the towers it was often difficult to make the men at the base understand signals from those above. This became a source of real danger in hoisting heavy stone, as the top of the tower was frequently wrapped in mist and hidden from the engineman below. The wind and other noises would often drown the voice, and at times a sharp whistle was heard but faintly. Bell wires were tried, but were liable to become deranged, so that they were not dependable. By great watchfulness cases of overwinding the hoisting rope were rare, and no serious results followed. Flag signals were the safest. With the flag held well out and the motions positively made, the signals, if seen at all, were unmistakable. A whistle was used as an additional safeguard.

Most of the fatalities occurred by men falling from a height —losing their balance or being dragged by falling equipment. One worker instinctively held on to his wheelbarrow as it slipped from a gangplank, and man and barrow went plunging through space. The men grew accustomed to working at great heights, and soon lost their caution. Several workers on the towers fell to their deaths by a careless misstep. The force of the wind near the edge of the tower was often very great and, coming in sudden gusts, it created an additional hazard. This is the one danger that bridgemen dread. A man would brace himself by leaning against the wind, only to lose his balance when the pressure suddenly changed. This happened more than once with fatal results.

5

Each of the two main towers took five years to build—one year to sink and fill the foundation, and four years more to complete the masonry. The Brooklyn tower was thus finished in June, 1875, and the New York tower—started a year later—in July, 1876.

There was a feeling that the work should have been finished more quickly. But solid construction that is to endure cannot be rushed. There were some delays from pecuniary embarrassments and legal obstructions, over which the engineers could exercise no control. For a large part of the time, however, the utmost capacity of the docks and piers was taxed to do what was done.

During the progress of the work popular jealousy of a private company controlling so much public money began to make itself felt in various ways. It had, moreover, become apparent that a much larger sum than the private company had been able to raise would be required to carry the bridge to completion. At the same time, the work was so well advanced, the plans and the methods of construction were so firmly fixed by what had already been done, that its friends now felt willing to resign the great enterprise entirely to the two cities, and prepared a bill to that effect. Following a suspension of construction due to a lack of finances, this bill was passed by the legislature in May, 1875, and was accepted by the city governments.

Under the charter thus amended the bridge ceased to be a private project and became public property, two-thirds to be paid for and owned by the city of Brooklyn, and one-third by the city of New York. The payments made by the private stockholders were reimbursed, with interest, and their title extinguished. The Engineer and his staff, as well as the principal working members of the directorate, retained their positions, so that the work remained, after all, a unit from beginning to end.

6

At a distance shoreward from each tower a massive masonry anchorage was constructed. Designed primarily to resist the pull of the four great cables, these anchorages also form a part of the land approaches to the suspended floor of the bridge. The construction is solid, with the exception of two arched passageways running through the masonry at street level.

In the heel of each anchorage—the end farthest from the tower—four giant anchor plates are buried, one for each cable. These anchor plates, located above tide level, are held down by the enormous dead weight of solid masonry piled upon them. Rising from the anchor plates, and likewise imbedded in the masonry, powerful chains of wrought-iron bars are carried upward and forward in a curving line to the front of the anchorage, where firm connection is made to the nineteen strands composing each cable. No part of the wire cable is buried in masonry or concrete. The two Roeblings, intuitively, knew better. Such imbedded construction has been used by later engineers, with disastrous results from the inevitable corrosion.

The anchorage on the New York side was located in the block bounded by Cherry, Dover, Roosevelt, and Water streets, a district of historic associations. Among the parcels of property included was No. 1 Cherry Street, where George and Martha Washington lived from 1789 to 1790, when New York was the capital of the United States. A tablet on the anchorage marks the site of this first presidential mansion. On the first block of Cherry Street directly north of the anchorage there were still standing houses almost as old. No. 23 was the haunt of privateers during the war of 1812. Beside it stood other old-timers with ornamental doorways and glimpses of slenderly banistered staircases. Prosperous skippers inhabited these houses once, and later, the editors of the publishing houses in Franklin Square, together with their more prosperous contributors. Now the district is "Little Spain." The modest shop windows display enormous loaves of bread, tins of olive oil, and brick-red ollas. The present inhabitants are probably entirely unaware that the sites of their crowded tenements are haunted by the ghosts of ancient buccaneers and by the spirits of literary celebrities of a bygone era.

In May, 1875, the workmen began pulling down the warehouses and old rookeries which covered the ground to be occupied by the anchorage. The site was found to be on the edge of a buried swamp. To keep the water down during the progress of the excavation, three steam pumps were required, removing a steady flow of six hundred gallons a minute. Old building foundations, mud, and refuse had to be dug out, and the excavation was continued until a uniform bottom of clear sand and gravel was exposed over the entire surface. So rapidly was the work prosecuted that, in three months from the start, the excavation was finished and the solid foundation of timber and concrete was in place. As soon as two courses of limestone had been laid, the four huge anchor plates were accurately set. From these anchor plates the successive links of the anchor chains were erected as fast as the masonry was built up to hold them. To speed the work, the masons worked in three shifts, day and night. Progress was rapid until the end of the year, when suddenly the work was halted—funds were exhausted.

Colonel Roebling hopefully stated that the early summer would find the anchorage finished "if New York City responds in furnishing the money for its share of the cost." In May work was resumed, and by midsummer the masonry was completed, ready for cable making. The monumental stone structure, occupying a city block and rising, virtually solid, to a height of nine stories, together with the powerful anchor chains accurately set and imbedded in its sixty thousand tons of masonry, had been built in a total working time of only twelve months. The Brooklyn anchorage, completed in the preceding year, had taken twice as long.

All the weight and solidity and massiveness of the bridge are in the towers, the foundations, and the anchorages. The rest, for the most part, is wire—wire, slender by contrast, spiderwebbed against the background of the sky, but endowed, through the magic of Roebling's art, with strength to match the time-defying resistance of the masonry.

7

The final preparations for cable making had to wait for the delayed completion of the New York anchorage. As soon as each anchorage was ready, the large area on top was immediately taken over for the cable-spinning machinery.

Through the summer of 1876 the preparatory work for cable spinning was vigorously carried forward. The machinery to operate the traveling ropes was made on the same general plan as that which the Roeblings, father and son, had developed and used so successfully in making the cables for the Cincinnati bridge, with some improvements and changes to meet local requirements. This time, four cables were called for instead of two, and of much larger section. The installation was in duplicate, for making both pairs of cables at the same time, so that two years might be saved in the stringing. The machinery that had to be set up included giant driving wheels, guiding wheels, and tightening pulleys, as well as sliding beveled gears, countershafts, clutches, levers, pinions, and segment gearing—all of which had to be designed and prepared in advance.

Before the wire spinning could start, a variety of temporary cables had to be strung from shore to shore: the traveler ropes, carrier cables, cradle ropes, footbridge cables, and handrail ropes. In addition two "storm cables" were provided, extending in an arch curve from tower to tower below the footbridge and serving as inverted cables to prevent the light footbridge from being lifted by the wind. All this equipment, including the machinery, was planned and calculated by Colonel Roebling and was minutely specified in notes and drawings which he prepared. The entire procedure for cable construction was worked up by him and was covered in memoranda which he laboriously wrote out for the instruction of his staff.

It is of interest to note that all of these wire ropes were made at the plant of John A. Roebling's Sons at Trenton, after competitive bidding.

8

Before the cables were in place, New York and Brooklyn stared up at the river-wide space between the bare towers and wondered by what wizardry a bridge could ever be strung across it. But the engineer had it all planned.

The difficult and dangerous problem, how to get the first wire ropes across the river and over the streets and buildings from one anchorage to the other, was safely and expeditiously solved. The two "traveler ropes" were the first to be taken over. With the giant reel on a scow in front of the Brooklyn tower, the free end of the rope was hoisted over the top of the tower and then carefully hauled to the anchorage, some nine hundred feet away. Fenders and supporting trestles had been erected at various intermediate points, and men were stationed on the roofs of the intervening buildings to guard them from injury as the wire rope was hauled over them.

All being ready, the scow bearing the reel was then towed across the river by two steam tugs, the corresponding length of steel rope being meanwhile paid off from the reel and allowed to sink to the bottom of the river. In eight minutes the dock at the foot of the New York tower was reached. There the wire rope was raised to the top, drawn down on the land side of the tower, and fastened to the grooved drum of the stone-hoisting engine at the tower base.

The rope now had to be raised from the bottom of the river to its permanent position at an elevation of some two hundred feet above the tide. It was at first supposed that this would have to be done on a Sunday morning, when there were few vessels in the way; but by frequent observations the engineers learned that they could get a few minutes of clear river on weekdays, and it was resolved to make the attempt as soon as everything was ready.

On August 14 the attempt was made. Large crowds were drawn to the scene by a feeling of the historic importance of the event. Engineers Martin and McNulty took positions on the New York tower at roadway level, to watch the craft in the river and to signal when the river was clear. Engineer Collingwood clambered up the hoisting frame on top of the tower to note the sag of the rope, with William Brown, carpenter, alongside to pass signals down to the operator of the hoisting engine. The paralyzed Chief Engineer watched the operation through his field glasses, from the window of his sickroom.

A steam tug was stationed in the river to wave all vessels away. On account of the immense traffic, it was a long and tedious wait before the river was clear enough. At last Collingwood heard from Martin the welcome words, "Go ahead!" and passed them to Brown, who signaled the engineman. At the same time a cannon shot was fired as a signal to shipping and to the thousands of spectators who lined the shores. In a few seconds the rope began to move; there was a ripple around it in the water; it began to draw away from both shores toward midstream. Faster and faster the space of clear water between the two visible parts narrowed, and in four minutes, with a sparkling *swish*, the dripping rope swung clear of the surface of the water, amid the cheers of the spectators on the wharves and ferryboats, and the shouts of the workmen on the bridge. Up went the rope, and in a few minutes it hung clear above all masts, to form the first connecting link between the cities of Brooklyn and New York.

Later in the afternoon the second rope was raised in the same way. With both ropes firmly secured on the top of the tower, the two ends were then carefully hauled over the streets and roofs to the New York anchorage, much in the same way as this had previously been done on the Brooklyn side. The free ends were next spliced together, around the driving and guiding wheels at the anchorages, thus forming one endless rope, capable of being worked to and fro from anchorage to anchorage, to draw loads. In this way, the first physical connection across the East River—a traveling cableway—was made. The rope was arranged to run over wood-lined wheels on top of the towers, and although it saw much hard service during the next two years, it bore no perceptible marks of wear.

9

To accomplish the next operations would require men to work while hanging on this slender aerial all the way from tower to tower. In order to give the workmen an example of confidence and courage, E. F. Farrington, the master mechanic, now undertook to make the first passage of the line.

On Friday afternoon, August 25, the running gear for the endless traveler rope was ready. A boatswain's-chair, consist-

OF GRANITE AND STEEL

ing of a small board for a seat, slung by short ropes uniting in a ring overhead, was secured to the traveler rope at the Brooklyn anchorage, and Farrington took his seat on the suspended board for the first trip over the line of the future bridge. As he looked down from his high starting point, he saw the housetops beneath him black with spectators, the streets far below seemingly paved with upturned faces, the ferryboats packed and the New York shore crowded with a thrilled audience. When he gave the signal, the engine at the Brooklyn anchorage was started, and the endless cable—6,800 feet in length—began to revolve around the drums. Then, as he swung out into space, with the whirring rope undulating like a flying serpent through the air, the boom of cannon far below announced to the world that the first human crossing of the new link between the two cities was being consummated. A million people craned their necks from the streets and docks and housetops and boats along the river, and swallowed hard at their hearts. Away went the speeding rope, looking like a spider's thread to the watchers below, bending and swaying with the human weight that rode its pendent curve. The crowds gasped as they watched the rider, sweeping first downward toward the housetops till the deepest sag his weight could give the slender rope was passed, and thence soaring sharply upward to the top of the first high tower in his course. Here he gave a signal to slow the rope nearly to a stop, while the men on the tower, cheering, lifted the seat and its rider over the parapet, supported both across to the other side, and launched them off the dizzy height again. Again the cannon roared, and the thousands of spectators swung their hats and cheered with wild excitement, while all the steam whistles on land and water shrieked their uttermost discords. Some of the ferryboats stopped in the middle of the river, and the ladies on the boats, on the piers, and on the housetops shrieked their cheers and wildly shook their handkerchiefs, while men and boys swung their hats and shouted until they were hoarse. Farrington, obviously enjoying his ride, waved his red bandanna in the breeze. Near the middle of the crossing he clapped his hands and yelled joyously to the crowds below. As he neared the New York tower, the wire cable was bent so that the trip was almost perpendicular, and Farrington stood up in his seat and was thus safely carried to the top. As he set foot on the masonry, the first man to have crossed, he was greeted with terrific cheers. Once more he was launched into space, and the last lap, from the New York tower to the anchorage, was completed. At the end the aerial pioneer had to elude the hysterical crowds that wanted to carry him through the streets in triumph.

Farrington had also been the first to swing across on the Roebling cableways at Niagara and Cincinnati, and now the first crossing of the greater span over the East River climaxed his proud list of public triumphs.

The two parts of the other traveler rope were taken across by being lashed to the first one and hauled over by the steam-operated driving machinery. After being taken over, the ends were secured on the two anchorages, and the lashings were cut loose by men seated in boatswain's-chairs which were arranged to slide on the first traveler rope. With the aid of these first ropes, all of the remaining ropes and cables were taken across and installed.

10

The "cradles," or cross bridges, were soon finished. They were made of oak, combining lightness and strength, with open floor construction to allow free passage of the wind and thus reduce oscillation. They were put together on the ground, hoisted to the tower tops, and then slid down the cradle ropes to their respective positions and fastened there. The cradles thus formed five transverse platforms where men could stand to control and regulate the wires during the stringing of the cables.

The footbridge, extending from one anchorage to the other, over the tops of the towers, provided access to the cradles. It was made of oak slats laid directly on the footbridge ropes, with spaces between the slats for the wind to pass. This floor was all put together in sections, and was laid as soon as the storm cables were attached underneath.

The footbridge arrangement, designed and chosen by Colonel Roebling for this crossing, was an improvement over the construction which he had helped his father install over the Ohio at Cincinnati. Instead of being suspended along a straight line at the future bridge roadway level, it was hung in a free curve sixty feet higher, at the level of the cradles and strands. From a low footbridge, the cables could have been reached only by means of long vertical rope ladders, difficult and dangerous to climb; moreover, with the guys and storm cables underneath, it would have obstructed navigation. The high construction was fully 210 feet above the river, and thus gave generous accommodation to shipping.

The Colonel, recalling experiences at Cincinnati, did not hesitate to express his belief that this frail temporary structure might be disabled more than once by violent gales before the main cables were completed. Profiting by that earlier experience, he provided safeguards for increased resistance. The principal security of the "catwalk" against the wind was the pair of inverted storm cables, assisted by a number of underfloor stays anchoring the river portion to the tower masonry below. "All waves must be checked," he explained, "before they reach the towers, where alone the ropes can be injured. This is done by the underfloor stays, and by securely fastening the ropes to the masonry. The inverted parabolic storm cables resist uplift and prevent the footbridge being carried away bodily."

Washington Roebling was more than an apt pupil of his father. He had not only grasped the concepts originated and developed by the pioneer master builder, but he also extended their application so as to improve on his father's designs. The footbridge over the Ohio, in its much briefer service life, had thrice been wrecked by the wind; the more advanced design over the East River, although longer in span and higher in elevation, withstood exposure to storm and gale through the two years it was in use, without any damage.

II

In 1867, in his first and only report on the proposed East River bridge, John A. Roebling had discussed the use of *steel* wire for the cables, as a means of keeping their size from being unwieldy. Now, nine years later, his son and successor had to make the momentous decision. Cables of steel wire would be an untried experiment. After making various tests and carefully weighing all considerations, the Colonel decided to take the courageous step. He had become convinced that, although iron had always been used in the past, steel was "the metal of the future," and that it ought to be used in the new record-breaking span.

At that time it was not known what quality could be obtained in the new material, and whether it could be made sufficiently uniform to answer the purposes required. Specimens of steel wire were therefore procured from the leading foreign and American manufacturers, and a series of tests was made to determine what requirements should be specified in contracting for it. Following these pathfinding studies and experiments, Colonel Roebling specified that all wire for the bridge "must be made of steel," and that it must have a tested strength of 160,000 pounds per square inch, nearly twice the strength of the charcoal iron wire that his father had used in the great cable spans that preceded this work.

The announcement of this revolutionary change in cable material was not received without lengthy controversial discussions in the profession and in the iron industry; but it has passed into history that bidders were found who undertook to produce the steel wire as specified, and it was furnished. This conspicuous adoption of the new structural material played a historically important part in establishing steel in engineering acceptance. After the success of this first application of steel wire for cables, iron wire was never again used for suspension bridges.

Another significant innovation in the cable material for the Brooklyn Bridge was the adoption of *galvanized* wire. As the cables were to be suspended in a salt atmosphere, Colonel Roebling wisely decided that galvanizing was the only sure safeguard against rusting of the wires—a decision that contributed largely to the resistance of the structure against the corrosive attack of the elements through the years. The coating was applied by running the wire through a bath of molten zinc. The protection of the full surface of every steel wire with an unbroken coating of zinc added to the cost, but the longevity thereby secured was worth many times the relatively small additional investment. This truth has been painfully learned by engineers, in more recent years, when they thought they knew better and tried to economize by omitting the galvanizing.

One of the draftsmen in Colonel Roebling's organization at that time was a young engineer by the name of Lefferts L.

Buck. Twenty years later the second bridge to span the East River, the Williamsburg Bridge, was under construction, and the chief engineer was—Lefferts L. Buck. As an economy measure, he decided to dispense with galvanizing the wires. If he were alive today, he would be one of the first to admit that his decision was a mistake. He had received his initial training under a master builder of remarkable engineering instinct, but the apprentice had failed to learn the lesson which the master had intuitively grasped and emphatically recorded for the guidance of his contemporaries and of those who came after him.

Colonel Roebling's specifications were unique; they not only specified, but they also explained, instructed, and illuminated. Instead of merely stating peremptorily that the wire must be galvanized, he gave the reasons for the novel requirement so that the necessity of a thorough compliance would be appreciated. "The cables of the East River Bridge," he wrote, "are suspended directly over a salt-water stream, and are, in addition, exposed to the salt air of the neighboring sea shore. Experience has shown that the ordinary means of protection, such as paint, oil or varnish, which would be ample in the interior, are totally inadequate to prevent rusting in localities near the coast. The only certain safeguard is a coating of zinc, which acts by absolute air-tightness, as well as by its galvanic action, and is not easily abraded. . . . The galvanizing must be done throughout in a thorough and perfect manner; each ring will be inspected in this regard by the inspector, when he tests the wire. All rings will be rejected which show spots imperfectly covered. . . . The galvanizing must be of uniform thickness, and must not scale off, or show any cracks when the wire is bent."

12

In September, 1876, as the time approached for securing bids on the wire for the cables, the Executive Committee, apparently influenced by a desire to avert any possible public criticism, adopted a resolution submitted by Mr. Hewitt: "That bids from any firm or company in which any Trustee, Officer or Engineer of the Bridge has an interest will not be received or considered, nor will the successful bidder be allowed to subcontract any portion of his contract with any such firm or company without the consent of the Trustees."

The adoption of this rule placed Colonel Roebling in a difficult position. The fact of his being the engineer of the bridge was going to penalize his brothers by arbitrarily disqualifying the cable wire manufactured at the family's Trenton plant; and it was going to handicap the project by eliminating a most likely bidder for this important material. Determined that others should not suffer on account of his connection with the structure, the Chief Engineer made his personal decision. The following month he sold out his interest in John A. Roebling's Sons—three hundred shares of the capital stock, valued at $300,000. The company was now free to bid on the wire for the great cables.

Two months later the bids and samples of the wire proposed to be furnished were received from eight concerns in the United States and in England. The quoted prices ran up to fourteen cents a pound. The lowest quotation was that of John A. Roebling's Sons, of Trenton: 6¾ cents a pound for Bessemer steel wire. The next lowest bid was that of J. Lloyd Haigh of New York: 8.7 cents a pound for crucible cast steel wire. The Roeblings' bid for this grade of wire was nine cents a pound.

After some deliberation the Executive Committee voted that the cable wire must be of crucible cast steel. And the contract—for 3,500 tons of wire—was awarded to Haigh.

13

In the spring of 1877 the footbridge was completed—a narrow, swaying plankway, only four feet wide, suspended high above the level of the roadway of the future. Swung in its lofty pendent arc above the river, it connected New York with Brooklyn as by an aerial highway.

The situation was an airy one, and in a breezy neighborhood. The wind came not only over the sides, but also through the floor, whose cracks were more than half as wide as its planks. This "porousness" gave the wind less hold and thus reduced oscillation.

Although intended only for the use of the workmen during the stringing of the cables, the catwalk excited the curiosity of

ARCHES IN THE APPROACH TO THE BRIDGE

THE BROOKLYN BRIDGE

many adventurous persons who eagerly sought the privilege of crossing the slender and apparently frail structure. From the hour this temporary footpath was opened, applications began to pour in from thousands of people, of every age and condition in life, for permission to cross from tower to tower. The passage was safe enough to persons of reasonable caution and strong nerves; but those who were liable to dizziness had all they could do to hold on when they looked down and saw the swaying panorama of shore and river far below them. Some who started boldly from the top of the tower became so terrified when they found themselves suspended in mid-air above the river that they were paralyzed by fright and they could neither go on nor go back without assistance from the workmen. It became necessary to take men from their work to watch these travelers, lest they fall from the high pathway through fright or nervousness. Then there were some visitors, more daring and exuberant than others, who fancied it a jolly lark to skim over with a hop, skip, and jump, pausing occasionally to indulge in a jig, and thus jarring the catwalk enough to slow up work, besides tending to injure the walk. Again, there were many who wanted to linger far beyond the stipulated time, and who even demanded the company of some of the engineers or foremen to answer all manner of questions, to identify points of interest in both cities, and to explain the technical details of the great enterprise—all of which would have required the thought and talk of a month.

During the early part of May the number of passes averaged over seventy a day, and the number of applications kept increasing. That there were no accidents was due entirely to the strict watch that was kept over the visitors, for there were hundreds of instances of narrow escapes.

All of this proved to be such an annoyance and such an interruption, to say nothing of the liability, that it was decided not to issue any more passes to those who wanted to cross merely for the sake of the panoramic view or for the excitement of the trip. The engineers determined that the labor of the men should no longer be interfered with; and strict orders were finally issued that no one would be allowed to cross except those actually engaged on the work.

14

A visitor privileged to watch the successive operations of the cable spinning would have found the whole spectacle a fascinating experience. On top of the Brooklyn anchorage was the spinneret from which the great cobweb emerged thread by thread. Here the large drums, each designed to hold ten miles of wire, were mounted in upright frames—eight drums in a row for each cable. The large rings or coils of wire, as they came from the factory, were mounted on reels behind the drums; and between each reel and drum a man was stationed to guide the wire through a piece of oiled sheepskin held in his hand, thus giving it a final coating of oil, the last it was to get as a single wire before entering into the cable strand. Brakes were provided to regulate the speed of the drums as the wire was run out.

For the cable spinning one of the first essentials that had been put in place was the traveler rope. On each half of this endless rope was hung a "traveling wheel," resembling a bicycle wheel with the tire missing. Over this grooved wheel a bight of wire would be slung.

Then the large driving pulley of the traveler rope would begin to turn, and the attached traveling wheel would start on its outward journey from the Brooklyn anchorage. The tinkling of an attached cowbell gave warning to the men along the route as the wheel approached. The wire gradually unwound from its feeding drum as the spinning wheel went on its aerial jaunt over the towers, and finally came to rest at its destination, the New York anchorage. There the men would grab the bight of wire, slip it out of its carrying groove, and put it around a heavy iron shoe—the "strand shoe"—attached to the anchor chain. A double wire had thus been carried across the river from anchorage to anchorage.

The motion of the endless rope was then reversed, and the empty wheel would start its homeward trip to Brooklyn, while its companion wheel, carrying a bight of wire for the second cable, started its outward journey from the home anchorage. And thus the operation continued, a filled wheel constantly going out and an empty one returning, each outward trip adding two wires to one of the cables.

After each wire was drawn across, it had to be brought to the same length, sag and curvature as the wires preceding it. Flagmen, stationed on the three cradles between the towers, signaled to the men on top of the tower, or down on the anchorage, the amount of raising or lowering required, and the wire would be pulled up or slackened until its sag accurately matched the initial guide wire or the wires already strung.

The spinning of the wires from anchorage to anchorage would be continued in this manner until a strand was completed for each of the two companion cables that were being simultaneously strung. Thus each wire is continuous in its strand, like the yarn in a skein, traveling eternally from Brooklyn to New York and back—up over the top of one tower, down in a long sweeping curve above the tideway, up over the other tower and down to the anchorage, where the cast-iron shoe around which it is wound is gripped by the everlasting clutch of the huge cement-bound mass of masonry. The wire in each skein is a million feet long—nearly two hundred miles—and its accurate stringing at dizzy heights over the tower tops, almost endlessly back and forth from anchorage to anchorage, was a test of the ingenuity and patience of man.

There is no spiral or twist of the wires in these ponderous cables, as there is in wire rope. Every stretch of wire lies flat and separate and parallel. When all were in place, they were laboriously bound together—first all the wires in a strand, and then, when all the strands were up, the whole symmetrical aggregation was compacted and bound into cylindrical form. In this final operation the strands gave up their separate identity and combined their individual strengths to make a single unified cable of integrated strength.

15

The first cable wire was drawn across the river on May 29, 1877, and from that time on cable making was continuous. It went on daily, in summer's heat and winter's cold, and again through another spring, summer, autumn, and winter, until the four cables were finished.

After each strand of nearly three hundred wires was laid, it was bound into a compact bundle by wrapping several turns of soft wire around it at short intervals throughout its length. The purpose of these "seizings" was to keep the strands distinct during the cable stringing and to prevent crossing or chafing of the wires. The operation was performed by men who went out on hanging platforms called "buggies," suspended from overhead trolley wheels which rolled along the strands on which the work was being done.

While the strand was being made, it hung at an elevation considerably above its ultimate position in the cable—actually fifty-five feet higher at mid-span. The tension in the wire in this initial position of the strand, with its flatter sag, was thereby nearly doubled. This tested the strength of the wire, took out all waves and bends, and led to the easier detection of any defective wire or splice. After the seizing was done, the strand was let out at the anchorages and carefully lowered to its correct final position, previously determined by calculation.

The strand laying went on through almost all kinds of weather—a stop being made only when ice on the wires or a strong wind blowing across the line of the bridge prevented the proper regulation of the individual wires.

Sun and wind played pranks with the work in the summer; the deflection of the strands varied an inch for every three degrees of temperature. In the winter, snow and ice coated the wires, the catwalk, and the running gear, adding to the difficulties and the dangers of the work.

As a protection against the freezing winds for the men working out on the land and river spans, little shelters made of boards were set up on the cradles. Atop the towers boardings were put up and small stoves installed for the relief of frozen hands and shivering bodies.

16

In July, 1878, three months before the cable stringing was to be completed, Colonel Roebling reported his discovery of frauds practiced by the contractor for the cable wire. From the first the engineer had found it necessary to test every ring of wire made by this contractor. The inspections were made at Haigh's works in South Brooklyn. Attempts to bribe the inspectors were anticipated by Roebling and were easily forestalled. Then it was noticed that the pile of rejected wire, instead of increasing as it should from the constant rejections, was growing smaller. A watch was therefore set, and the trick was discovered.

A wagon load of eighty rings of accepted wire, as it left the inspector's room with his certificate of acceptance, instead of going straight to the bridge, was driven to another building where it was quickly unloaded and replaced by a load of rejected wire, which then went to the bridge with the same certificate of inspection. Thus rings of wire previously rejected were being surreptitiously palmed off as accepted wire. Tests clinched the fact that a switch had been made.

Two days later the contractor's men were caught again, unloading the good wire and filling up with rejected wire. The good wire taken from the wagon was being transferred back to the inspection room, and thus repeatedly palmed off as new wire in order to obtain additional certificates of acceptance for the same routine of substitution.

The distressing aspect of this affair was that all the rejected wire that had come to the bridge had been worked into the cables and could not be removed without undoing more than a year's work. It was impossible to tell how long the contractor had been practicing this deception. Colonel Roebling judged that it had probably been going on for six months—since January, for at that time the quality of the wire had suddenly declined. According to the inspector's books the total quantity of the wire rejected, since the beginning of the work, amounted to nearly five hundred tons. A substantial part of this had gone into the cables. "How to ascertain the total quantity is a difficult task," Roebling observed in his report. "An engineer who has not been educated as a spy or detective is no match for a rascal. This is the first instance of deliberate fraud which has come to my notice during the nine years that I have been the engineer of this work and covering an expenditure of $8,000,000."

The bridgebuilder now had a difficult decision to make. Condemnation and rejection of the cables already completed would involve the loss of a year or more of precious time. After much thought and investigation Roebling ruled that the contractor must, at his own expense, place additional good wires in the cables to make up for the calculated deficiency in the strength of the wires previously erected.

17

In John Roebling's previous suspension bridges, the cables never contained more than seven strands. The East River bridge cables were the first to be formed of nineteen strands. The number seven or nineteen is chosen so that the strands can be arranged in a compact hexagon which can then be squeezed into a still more compact cylindrical cable. Between seven and nineteen there is no number which makes this possible.

The change from seven to nineteen strands introduced a new construction problem. It necessitated a two-step squeezing operation, in which the seven central strands were first compacted, and the twelve outer strands were finally consolidated with the central core in a separate squeezing and binding operation.

For the squeezing and seizing of the inner core of seven strands—and again later for the final squeezing and wrapping of the completed cables—three or four workmen occupied the perilous swinging buggy. On this dizzy aerial perch the bridgemen worked almost at constant risk of their lives. Yet no accidents from falls occurred.

In fact, the construction of the great bridge strikingly demonstrated how men can become adjusted to living and working under abnormal conditions—whether in the dense atmosphere of compressed air in the caissons below the bed of the river, or on the unguarded swinging platform which traversed the lofty cables. The cable workers were perfectly at their ease, and often tempted fate by walking out on the cable, scorning even the safeguard of the slight handrope that was provided. The workmen, as well as the engineers, walked along the cables unconcernedly, and even trotted down the inclines, barely touching the handrope. One of the men, when questioned by an amazed reporter, laughingly remarked that working on the buggy and getting in and out in that perilous fashion had one advantage at least, and that was that tools were safe from predatory visitors; dishonest strangers and souvenir hunters would not be tempted to venture from the swaying catwalk to the still more dizzy, narrow, and often greasy core of cable strands.

Despite the hazardous work aloft, everything proceeded smoothly, without accident, until one day in June, when the fifteenth strand of the north cable had just been completed and was in process of lowering by "letting out" at the New York anchorage. Suddenly—by the unfortunate fouling of a holding rope, causing it to break—the great strand broke loose from its adjusting tackle, carrying with it the large cast-iron shoe and its ponderous attachments. As the end swept from the anchorage, it dashed off several of the men at work there, and then, with a frightful leap of nine hundred feet through the air, grazing the houses and peopled streets below, it landed in the bridge yard close under the New York tower. But only for an instant. The great weight of the strand mid-stream whizzed the free end up the tower and over the top, with terrifying and increasing speed, until the whole strand madly plunged into the river, narrowly missing the ferryboats that were plying, crowded with human freight, below the line of the bridge. The consequences might have been even more disastrous if any shipping had been hit by the strand as it lashed wildly through space. The cost of this accident was two bridgemen killed, three workmen seriously injured, and the loss of the strand. All the work of spinning the strand had to be done over again. The lost strand was cut into lengths of four hundred feet by divers and fished up from the bottom of the river, in order to salvage the wire.

Four months later the spinning of the cables was completed. On October 5, 1878, the last wire was run across.

The next operation was to draw all the strands of a cable together to a true cylindrical form, by means of powerful circular squeezers, and to apply the cable wrapping. The wrapping wire was put on under considerable tension, using a rotating machine operated by men in the traveling buggy slung from the cable. It was calculated that, as a result of the compacting and wrapping, the friction between the wires was so great that it would be impossible to pull out ten feet of any cable wire, even if cut at both ends.

18

Early in 1878 a severe northeaster hit the city, destroying shipping and structures along the waterfront. The great unfinished span over the East River stood up against the gale and came through unscathed.

But soon the bridge had to face a storm that was more difficult to weather—a rising storm of public opposition. Already the limit of $13,500,000 fixed by the legislature had been exceeded, and yet the work was far from completion. As a natural consequence the undertaking had aroused the strenuous opposition of influential citizens, who insisted that no more of the city's money be "squandered on the bridge." A taxpayer's suit was filed, seeking an injunction to stop the work by forbidding any further expenditures.

Prominent in this attack was the "New York Council of Reform." This group, in public meetings and in print, assailed the bridge on every conceivable ground, especially for its inability, when completed, "to withstand the force of storms such as that which has just made such havoc along our coast and in neighboring cities."

The structure was illegal, they claimed, because it obstructed navigation. "Since the commerce of this city is its life, and has a State and national importance, no such injury to it can be tolerated." The Federal government would certainly object to the obstruction. "The bridge will be judically condemned, and it is therefore a criminal waste to spend any more of public money upon it."

Finally, the objectors declared, the structure as designed was unsafe: "This is wholly an experimental bridge. It is the highest and longest in the world. The history of suspension bridges in this country and in Europe shows their most dangerous exposures to be that to storms, producing oscillations and ruptures. Five of the largest suspension bridges in this country, and several in Europe, have been destroyed within a few years after their erection in this manner. . . . If an eddy of air were to strike the bridge from underneath with greater force than its own weight, it would be lifted, to crash back again with its destructive momentum of thousands of tons."

In conclusion, the opposition insisted that it would be foolish and wicked to spend more money on "a bridge that is not called for, cannot serve the purposes for which it is built, very seriously damages a large part of the commerce of this harbor, taxes the financial ability of these two cities to their utmost, and cannot fail either to be taken down by the mandate of the courts or demolished by the winds."

The controversy and the litigation dragged on for months, with volumes of testimony and arguments. Construction proceeded under a handicap, with available funds approaching exhaustion.

Finally, in November, 1878, work on the bridge had to stop for lack of funds. The trustees had applied for a mandamus to compel the authorities of the city of New York to pay the last two installments required of them under the appropriation originally authorized, but this application was denied by the courts. All work on the approaches was stopped; the delivery of stone was suspended; the cables were left unfinished; and the engineering force was reduced.

When Colonel Roebling was directed to suspend all construction not absolutely essential to preserve the work already done, he was asked to send an estimate of the cost of the necessary protective measures. His estimate was $50,816 to finish the cables and to keep the engineer corps on half pay for six months. "I need all my assistants when the work is resumed," he wrote. "To disband them is very easy—but to replace them with new men and drill these into a proper state of efficiency would be no small task."

In the spring, after six months of shutdown, some funds were made available and the work was resumed. The following winter a bill was prepared for submission to the legislature to appropriate the money needed to complete the bridge.

The uneasiness of the public in regard to the bridge situation was fed by rumors of dissension among the trustees. Stories were circulated to the effect that some of the officers were lukewarm to the project and secretly desired to prevent its completion. A reporter quoted one of New York's representatives on the board as having remarked, "Our city is not a jealous city, but then to ask it to build a bridge in order to send its trade to a neighboring city is asking a good deal even from the best of natures."

This alleged quotation aroused a storm of indignant protest. The trustee was charged with selfish and short-sighted provincialism. His supposed attitude reminded one editor of a parallel case—that of a prominent Kentuckian, who, in discussing a proposed railroad to be built across the state line, expressed himself "in favor of building railroads *into* Kentucky, but averse to building them to go *out* of the state."

This unfortunate discussion only deepened the factional split between the trustees from the two cities. Some of the New York representatives characterized the project as an extravagant and nefarious scheme "conceived in iniquity" for the profit of the politicians and of those privately interested in Brooklyn real estate. As one incensed opponent described it, the bridge was being built "to drain the resources of the city of New York in order to fertilize the sandy wastes of Long Island."

19

Originally, bids for the suspended structure had been asked on the basis of wrought iron, because at that time it was questionable whether proper shapes and the desired quality could be produced in steel. Within a year, through continued investigations and tests made for the Chief Engineer by Colonel Paine, these difficulties were overcome. Thereupon, in order to secure increased capacity of the bridge from the gain in strength by substituting steel, Colonel Roebling decided to make the change. With the approval of the trustees he proceeded to call for new bids, this time for *an all-steel bridge*. The field was a new one, and the engineer had to make use of the latest knowledge on the subject, and at the same time to proceed with caution and make use of every possible check.

The work proceeded. From the streets below, the dangling suspenders, from which the floor was to be hung, looked like spider threads. They were, however, stout ropes of steel, able to sustain a weight of more than fifty tons each, or five times the heaviest load likely ever to fall on them. They were attached to the cables by wrought-iron straps. On the lower end of each suspender rope were threaded stirrup rods, by means of which the steel floorbeam could be raised or lowered to regulate the grade of the floor.

The floorbeams near the towers were attached to their proper suspenders and swung off. A track was then laid on the floor, and the succeeding beams were carried out to their positions on platform cars. These floorbeams were unlike any ever before used. Instead of using ordinary beam sections, each floorbeam was built up in the form of a triangular lattice girder, thus affording greater strength with less weight.

By September, 1881, the floorbeams were nearly all in position when work was suddenly again suspended. The stoppage gave rise to new questions and surmises among the general public. Had the project finally struck an insurmountable snag? Had the engineers, at this eleventh hour, discovered that their design was all wrong? Would they have to take the span down and start all over again?

To set these doubts and conjectures at rest, the president of the board issued a statement presenting a frank explanation. The interruption in the work was occasioned by delays in receiving the material from the foundries, due to difficulties encountered in producing and shaping the steel. No structure of this size had ever been built of steel. The contract was unprecedented in its nature. The trusses called for steel in shapes and sizes never produced before. "The contractors supposed, as did their competitors for the contract, that their plant, which was calculated for iron, would answer for steel, and they made their estimates accordingly. But their machinery was inadequate; it broke down and was necessarily abandoned." Special machinery had to be constructed at great expense, more than wiping out any profit on the job. "The contractors have gone on honorably with their work, notwithstanding the advance in the cost of iron, and since they have got their works in shape they have finished one-half of the whole contract."

The mechanical and engineering difficulties of rolling the specified steel shapes were finally surmounted, and the erection of the suspended steelwork of the bridge was pushed with the utmost speed.

The erection of the floorbeams was completed on December 10. The following day eleven of the trustees, escorted by Mrs. Roebling and the assistant engineers, walked across the span on a footpath of planks. This was the first official crossing of the bridge at roadway level. The crippled Chief Engineer would have given a great deal to be there; his wife made the trip in his place.

20

Early in October, 1881, the Executive Committee reported to the trustees that about one thousand additional tons of steel would be needed as a result of changes in design. At once many questions were raised, and a resolution was passed requesting the Chief Engineer to explain the increase. At the following meeting his reply was presented. He had been authorized, he explained, to strengthen the structure so that a train of heavy railroad cars might be run over it with absolute safety. To provide the headroom required for Pullman cars, all four middle trusses had to be raised, and the posts required extra bracing. "We will now have a bridge so rigid," he concluded, "that the ordinary traffic will have no visible effect on it."

When this communication was read to the trustees, Robert B. Roosevelt jumped to his feet, demanding to know who had authorized the change. James Stranahan immediately rose to reply. He accepted full responsibility. "I am glad the thing was done," he added, "and I think every member of the board may rejoice in the fact that the capacity of the bridge is now equal to the passage of any train of cars that may pass over any of the roads in the country."

The thing was done. Although we do not know whom to thank for the increased capacity and resistance, we can now be grateful for the fact that, between the trustees and the Chief Engineer, they built better than they knew—certainly better than their contemporaries comprehended.

As soon as the public learned that one thousand additional tons of steel would be required, a new storm of criticism arose. To the man in the street the conclusion was obvious: either the engineers had proved their incompetence by making their earlier design inadequate, or they were now extravagantly proposing a needless increase of material at public expense. It was evidently a case of either "blunder" or "plunder." The New York *Times* joined the popular clamor and rushed to the attack. In the strongest editorial language it excoriated the "stupidity" of the engineers and challenged the safety of the bridge under the unforeseen tonnage of steel now to be added to the span's burden.

In answer to this attack *Engineering News*, the country's leading construction journal, came out with a vigorous editorial ably defending the engineering of the project:

> The cost of a few additional tons of steel, over which the shortsighted are howling as an evidence of mistaken calculation, is in reality but a healthy outgrowth of expanded ideas of the importance of this structure. In the fourteen years which have elapsed since the bridge was designed, enormous strides have taken place, and are now taking place daily, in the volume, weight and character of railroad intercommunication. Such things, for instance, as Pullman cars, weighing over thirty tons, had never been thought of at that day. The bridge, when designed, was in advance of the highway bridges of the day, but its purpose was but that of a highway bridge. City travel, as then conducted, required cars of but two or three tons weight. Without any disparagement to the talents of the elder Roebling, we believe that it is entirely owing to what the *Times* is pleased to consider as the stupidity of the present engineers, that the structure will be capable of withstanding all the strain which by any possibility it can be subjected to; and when completed we are confident it will reflect the greatest credit upon all who have been instrumental in perfecting it, with an eye to its future needs and in disregard of the clamors which it might provoke.

Seeds of doubt and suspicion concerning the capacity and safety of the structure were growing into rumors, and finally into open charges. Laymen, editors, and even engineers, were suggesting—first in whispers and then in shouts—that the span was being overloaded with a dead weight of steel and would never have strength enough left over to carry all of the ambitious highway and railway loading that was contemplated. A typical news item in this vein appeared in the Chicago *Times*. The story now included a new twist: the suggestion that a third pier would have to be built under the middle of the span to support the added weight!

At a meeting of the trustees in December a resolution was introduced by Robert B. Roosevelt, asking the Chief Engineer to report "on the total weight that the cables and the suspended structure will be able to bear, and what will be the weight of the suspended structure when completed, with tracks and all, ready for use by foot passengers, vehicles, and cars."

Within a month Colonel Roebling submitted his calculations and report. The cables and stays had ample strength to carry all of the proposed loading, with a margin of safety of four. "The cables are strong enough to pull up the anchorages," he added, by way of graphic emphasis.

That the bridgebuilder was conservative in his conclusion of abundant reserve strength is proved by the fact that, after more than a generation of service, the great structure has actually carried, with perfect safety, a total useful load three times as great as that for which it was designed. It has been an open secret for years that the Brooklyn Bridge has been overloaded beyond its figured strength, but Washington Roebling knew that it would be.

A careful study of the operations of the Board of Trustees would have revealed to an inquiring mind at the time, as it more clearly reveals in retrospect, a point that explains both the enduring quality of the achievement and the atmosphere of suspicion that it aroused. It is this: There was some one individual connected with the work, possibly two acting separately but toward an identical objective, who—with a foresight which could not have been understood, had it been revealed—worked persistently, from beginning to end, to bring it about that the Brooklyn Bridge should, by its capacity to meet all demands upon it, be something more than a limited and transient convenience between two cities. To this individual or pair of individuals, satisfied that the future would fully justify the outlay, the initial cost was a secondary consideration. The primary and compelling goal was to produce a structure of enduring power, to give a full measure of usefulness for generations to come.

In the annals of the project, the evidence suggests that two men had this vision—James S. T. Stranahan and Washington A. Roebling. Each was willing to brave popular distrust and personal attack in working toward an inspired goal. Whether they worked separately or in acknowledged understanding will, perhaps, remain an unsolved mystery.

21

The whispering campaign of criticism and doubt kept gaining in volume and vehemence. The enemies of the project were now "hitting below the belt"—the physical disability of the Chief Engineer was made a focal argument in the attacks.

Toward the end of June, 1882, the trustees passed a resolution requesting Colonel Roebling to be present at the next board meeting. He replied in writing:

I am not well enough to attend the meetings of the Board, as I can talk for only a few moments at a time, and cannot listen to conversation if it is continued very long. My physicians hope that living out of doors and away from the noise of a city may lessen the irritation of the nerves of my face and head. I am now able to be out of my room occasionally. I did not telegraph you before the last meeting that I was sick and could not come, because everyone knows I am sick, and they must be as tired as I am of hearing my health discussed in the newspapers.

I believe there is not a day that I do not do some sort of work for the Bridge. . . . My assistants do the work assigned them with perfect confidence that they can always refer to me for any advice or assistance they need. The work to be done this summer is very plain routine. There is nothing to be done that has not been done before, and the interests of the Bridge do not suffer in any way from my not being there. If the Edgemoor Iron Company will furnish the steel as fast as we need it, the work will go on all right and without delay.

It is very important that authority be given me at your next meeting to advertise for the iron for the elevated termini, the spruce planks for the floors, the steel rails for the tracks on the Bridge, and the paint. Specifications for all these things are now ready, and they should be obtained as soon as possible, so that they may be ready when we want to use them.

I shall be most highly honored to be present at meetings of the Board as soon as I am well enough to be of any use there.

The underground campaign went on, and soon it took the form of an attempt to displace Colonel Roebling as chief engineer of the bridge. Through twelve years of trials and tribulations he had carried the project forward successfully; and now, after all his work and sacrifice, it was proposed to take the work away from him only nine months before its completion. At a meeting of the trustees in August, Seth Low, then mayor of Brooklyn, introduced the following resolution:

Whereas, The Chief Engineer of the bridge, Mr. W. A. Roebling, has been for many years and still is an invalid; and

Whereas, In the judgment of the board the absence of the Chief

Engineer from the post of active supervision is necessarily in many ways a source of delay, therefore,

Resolved, That this board does hereby appoint Mr. Roebling Consulting Engineer, and Mr. C. C. Martin, the present First-Assistant Engineer, to be the Chief Engineer of the New York and Brooklyn Bridge; and

Resolved, In so doing the board desires to bear most cordial testimony to the services hitherto rendered by Mr. Roebling, and to express its regret at the necessity of making the change at this time.

After some discussion it was decided to lay the matter upon the table until the next regular meeting, to be held in September.

As soon as the crippled engineer heard of this proposal to take his bridge away from him, he decided to fight. He was physically incapacitated, but he had kept his fighting spirit. He started to dictate a statement to his wife. After many starts he found what he wanted to say. Without asking for sympathy, he told what he was doing on the bridge and why he should not be displaced. It was an appeal to the engineering profession to prevent an injustice.

The American Society of Civil Engineers was meeting at the time. Mrs. Roebling went to the meeting and secured permission to read her husband's statement. It was the first time that august body had been addressed by a woman. The drama of the situation, the simple eloquence of the appeal, and the impressive presentation produced an immense sensation. The backing of the engineering profession was secured, and the confidence of the public was won.

22

While the bridgemen were spinning the wire cables and suspending the steel spans high above the river, large forces of laborers and masons were busily at work at the two ends of the bridge, clearing the way and extending the line of the structure through the blocks of city streets. The landmarks of a bygone age, historic old houses on the waterfronts of both cities, vanished silently, and where they had stood, by and by there grew piles of brick and granite to form the approaches.

For five years, from 1877 to 1882, work on the building of

the approaches went on. The masonry construction was of solid character. In the midst of so much sham and superficiality in the architecture of those days it was worth something to have such a fine example of thoroughly honest design and execution. No problem in brick or stonework was dodged, and everything was built to endure.

On the New York side the dominant impression is that of the vaulting arches, growing higher and higher as the hill falls and the bridge roadway rises to its jumpoff. The openings of these arches have been walled with brick to make warehouses, mostly for hides. For more than a century the hide and leather business has centered around Cherry Hill. Amidst the odd cave shops, grown ripe with use, the viaduct of stone arches completes an old-world picture reminiscent of Rome—a simple beauty of proportionate building, weathered by time.

As the bridge neared completion, public interest in its progress became livelier, and the finishing touches absorbed the attention of all who had occasion to cross the East River ferries. A force of painters, working high above the water, were busy laying coats of white and buff on the trusses, cross girders, and braces. To the people below it seemed as though the painters, with their brushes and cans of paint, were walking on threads as they made their way along the narrow steel members. But the most fascinating sight was the riggers, slung sailor fashion in their boatswain's-chairs at the intersection of the gray network made by the slanting stays and the vertical suspenders. Each rigger was busily lashing the stays and suspenders where they crossed, until the diamond-shaped spaces of the open mesh were perfect in form. These men were all regular sailors, and felt at ease working aloft. Seen from a ferryboat at a distance, they looked like spiders spinning their webs under the cables.

It was at last evident that the end of the great work was at hand, and that it would not be many weeks before the great tide of travel and traffic would begin to flow to and fro across this bridge.

The formal opening was now set for May 24, and the program had been arranged. Ten thousand invitations had been sent out to various parts of the country, to national and state officials and to other distinguished guests.

One large element in the population refused to join in the

festivities. The keen susceptibilities of the Irish were offended because the opening date happened to be the Queen's Birthday. They found in this coincidence apparent evidence of a too great Anglo-American friendship. Talk was even heard of putting dynamite to the structure in remonstrance. Kingsley patiently explained that, when the date was chosen, none of the committee knew that it was Queen Victoria's natal day. It was too late, however, to recall the invitations and to change the arrangements for the official ceremonies.

23

The great work was finished. It had cost over twenty lives, and had taken more than thirteen years to build.

The two master builders had paid their price—one with his life, the other with a crippled body.

The highest and the humblest workmen had been influenced by the spirit of courage and devotion in which the work had been started and carried forward.

Orators and writers paid their tributes to the significance of the achievement. It was hailed as a mighty engineering feat—the longest span in the world—the longed-for link between New York and Brooklyn—the promise of a "Greater New York."

About its beauty the spokesmen of the day said little. The structure did not represent the contemporary style in architecture. It had not a single battlement, a single inset of red brick or white marble, a single scroll of cheap ironwork, or a single panel of gilded decoration. But later generations clearly see what the Roeblings wrought. In an age of superficial embellishment and pretentious atrocities the two bridgebuilders gave no thought to the prevailing fashion in design. They were not architects, but workmen seeking only to do an honest and sturdy job. How the thing was going to work and how to make it last were their chief concern. So they built honestly and soundly. And sound building is beautiful building.

Except for those little devices by which a good builder strengthens the main lines and hides the rough spots, they seem to have given no conscious thought to looks. There must be approaches to lift the span up in the air to clear the masts on the river below. The Roeblings lifted them, therefore, over a series of arches whose joints they finished off with cushions of smooth stone. At the point where the span rises and flies across the river, there must be towers with double arches to let the traffic through. Simply because Gothic arches suited their engineering plan, the Roeblings built Gothic arches. The tops of the towers they finished naturally and simply in plain heavy stone. That is all—not an inch of carving, not a single concession to ornamentation. No conscious art; merely the honest work of master craftsmen. But in this fine and sincere workmanship the builders, father and son, recaptured the simple and natural beauty that endures. The Brooklyn Bridge remains the finest monument of their age.

No inspired builder of a medieval cathedral brought to his work a greater singleness of purpose, a more selfless devotion, than the two Roeblings lavished on their master bridge. The best in their characters went into its building.

Centuries hence men will still marvel at the gigantic towers—more wonderful than the pyramids—standing like sentinels above the busy waterway, with the everlasting sea beating against their mighty bases, and with their granite masonry rooted deep below the rushing tides on the rock which those tides had not touched for untold ages.

There is timeless strength in those towers, and poetry in the cable-borne span. The two are harmoniously joined. Between the two pierced granite towers the arching roadway slowly sweeps upward to meet the swift downward sweep of the cables. These curves and proportions were not accidents. It is no accident that the Brooklyn Bridge still remains the most satisfying esthetically of all great bridges—because its two master builders were artists at heart. Of granite and steel and dreams, the Bridge was built.

The father had dreamed the Bridge. The son, with gallant fighting spirit, had carried the dream to fulfillment. The empty gap his father had planned to span was spanned. Water, earth, fire, and the powers of darkness had been completely confounded.

People spoke of the Bridge with wonder, and with a certain measure of awe. On one point all authorities of the time agreed. The finished work was universally acclaimed as "the triumph of the science of engineering." The seemingly impossible had been accomplished. Said the editors of the *Scientific American:* "The practical completion of the grandest piece of engineering the world has yet seen necessarily attracts attention, not only in the immediate vicinity of the work, but throughout the civilized world; not only from curious sightseers but from those who labor for the advancement of their fellows and who rejoice in the success of a stupendous undertaking."

But the Bridge was more than a triumph of engineering. It was also a triumph of the human spirit.

CHAPTER XX

THE BRIDGE IS DEDICATED

MAY 24, 1883.
It was a glorious spring day. Dawn revealed the twin cities resplendent with waving banners and streaming colors. From windows and roofs, from the forests of shipping along the wharves and from the vessels riding the sparkling waters of the bay, flags were floating proudly in the breeze; while high above all, from the massive, time-defying towers of the Bridge, the Stars and Stripes signaled to the world—from the gateway of the continent—the arrival of the long-awaited day.

The Bridge was completed. The dream of years was at last a reality. Once more the unconquerable spirit of Man had triumphed. Of faith and courage, a masterpiece had been wrought.

A spirit of tense excitement was in the air, a sense of high significance, a feeling of joyous celebration.

Almost before the sun was up, the thoroughfares of both cities had put on a festival appearance. Shops and offices were closed and all business was suspended. The streets and avenues were a maze of bright and stirring colors.

Tens of thousands of men, women, and children, in their Sunday best, surged through the flag-draped streets. Bands played. Street hawkers did a lively trade in souvenirs of all descriptions. Incoming trains continued to bring thousands of eager visitors to witness the celebration.

The crowds began to flow in the direction of the great span. Young and old, on foot, in carriages, stages, and horsecars, moved toward the center of the day's festivities, and there sought vantage points to see the show.

The scene along the river front was unique and beautiful, with hundreds of vessels, including five ships of the North Atlantic Squadron at anchor below the bridge, decorated from stem to stern. In the bright spring sunshine the rippling reflection from every ferryboat, warship, and tug, and from the hundreds of pleasure craft, made the waters of the river run with colors, like a dye vat.

And above this sparkling scene was the great vaulting span—the embodiment of a city's dream in granite and steel.

2

Far uptown in Manhattan, where Broadway cuts through Fifth Avenue at Twenty-third Street, a large crowd was packed around the Fifth Avenue Hotel, waiting for a glimpse of the President of the United States, General Chester A. Arthur, and the Governor of New York, the Honorable Grover Cleveland. The "Dandy Seventh" Regiment, wearing its famous gray and white dress uniforms, marched into view and lined up on the avenue. Colonel Emmons Clark, mounted on a spirited horse and with saber flashing, was in command.

Presently the distinguished guests appeared, the President accompanied by the members of his Cabinet, and the Governor with his staff. They entered their open carriages, the officers gave their commands, the band began to play, and the parade started. Down Broadway the procession headed, with flags flying, drums beating, horses prancing, and the proud columns swinging along in beautiful precision. To the stirring rhythm of the regimental band the parade moved briskly along the flag-lined route between the massed ranks of thrilled and cheering spectators. The President raised his silk hat, the Governor waved his hand, and the crowds cheered.

In quick time the procession had passed two miles of the famous thoroughfare and reached City Hall Plaza. There it was joined by Mayor Franklin Edson and other New York City officials. The columns wheeled left, with sabers and gunbarrels catching the sun, and the glittering parade approached the entrance to the world's latest wonder—the East River Bridge.

Grandstands had been erected on the roofs of the adjoining buildings—the Harper Brothers Publishing House, the *Staats-Zeitung* Building, the Richard K. Fox Building (where the famous *Police Gazette* was published, the only building in New York boasting gilded fire escapes)—and on all other available structures near the New York end of the bridge. The ladies were dressed in their finest; they sat in the sun under gay parasols, with their silk-hatted escorts near them, waiting for the parade to arrive and for the dramatic moment when the new President and the Governor would enter upon the great span.

The presidential party with its military escort proceeded over the rising arches of the New York approach, and then over the first span to the New York tower. Here the procession was met by officials and a military guard from Brooklyn. President Arthur and Governor Cleveland descended from their carriages, to walk over the bridge. The band struck up, and between lines of brilliantly uniformed troops, the procession passed through the lofty portal of the tower and entered the roadway of the main span.

This was the moment for which the crowds had been waiting. As the officials set foot on the span high above the river, the historic event was proclaimed by the thunder of guns. The heavy cannon at Castle William on Governors Island thundered out across the harbor, the artillery at the Navy Yard in Brooklyn boomed out over the river, and the battery of guns on the summit of Fort Greene took up the reverberating salute. Boats set their horns and steam sirens shrieked in chorus. And a million men and women cheered.

The ships of the Fleet, from their station below the bridge, joined in the thundering ovation. The yards of the men-of-war were manned, their colorful strings of signal flags snapped in the breeze.

As the official procession moved out in triumphal progress along the span, the air was filled with a tremendous flood of joyous sound from ships and docks and factories and from the streets below—the clanging of bells, the roaring of a thousand steam whistles, and the frantic cheers of the crowds—while sounding from afar, clear and distinct above the clamorous din, the silver chimes of Trinity rang out over the river.

3

At the Brooklyn end of the bridge, in the vast enclosed terminal where the exercises of the day were to take place, a great assemblage was awaiting the approaching procession. The interior was bright and colorful with decorations—azure hangings of satin bordered with silver, and the coats of arms of the states amid a forest of flags. The Twenty-third Regiment of the National Guard appeared for the first time in the new state service uniform. The United States Marines added to the colorful spectacle. The building was thronged. Invitations had been at a premium. Ten thousand ticket holders had sought to crowd into a space designed to seat two thousand. Thousands of others were trying to break down the police and military barriers in order to hear the speakers. Representatives of the bench, the bar, the pulpit, and the press, prominent men from every section of the country, senators and representatives, governors and mayors and leading citizens filled the huge improvised auditorium. Conspicuous in one section of the hall were the officers of the Army and Navy in their brilliant uniforms.

When the presidential party and their escort entered the hall they were greeted with thunderous cheers. They occupied seats directly in front of the stand erected for the speakers. Cyrus W. Field, John Jay, and other distinguished guests were seated in adjacent sections. One section was reserved for the engineers and workmen who had built the bridge.

4

During these ceremonies a lonely man—paralyzed, crippled, and racked with pain—sat at an upper window of his home on Columbia Heights, viewing the scene from afar. He saw the crowds surrounding the bridge approach, and the host of more privileged citizens pouring into the reserved space. Through his field glasses he saw the distinguished procession coming over the bridge—the President of the United States, the Governor of the State, and all the other notables, together with the glittering military escort. His gaze was riveted on the scene. The breeze carried the cheers of the crowd and the strains of the band, but he could not hear the glowing tributes of the orators.

His throat was choked, and he could hardly keep back the tears; for this was the great moment of his life. Through long and weary years he had been enduring pain and fighting on for this consummation. This day gave meaning to his life, and to his father's before him. For this was his father's Bridge—and his Bridge. The father had dreamed the dream, the son had wrought the dream. Both had lived, and battled, and sacrificed, that this Bridge might be built, and both had given their last supreme effort to this achievement.

Beside the crippled man at the window stood his wife, who had been his ministering angel through the years of pain and struggle. She had been his eyes, his hands, his feet—recording his notes, carrying his instructions to the job, and bringing back reports of its progress. When others had tried to displace him just as the work was drawing to completion, she had eloquently presented his plea to stop the injustice. Brave and loyal and tender, she had given her strength to carry her husband through his soul-straining ordeal.

5

If the paralyzed engineer watching the celebration through his field glasses could have heard the words of the speakers, his throbbing heart would have beat faster. For there, at the scene of dedication, standing before the vast assembly of citizens, an honored civic leader whom they had selected as their spokesman—the Honorable Abram S. Hewitt—was eloquently recording Humanity's indebtedness to the Builders of the Bridge:

"When we turn to the graceful structure at whose portal we stand, and when the airy outline of its curves of beauty, pendent between massive towers suggestive of art alone, is contrasted with the over-reaching vault of heaven above and the ever-moving flood of waters beneath, we are irresistibly moved to exclaim, 'What hath *man* wrought!'

"Man hath, indeed, wrought far more than strikes the eye in this daring undertaking, by the general judgment of engineers, without a rival among the wonders of human skill. . . .

"But the Bridge is more than an embodiment of the scientific knowledge of physical laws. It is equally a monument to the moral qualities of the human soul. It could never have been built by mere knowledge and scientific skill alone. It required, in addition, the infinite patience and unwearied courage by which great results are achieved. It demanded the endurance of heat, and cold, and physical distress. Its constructors have had to face death in its most repulsive form. Death, indeed, was the fate of its great projector, and dread disease the heritage of the greater engineer who has brought it to completion. The faith of the saint and the courage of the hero have been combined in the conception, the design and the execution of this work.

"Let us, then, record the names of the engineers who have thus made humanity itself their debtor for a successful achievement, not the result of accident or chance, but the fruit of design, and of consecration to the public weal. They are: *John A. Roebling*, who conceived the project and formulated the plan of the Bridge; and *Washington A. Roebling*, who, inheriting his father's genius, and more than his father's knowledge and skill, has directed this great work from its inception to its completion. . . .

"During all these years of trial, and false report, a great soul lay in the shadow of death, praying only to stay long enough for the completion of the work to which he had devoted his life. I say a great soul, for in the springtime of youth, with friends and fortune at his command, he braved death and sacrificed his health to the duties which had devolved upon him, as the inheritor of his father's fame, and the executor of his father's plans. . . .

"But the record would not be complete without reference to the unnamed men by whose unflinching courage, in the depths of the caissons, and upon the suspended wires, the work was carried on amid storms, and accidents, and dangers sufficient to appall the stoutest heart.

"One name, which may find no place in the official records, cannot be passed over here in silence. With this Bridge will ever be coupled the thought of one, through the subtle alembic of whose brain, and by whose facile fingers, communication was maintained between the directing power of its construction and the obedient agencies of its execution. It is thus an everlasting monument to the self-sacrificing devotion of woman. The name of Mrs. Emily Warren Roebling will thus be inseparably associated with all that is admirable in human nature, and with all that is wonderful in the constructive world of art."

One after the other the following speakers added their words of acknowledgment and praise. From one who had been intimately identified with the bold enterprise from its inception, the Honorable William C. Kingsley, the hushed audience heard the story of the planning and the building, the heroism and the sacrifice:

"With one name, this Bridge will always be associated—that of *Roebling*. At the outset of this enterprise we were fortunate to secure the services of John A. Roebling. . . . While testing and perfecting his surveys . . . the man who designed this Bridge lost his life in its service. He was succeeded by his son, Colonel Washington A. Roebling. . . . Down in the earth, and under the bed of the East River . . . within the caissons . . . always on hand at the head of his men to direct their efforts and to guard against any mishap or mistake which might have proved disastrous . . . Colonel Roebling contracted the mysterious caisson disease. . . . For many long and weary years this man, who entered our service young and full of life, and hope, and daring, has been an invalid, confined to his home. He has never seen this structure as it now stands, save from a distance. . . . But every step of its progress he has directed. . . . Colonel Roebling may never walk across this Bridge, as so many of his fellowmen have done today, but while this structure stands all who use it will be his debtors."

6

The celebration was continued in both cities throughout the day and far into the night. Thousands upon thousands filled the streets and crowded the river front—gazing at the marine spectacle and the wondrous span arching overhead.

After the ceremonies a procession, headed by the President, the Governor, the speakers of the day, and the trustees, marched through the streets of Brooklyn to the residence of Washington Roebling—to do honor to the man who had risked and lost so much to make this day possible, who had given his health, his strength, and his career to bring his father's dream to completion. The silk-hatted and frock-coated dignitaries entered the modest dwelling, to clasp the hand of the paralyzed bridge-builder and to voice the nation's tribute, while the crowds outside cheered themselves hoarse. Hearts were touched, and eyes grew moist.

From the shouting newsboys everywhere the public was eagerly buying copies of the second printing of the souvenir edition of the Brooklyn *Eagle*. All circulation records had been broken—250,000 copies had already been sold. The newspaper heralded the completed bridge as "*the Eighth Wonder of the World.*" "Eighth in point of time, but not in significance," the editors hastened to add.

7

Meanwhile, with darkness falling, the city was undergoing a transformation. Torches and bonfires and bright Chinese lanterns were strung like fireflies.

Soon night came, with the stars crystal clear. And then, when the darkness was complete, a pyrotechnic display of unprecedented magnificence was staged from the illuminated span and towers of the bridge. All boats had been cleared from immediately below the span. The streets and housetops were packed, and the river was blocked with craft of every description carrying legions of thrilled spectators. Every conceivable vantage point was crowded—the tops of tall buildings, the roofs of warehouses along the waterfront, wharves and piers, tugs, lighters, excursion boats, ferryboats, sailboats, and barges. Bands were playing.

At eight o'clock the first fuse was touched off, releasing fifty giant rockets, and a few seconds later the two cities lay under a sparkling shower of gold, blue, red, and emerald fire. From both shores burst fountains and jets and tremendous geysers of varicolored flames and living cascades of brilliance eclipsing the stars.

Countless thousands witnessed the unforgettable spectacle. They saw the span flooded with the miracle light of eighty powerful *electric* lamps strung along the arching roadway! And above the Bridge they beheld the flight of numberless bombs and rockets, together with great showers of gold and silver rain, and Niagaras of fiery sparks and floating stars. Set pieces sent out parachutes from which hung colored balls of fire, or bombs which shook the sky with their detonations. The dazzling fireworks were visible for miles, and illuminated the sky so that it was as light as day.

At nine o'clock the span shuddered under the impact as five hundred more rockets, all fused together, were set off in breathtaking explosion. From the safety of their slips the ferryboats and other craft blasted the brilliance of the night with their steam whistles. The official display was over—a spectacle that was to be vividly remembered through life by all who had witnessed it.

Local illuminations and fireworks displays continued. The streets of Brooklyn were a mass of colored fire. Morning found thousands still on the roofs of the city's buildings, tired but happy.

8

Shortly before midnight the Bridge was thrown open to the eager public. For hours thousands had jammed the approaches at both ends of the new structure, waiting for the chance to cross on foot or by coach. At the signal there was immediately a dense rush of pedestrians, all anxious to be the first to cross. One man achieved the distinction of being the first to pay the one-cent toll and to pass through the barrier. But the second in line was an amateur mile runner; he sped across the span ahead of the pack of racing rivals and won lifelong glory as the first to reach the New York end. The excited crowd followed behind the panting runners. Thousands—250,000 in the first twenty-four hours—seized the opportunity to set foot on the magic span that had fascinated their imagination through the years of its building. And as they crossed over the lofty roadway, hung from its powerful strands of steel between the two sentinel towers, they felt the spell of the Bridge as millions of others were to feel it in the generations that followed.

Because certain intangibles of the human spirit were built into the Bridge, it has remained a powerful work of the imagination. To those who have felt its spell as they walked across its arching span, and to those who have looked up to its great towers and the tracery of its cables from the shadows below, the Bridge has been an object of fascination, of romance, of dreams.

Such a composition is not wrought by chance, nor is it a mere synthesis of engineering formulas and mechanical processes. The souls of men had been poured into its creation.

9

Twenty-five years later a monument is being unveiled to the master builder, John A. Roebling. Many thousands are gathered to honor the memory of the man. An orator is speaking.

The vast audience is hushed, as his words find an echo in every heart. In inspired eloquence, linking the magic of the Bridge with the personality and genius of its creator, the speaker pays Humanity's tribute:

"Yes, Brooklyn Bridge is beautiful. All the latent poetry of the mathematician—and in its highest reaches mathematics becomes divinest poetry; all the aestheticism of the architect; all the musician's sensitiveness to harmony; all the mysticism of an idealist philosophy; whatever of faith, feeling, reverence John Roebling cherished in his heart, was here voiced like a ringing cry. As if conscious of his pending doom, his genius stands embodied in this form—an aspiration visible—a soul's bid for immortality."

EPILOGUE

TO THE SECOND EDITION

To repay my debt of inspiration to the Brooklyn Bridge and its builders, I took five years out of my professional life to write this book—

> "To tell the Saga of those stirring days,
> To tell the fight against terriffic odds,
> To tell the way in which mere men can raise
> A Monument well worthy of the Gods!"

I put everything I had into this work—"whatever of faith, feeling, reverence" I cherished and all of the romance, drama, and poetry that I felt. I have been repaid for this labor of love by its heart-warming reception by old and young, and by the assurances I have received that this story of the human spirit will live on as an enduring inspiration to future generations.

One accomplishment of this book has been to redirect attention to the grand old Brooklyn Bridge as one of the most inspiring monuments of the age. In recent years, this world-famous structure has been little more than a majestic monument—not because of any inadequacy of strength but for the simple reason that the traffic needs of the Automobile Age could not possibly have been foreseen eighty years ago when the bridge roadways were planned. In 1948, the municipal authorities decided to have the bridge reconstructed to meet modern traffic needs, and the engineering of this task was entrusted to me. I was engaged to make the engineering studies and to draw the plans for rebuilding the structure—the suspended roadways, the stiffening trusses, and the approaches—so as to increase the capacity and the accommodations of the bridge from two lanes to six lanes of modern highway traffic. This engineering engagement gave me a particular thrill of satisfaction. The world-famous structure that had inspired me in my youth had come back to me at the height of my career. The crowning work of the pioneer master builders had been entrusted to my professional skill and care for lengthening of its life and enlargement of its usefulness. It was another dream come true.

When people expressed concern about the proposed reconstruction and modernization of the beautiful old span, I was glad to reassure them that the appearance of the Bridge would not be changed. To me, Brooklyn Bridge is sacred.

Another accomplishment of this book has been to direct attention to the heroic contribution of a woman in the building of the Bridge, symbolizing the part that women silently play in the inspiration of all great achievements. Arrangements have been made for the erection of a memorial tablet on the Bridge, bearing these words:

THE BUILDERS OF THE BRIDGE

Dedicated to the memory of
EMILY WARREN ROEBLING
(1843–1903)

With faith and courage, she helped her stricken husband

COL. WASHINGTON A. ROEBLING, C.E.
(1837–1926)

to complete the construction of this Bridge
from the plans of his father

JOHN A. ROEBLING, C.E.
(1806–1869)

who gave his life to this Bridge.

———

"Back of every great work we can find the self-sacrificing devotion of a woman."

———

This tablet erected 1951 by
THE BROOKLYN ENGINEERS CLUB
with funds raised by popular subscription.

New York, 1950.   D. B. STEINMAN.

BIBLIOGRAPHY
PRINCIPAL SOURCES AND REFERENCES

BIBLIOGRAPHY

PRIOR BIOGRAPHIES

Auener, Wilhelm, *Johann August Roebling*, Mühlhausen, 1931.
Schuyler, Hamilton, *The Roeblings: A Century of Engineers, Bridgebuilders and Industrialists; Story of Three Generations of an Illustrious Family, 1831–1931*, Princeton University Press, 1931.
Stuart, C. B., *Lives and Works of Civil and Military Engineers of America*, Van Nostrand, 1871.

BIOGRAPHICAL SKETCHES AND REFERENCES

Bentley, J. C., "Roebling: Master Bridge Builder," *Manufacturers' Record*, Vol. 100 (1931), pp. 18–20.
"A Biographical Sketch of John A. Roebling," *The International Steam Engineer*, July 1923.
Bishop, J. L., "John A. Roebling, of Trenton, New Jersey," *History of American Manufactures*, Vol. 2, pp. 578–79.
"The Brooklyn Bridge: A Short History of John A. Roebling and of the Company He Founded," *Trenton* Magazine, June, July, Sept., 1933.
Dictionary of American Biography, Vol. 16, pp. 86–89, 89–90, Scribners, 1935.
Federal Writers' Project, New Jersey, "The Romance of the Roeblings," in *Stories of New Jersey*, pp. 333–42. Barrows, 1939.
"Great Bridges and Their Builders," *American Railroad Journal*, Vol. 54 (1881), pp. 257–58.
"J. A. Roebling," *Beecher's Magazine*, Vol. 3 (Jan. 1871), pp. 129–34.
"J. A. Roebling Came to U. S. a Century Ago," Trenton *Sunday Times-Advertiser*, Aug. 2, 1931.
John A. Roebling. An Account of the Ceremonies at the Unveiling of a Monument to His Memory; Address by H. Estabrook, Roebling Press, Trenton, 1908.
"John A. Roebling, C.E.," *Engineering News*, Vol. 10 (1883), p. 246.
"John A. Roebling, C.E.," *Scientific American Supplement*, Vol. 16 (1883), pp. 6268–69.
"John Roebling Was Pioneer Bridge Builder," *The Herald-Advertiser*, Apr. 7, 1935.

Knorr, E. R., "Johann August Roebling, deutsch-amerikanischer Architect und Ingenieur," *Illustrierte Zeitung*, No. 1367 (1869).
Lee, F. B., "John A. Roebling," *Genealogical and Personal Memorial of Mercer County, New Jersey*, Vol. 1, pp. 46–47.
Lesser, A., "Johann August Roebling," *Weltpost III*, Leipzig, 1884.
"Mr. John A. Roebling," *Journal of the Franklin Institute*, Vol. 54 (1867), pp. 410–13.
The National Cyclopedia of American Biography, Vol. IV, pp. 404–06, James T. White, New York, 1897.
"The 'Roebling' Bridge," *American Society of Civil Engineers—Proceedings*, Vol. 51 (Aug. 1925), pp. 259–60.
"Roebling, Master Bridge Builder," *Manufacturers' Record*, Oct. 1, 1931.
Steinman, D. B., and Watson, S. R., *Bridges and Their Builders*, Putnam's, 1941. Chapter 11, "The Roeblings and the Brooklyn Bridge," pp. 207–49.
Strawn, Arthur, "Sires of the Big Bridge," New York *Herald Tribune*, Oct. 11, 1931.
Van Deusen, E. A., *Biography of John A. Roebling*, Lefax Filing Index.
von Treitschke, H., *Deutsche Geschichte des neunzehnten Jahrhunderts*, Vols. III, V, Leipzig, 1886.
Wandsleb, Alfred, "Bunte Bilder aus dem Leben der amerikanischen Röblings," *Mühlhauser Anzeiger*, Nos. 246, 250, 253, 256 (1926).
———— "Johann August Roebling," *Mitteldeutsche Lebensbilder*, Vol. 2 (1927), Magdeburg.
"Washington Bridge Due to Immigrant," Brooklyn *Eagle*, Oct. 26, 1931.

CHAPTER I: THE WALLED TOWN

Brinkman, Ernst, *Aus Mühlhausens Vergangenheit*, Mühlhausen, 1925.
Jordan, R., *Alt Mühlhausen in Bild und Wort*, Mühlhausen, 1925.
Kettner, Emil, *Mühlhauser Geschichtsblätter*, Mühlhausen, 1917.
Mühlhauser Stadtarchiv. (Genealogy of the Roebling family.)
Scrapbooks containing notes and letters regarding the genealogy of the Roeblings.

THE EDUCATION OF JOHN ROEBLING

Dieteleyn, J. F. W., *Auszug aus Naviers Abhandlung über die Hängebrücken*, Berlin, 1825. (Presumably covers Dieteleyn's lectures on suspension bridges during Roebling's attendance, 1823–26.)

BIBLIOGRAPHY 423

Dieteleyn, J. F. W., *Grundzüge der Vorlesungen über Strassen-, Brücken-, Schleusen-, Kanal-, Strom-, Deich-, und Hafenbau, gehalten an der Königlichen Bauakademie zu Berlin von 1824–1831*, 5 Vols., Berlin, 1832. (Dieteleyn's lecture notes on the engineering of highways, bridges, dams, canals, rivers, dikes, and harbors.)

Die Technische Hochschule zu Berlin, 1799–1924: Festschrift, Berlin, 1925.

Kühnemann, J. G., *Chronologische Uebersicht der merkwürdigsten Begebenheiten der allgemeinen Geschichte, von den ältesten zeiten bis auf das Jahr 1811*, Halle, 1811. (Copy in John Roebling's library. Believed to be his high school textbook in world history, including the history of America.)

Navier, C. L. M. H., *Rapport et memoire sur les ponts suspendus*, Paris, 1823. (Describes the contemporary beginnings of the suspension bridge art in England, France, and America.)

Original school notes and drawings by John Roebling.

THE CHAIN BRIDGE AT BAMBERG

Le Cocq, G. L., "Ponts Suspendus," *Encyclopédie Scientifique*, Vol. 1 (1911), p. 27, Paris. (The second chain bridge in Germany. Span 210 ft. Over the Regnitz. Replaced an old timber arch of 1808. The completion date is given as 1829, which would place it in Roebling's apprenticeship period after graduation.)

Mehrtens, G. C. (trans. by L. Mertens), *A Hundred Years of German Bridge Building*, Berlin, 1900.

CHAPTER II: THE EMIGRANT

Gemeinnütziges Unterhaltungsblatt, published at Mühlhausen by E. W. Roebling, 1831–40.

Krebs, A., *Die Gedächtnisfeier zu Ehren des Ingenieur Johann August Roebling am 11 Juni 1876*, Mühlhausen, 1876.

Miscellaneous information contributed by Roebling relatives in Mühlhausen, etc., 1931, etc.

Mühlhauser Magistratsakten, D16, Vol. 3, No. 1, commencing 1876, "Die Familie Roebling in Amerika" (with letters exchanged and newspaper excerpts).

Mühlhauser Wochenblatt, 1831 and 1832.

Original engineering notes, estimates, and sketches by John Roebling, covering his engineering apprenticeship in Westphalia, 1826–30.

Original letters received by John Roebling from correspondents, 1830, etc.

Original manuscript notebook of John Roebling on the history and government of the United States, 1830, etc.

Original notes of John Roebling giving lists of prospective fellow colonists and their respective subscriptions.

Roebling, J. A., *Tagebuch meiner Reise von Mühlhausen in Thüringen über Bremen nach den Vereinigten Staaten von Nordamerika im Jahre 1831, geschrieben für meine Freunde*. Eschwege, 1832.

—— (Trans. by Edward Underwood), *Diary of My Journey from Mühlhausen in Thuringia via Bremen to the United States of North America in the Year 1831*, Roebling Press, Trenton, 1931.

CHAPTER III: AMERICA, 1831

Berühmte deutsche Vorkämpfer für Fortschritt, Freiheit und Friede in Nordamerika, Cleveland, 1899.

Cronau, Rudolf, *German Achievements in America*, published by the author, 1916.

Hendricks, R. J., *Bethel and Aurora*, R. R. Wilson, 1933, pp. 1–11, 244–49. (The story of John Roebling's romance with Helena Giesy. The author knew her reverently in his youth, and married her niece. A discrepancy in the dates is acknowledged by him and may be attributed to his uncertain memory after the lapse of half a century.)

—— "Reverting to Roebling" (Editorial), *Daily Oregon Statesman*, Jan. 31 (?), 1924.

—— "What Might Have Been" (Editorial), *Daily Oregon Statesman*, Jan. 24, 1924.

Körner, G., *Das deutsche Element in den Vereinigten Staaten von Nordamerika, 1818–1848*, Cincinnati, 1880.

Liefmann, R., *Die kommunistischen Gemeinden in Nordamerika*, Jena, 1912.

Lohmann, M., *Die Bedeutung der Deutschen Ansiedlungen in Pennsylvanien*, Stuttgart, 1923.

Och, J., "Der deutschamerikanische Farmer," *Staatswiss. Diss.*, Freiburg, 1913.

Original letters to John Roebling from family and friends, 1831, etc.

Roebling, J. A., Letters to Ferdinand Baehr, 1831. (Bound volume of translation of the letters, in the possession of the Roebling family.)

—— "Opportunities for Immigrants in Western Pennsylvania in 1831" (from Roebling's letters to Ferdinand Baehr, etc.), *Western Pennsylvania Historical Magazine*, Vol. 18, No. 2 (1935), pp. 73–108.

BIBLIOGRAPHY

CHAPTER IV: THE SURVEYOR

Private papers, correspondence, copybooks, drawings, and notes of J. A. Roebling, 1831-48.
Roebling, J. A., "Some Remarks on Suspension Bridges, and on the Comparative Merits of Cable and Chain Bridges," *American Railroad Journal*, Vol. 12 (Apr. 1, 1841), pp. 193-96.
"Suspension Bridges, No. 1," *American Railroad Journal*, Vol. 4 (1835), p. 19.

FAIRMOUNT BRIDGE, PHILADELPHIA

Ellet, Charles, Jr., "Plan of the Wire Suspension Bridge About to be Constructed Across the Schuylkill at Philadelphia," *American Railroad Journal*, Vol. 10 (Mar. 1, 1840), pp. 129-33.
——— *Report and Plan for a Wire Suspension Bridge Proposed to be Constructed across the Mississippi River at St. Louis*, Wm. Stavely & Co., 1840.
"The New Bridge," *American Railroad Journal*, Vol. 14 (Apr. 1, 1842), p. 224.
Roebling, J. A., "The Cincinnati Bridge," *Engineering*, London, Vol. 4 (1867), p. 75.
"Suspension Bridges," *American Railroad Journal*, Vol. 18 (1845), p. 812.

CHAPTER V: THE INVENTOR

Dunbar, Seymour, *A History of Travel in America*, Vol. III, Bobbs Merrill, 1915.
"First Triumph of John A. Roebling, Father of the Suspension Bridge, Came When He Routed Hostile Combine and Put Wire Rope on Canal," Trenton *Advertiser*, Sept. 23, 1923.
Guide Book of the Central Railroad of New Jersey and its Connections through the Coal-Fields of Pennsylvania, Harper, 1864.
Information on Roebling's patents furnished by C. C. Sunderland, Chief Engineer of John A. Roebling's Sons Company, Trenton.
Original letters from Ellet, Young, etc., and Roebling's original copies of his own letters, 1839, etc.
"Roebling, Brooklyn Bridge Builder, Made First Cable in Tiny Shop at Saxonburg," Pittsburgh *Press*, July 28, 1929.
Roebling's notes, drawings, and correspondence covering his inventions and patent applications, 1834-48.

EARLY SUSPENSION BRIDGE PATENTS (JOHN A. ROEBLING)

"Anchoring Suspension Chain for Bridge," *Patent Office Report (U. S.)*, 1846, Vol. 1, p. 298, Patent No. 4710, Aug. 26, 1846.
"Apparatus for Passing Suspension Wires for Bridges," *Patent Office Report (U. S.)*, 1847, Vol. 1, p. 74, Patent No. 4945, Jan. 26, 1847.
"Application of Traveling Wheels for Spinning Suspension Cables," *Journal of the Franklin Institute*, Vol. 15 (1848).
"Improvements in the Wire Cable or Chain Suspension Bridge," *Mechanics' Magazine*, London, Vol. 48 (Jan. 22, 1848), pp. 91–92.

CHAPTER VI: THE BRIDGEBUILDER

Annales des Ponts et Chaussées, 1831, Vol. 1, p. 93; 1834, Vol. 1, p. 129; 1834, Vol. 2, pp. 157–72; 1844, Vol. 2, p. 89; 1850, Vol. 2, p. 294.
Berg, *Der Bau der Hängebrücke aus Eisendraht*, Leipzig, 1824.
A Century of Progress—History of the Delaware and Hudson Company, 1823–1923, Albany, 1925.
Construction of Parallel Wire Cables for Suspension Bridges, John A. Roebling's Sons Company, Trenton, 1924.
Johnson, C. A., "The Development of the Suspension Bridge," *Architectural Forum*, Vol. 49 (1928), pp. 721–28.
Mahan, D. H., "Monongahela Wire Bridge," "Wire Suspension Canal-Aqueduct-Bridge over the Allegheny River at Pittsburgh," *Civil Engineering*, 1846 and later editions.
Mehrtens, G. C., *Eisenbrückenbau*, Vol. 1, pp. 485–92, Leipzig, 1908.
Morandière, *Traité de la Construction des Ponts*, Paris, 1888.
Roebling, J. A., "The Wire Suspension Aqueduct over the Allegheny River at Pittsburgh," illustrated, *American Railroad Journal*, Vol. 18 (Oct. 9, 1845), pp. 631, 648–49.
——— "Wire Suspension Bridge over the Monongahela at Pittsburgh," illustrated with large lithographed plate of the designer's drawings, *American Railroad Journal*, Vol. 19 (June 13, 1846), pp. 376–77.
Tyrrell, H. G., *History of Bridge Engineering*, published by the author, 1911.
White, Joseph, and von Bernewitz, M. W., *The Bridges of Pittsburgh*, Cramer Printing and Publishing Co., Pittsburgh, 1928.

ALLEGHENY RIVER AQUEDUCT, PITTSBURGH

"Bridge Across the Allegheny," *The Civil Engineer and Architect's Journal*, Vol. 8 (June 1845), p. 195.

"Bridge Across the Allegheny," *The Practical Mechanic and Engineer's Magazine*, Vol. 4 (June 1845), p. 251.

Roebling, J. A., "The Wire Suspension Aqueduct over the Allegheny River, at Pittsburgh," *Journal of the Franklin Institute*, Third Series, Vol. 10 (1845), pp. 306–09, illustration p. 307.

────── "The Wire Suspension Aqueduct over the Allegheny River at Pittsburgh," *Mechanics' Magazine*, London, Vol. 44 (Jan. 17, 1846), pp. 36–39.

MONONGAHELA SUSPENSION BRIDGE

Lindenthal, Gustav, "The Monongahela Bridge—Rebuilding of the Monongahela Bridge at Pittsburgh," *Engineering News*, Vol. 10 (1883), pp. 314–15; Vol. 11 (1884), pp. 251–53, 265–66; *Engineering*, London, Vol. 37 (1884), pp. 239–41, 371.

The Olden Time, Pittsburgh, Vol. 1, 1846 (Reprinted 1876).

"The Suspension Bridge," *American Railroad Journal*, Vol. 19 (Feb. 21, 1846), p. 126.

"Suspension Bridge over the Monongahela at Pittsburgh," *American Railroad Journal*, Vol. 19, (Apr. 4, 1846), p. 216; (June 13, 1846), p. 376.

Wickersham, Col. S. M., "Monongahela Suspension Bridge," *Scientific American Supplement*, Vol. 15 (1883), p. 6201.

────── "The Monongahela Suspension Bridge at Pittsburgh, Pa.," *Engineering News*, Vol. 10 (1883), pp. 243–44; *The Iron Age*, Vol. 31 (June 21, 1883), pp. 3, 5, 7.

DELAWARE AQUEDUCT, RONDOUT AQUEDUCT, LACKAWAXEN AQUEDUCT, NEVERSINK AQUEDUCT

Boynton, H. C., "Bridge Wire Tested After 75 Years," *The Iron Age*, Vol. 121 (1928), p. 400.

Roebling, J. A., "Suspension Aqueduct on the Delaware and Hudson Canal," *American Railroad Journal*, Vol. 22 (Jan. 13, 1849), pp. 21–22.

Schuyler, P. K., "Lackawaxen Suspension Bridge Rebuilt for Present Use," *Engineering and Contracting*, Vol. 69 (1930), p. 421.

Whitford, N. E., *History of New York Canals*, 2 Vols. (supplement to 1905 Report of the N. Y. State Engineer and Surveyor).

"Wire Suspension Bridges," *American Railroad Journal*, Vol. 22 (1849), p. 421.

CHAPTER VII: A LONELY BOYHOOD

Estabrook, H. D., *Address at the Unveiling of a Monument to John A. Roebling*, Roebling Press, Trenton, 1908.
Roebling, W. A., *Early History of Saxonburg*, Butler County Historical Society, Butler, Pa., 1924.
Sipe, C. H., *History of Butler County, Pennsylvania*, Vol. 2, Chapter 20, Topeka, 1927.

CHAPTER VIII: SO LITTLE TIME

Mumford, Lewis, *The Brown Decades*, Harcourt, Brace, 1931, pp. 97–100.
——— "The Buried Renaissance," *The New Freeman*, March 15, 1930.
Original manuscripts of John A. Roebling.
Roebling, J. A., "An Atlantic Telegraph," *Journal of Commerce*, New York, Apr. 20, 1850.
Roebling, J. A., C.E., "The Great Central Railroad from Philadelphia to St. Louis," *American Railroad Journal*, Special Edition, 1847; also Vol. 20 (1847), pp. 122–25, 134–35, 137, 138–41, 155–57.
"Roebling, the Philosopher," Trenton *Times*, Aug. 3, 1931.

CHAPTER IX: THE INDUSTRIALIST

Catalogues issued at various dates by John A. Roebling's Sons Company.
"City Magnificently Honors Memory of the Roebling Who Founded a Great Industry," *Daily True American*, Trenton, July 1, 1908.
Dieterich, G., *Die Erfindung der Drahtseilbahnen*, Leipzig, 1908.
John A. Roebling's Sons Company Joins in Celebrating Trenton's 250th Anniversary, 15 pages, John A. Roebling's Sons Company, Trenton, 1929.
Kerney, James, "Roebling's Greatest One-Family Industry in America," Trenton *Sunday Times-Advertiser*, Oct. 25, 1931.
Letters from John A. Roebling to Charles Swan, 1849 to 1865.

"Monument to Memory of Great Trentonian Unveiled," Trenton *Times*, June 29, 1908.

Moore, K. W., "A Century of Wire Rope Engineering—John A. Roebling's Sons Company Celebrates Its 100th Anniversary," *Trenton*, Vol. 17 (Sept. 1941), pp. 1–3.

Mumford, J. K., *Outspinning the Spider: The Story of Wire and Wire Rope*, Robert Stillson Co., New York, 1921.

Roebling, W. A., "An Inside View of a Great Industry," manuscript written 1919, published in Schuyler's *The Roeblings*, 1931, pp. 330–66.

"Trenton Pays Tribute to the Great Engineer: John A. Roebling. Monument to Memory of Great Trentonian Unveiled," Trenton *Evening Times*, June 30, 1908.

Trenton *State Gazette*, Aug. 21, 1848; Apr. 22, 1850.

Wire-Roping the German Submarine, John A. Roebling's Sons Co., Trenton, 1920.

CHAPTER X: SPANNING NIAGARA

Beecher's Magazine, Trenton, Jan. 1871.

"The Bridges of Niagara Gorge," *Scientific American*, Vol. 80 (1899), pp. 296–97.

Dow, C. M., *Anthology and Bibliography of Niagara Falls*, 2 Vols., Published by the State of New York, 1921.

Dunlap, O. E., "The Romance of Niagara Bridges," *Strand Magazine*, Vol. 18 (1899), pp. 430–33.

"The Principal Bridges of the World: Suspension Bridges," *Scientific American Supplement*, Vol. 86 (1918), p. 294.

Roebling, J. A., *Final Report of John A. Roebling, Civil Engineer, to the President and Directors of the Niagara Falls Suspension and Niagara Falls International Bridge Companies*, May 1, 1855, Lee, Mann & Co., Rochester, N. Y., 1855.

——— "Memoir of the Niagara Falls and International Suspension Bridge," in *Papers and Practical Illustrations of Public Works of Recent Construction, both British and American*, by John Weale, London, 1856, pp. 1–39.

Rogers, Dr. R. J. (Letter describing Roebling's treatment of cholera cases at Niagara), *New York Medical Journal*, Nov. 15, 1884.

NIAGARA FOOTBRIDGE (ELLET)

American Railroad Journal, Vol. 21 (1848), pp. 98, 212; Vol. 29 (1856), pp. 241–43; Vol. 34 (1861), pp. 468–69; Vol. 54 (1881), p. 178.

The Civil Engineer and Architect's Journal, Vol. 11 (1848), pp. 191, 288.
Mechanics' Magazine, London, Vol. 48 (1848), pp. 87–88, 473; Vol. 49 (1848), p. 332; Vol. 54 (1851), p. 73.

LEWISTON SUSPENSION BRIDGE (SERRELL)

Engineering News, Vol. 41 (Jan. 12, 1899), pp. 18–20, 24–25.
Engineering Record, Vol. 40 (1899), pp. 286–91).
Scientific American, Vol. 6 (1851), pp. 186, 225.
Serrell, E. W., "The Applicability of Suspension Bridges to Railways," *American Railroad Journal*, Vol. 26 (1853), p. 167.
———— "Lewiston and Queenston Suspension Bridge," *Appleton's Mechanics' Magazine and Engineers' Journal*, Vol. 2 (1852), pp. 137, 138, 168, 216.

NIAGARA RAILWAY SUSPENSION BRIDGE (ROEBLING)

American Railroad Journal, Vol. 18 (1845), p. 795; Vol. 19 (1846), pp. 36–37; Vol. 20 (1847), p. 153; Vol. 22 (1849), pp. 3–4, 21–22; Vol. 25 (1852), pp. 546, 711; Vol. 28 (1855), pp. 187, 402–03; Vol. 29 (1856), pp. 241–43; Vol. 34 (1861), pp. 468–69, 822, 902–03; Vol. 43 (1870), pp. 285–88; Vol. 47 (1874), p. 357; Vol. 54 (1881), pp. 257–58; Vol. 60 (1886), p. 138.
American Society of Civil Engineers—Transactions, Vol. 10 (1881), pp. 195–224; Vol. 17 (1887), pp. 204–12; Vol. 40 (1898), pp. 125–77.
Annales des Ponts et Chaussées (Memoires et Documents), Vol. 2 (1852), pp. 211–26; Vol. 1 (1861), pp. 202–03; Vol. 2 (1877), pp. 662–64.
Appleton's Mechanics' Magazine and Engineers' Journal, Vol. 2 (1852), pp. 193–95.
Barlow, P. W., "Observations on the Niagara Bridge," *The Engineer*, London, Vol. 10 (1860), pp. 283–85; *Journal of the Franklin Institute*, Vol. 41 (1861), pp. 16–22, 89–93, 160–65.
———— *Observations on the Niagara Railway Suspension Bridge*, London, 1860.
———— "On the Mechanical Effect of Combining Girders and Suspension Chains," *The Civil Engineer and Architect's Journal*, Vol. 20 (Oct. 1857), p. 323; Vol. 23 (Aug. 1, 1860), pp. 225–30.
The Civil Engineer and Architect's Journal, Vol. 9 (1846), p. 73; Vol. 10 (1847), p. 162; Vol. 12 (1849), pp. 107–10; Vol. 15 (1852), p. 396; Vol. 18 (1855), p. 160; Vol. 20 (1857), pp. 231–32, 323; Vol. 23 (1860), pp. 225–30, 317–19; Vol. 24 (1861), pp. 53–55, 118.

BIBLIOGRAPHY

The Engineer, London, Vol. 8 (1859), pp. 138, 179, 225; Vol. 9 (1860), pp. 170–71, 190, 192, 319; Vol. 10 (1860), pp. 187, 283–85, 292; Vol. 11 (1861), p. 280; Vol. 13 (1862), p. 60; Vol. 42 (1876), p. 96; Vol. 52 (1881), p. 136; Vol. 64 (1887), p. 481.

Engineering News, Vol. 8 (1881), p. 108; Vol. 10 (1883), p. 246; Vol. 18 (1887), pp. 416–19.

Engineering Record, Vol. 40 (1899), pp. 286–91.

Institution of Civil Engineers—Proceedings, London, Vol. 14 (1855), p. 459; Vol. 16 (1857), p. 459; Vol. 26 (1867), pp. 266–68; Vol. 40, No. 2 (1874), pp. 173–74; Vol. 54 (1878), p. 190; Vol. 65 (1881), pp. 389–90; Vol. 93, No. 3 (1887–88), pp. 510–12; Vol. 133, No. 3 (1897–98), p. 422.

Journal of the Franklin Institute, Vol. 15 (1848), p. 335; Vol. 29 (1855), p. 233; Vol. 40 (1860), pp. 361–72; Vol. 41 (1861), pp. 16–32, 89–93, 160–65, 237–38; Vol. 54 (1867), p. 410; Vol. 65 (1873), p. 79.

Lindenthal, Gustav, "Old and New Forms of the Suspension Bridge," *Engineering Magazine*, Vol. 16 (1898), p. 367.

"The Old Niagara Railway Suspension Bridge," *Public Works*, 1920, pp. 370–72.

The Practical Mechanic's Journal, Vol. 2 (1857–58), pp. 218–19; Vol. 5 (1860–61), p. 250.

"Predicted Fall of the Niagara River Suspension Bridge," *Holley's Railroad Advocate*, Vol. 3 (May 2, 1857), p. 8.

Railroad Advocate, Vol. 1, Dec. 30, 1854, p. 3; Mar. 10, 1855, p. 2; Mar. 17, 1855, p. 2; Vol. 2, June 30, 1855, pp. 3–4.

Railroad Gazette, Vol. 19 (1887), pp. 715–16.

Roebling, J. A., "Iron, and Iron Bridges," *The Engineer*, London, Vol. 10 (1860), p. 187.

——— "The Niagara Suspension Bridge," *The Engineer*, London, Vol. 8 (1859), p. 225.

——— "Niagara Suspension Bridge," *Railroad Advocate*, Vol. 1 (Mar. 17, 1852), p. 2.

——— "Passage of the First Locomotive over the Suspension Bridge over the Falls of Niagara," *Journal of the Franklin Institute*, Vol. 29 (1855), p. 233.

——— "Railroad Suspension Bridge at Niagara Falls" (Report to the Directors of the Niagara Falls International and Suspension Bridge Companies, *American Railroad Journal*, Vol. 25 (Sept. 11, 1852), pp. 578–80.

——— "*Report on the Condition of the Niagara Railway Suspension Bridge—1860*," Aug. 1, 1860; (Report Reprinted), *Journal of the Franklin Institute*, Vol. 40 (Dec. 1860), pp. 361–72.

Roebling, W. A., *A Reply to the Recent Criticism Made by Mr. Edward Wasell upon the Niagara Railway Suspension Bridge*, S. B. Leverich Co., New York, 1877.
Scientific American, Vol. 7 (1852), p. 409; Vol. 3 (New Series) (1860), p. 185; Vol. 36 (1877), pp. 249, 297; Vol. 40 (1879), pp. 335, 337; Vol. 43 (1880), p. 202; Vol. 45 (1881), pp. 31, 35, 36.
Scientific American Supplement, Vol. 3 (1877), p. 1073; Vol. 16 (1883), p. 6268.
Street Railway Review, Vol. 7 (1897), pp. 658–59.

CLIFTON BRIDGE AT NIAGARA FALLS (SAMUEL KEEFER)

American Railroad Journal, Vol. 40 (1867), p. 486.
The Clifton Suspension Bridge at Niagara Falls, Brundage, Niagara Falls, N. Y., 1872.
The Engineer, London, Vol. 23 (1867), p. 505; Vol. 26 (1868), p. 278.
Engineering, London, Vol. 4 (1867), pp. 203, 590; Vol. 5 (1868), p. 592; Vol. 7 (1869), pp. 268, 285, 300, 301.
Engineering News, Vol. 35 (Jan. 2, 1896), pp. 13–14.
Institution of Civil Engineers—Proceedings, London, Vol. 144 (1900–01), pp. 69–85.
Journal of the Franklin Institute, Vol. 54 (1867), p. 277; Vol. 55 (1868), p. 292.
Keefer, Samuel, "The Clifton Bridge, Niagara," *Van Nostrand's Eclectic Engineering Magazine*, Vol. 2 (1870), pp. 318–19.
Maw, W. H., and Dredge, James, *Modern Examples of Road and Railway Bridges*, "The Clifton Suspension Bridge, Niagara Falls," pp. 102–12, London, 1872.
Scientific American, Vol. 19 (1868), p. 213.
Strand Magazine, Vol. 18 (Nov. 1899), pp. 430–33.

KENTUCKY RAILROAD SUSPENSION BRIDGE (ROEBLING)

American Railroad Journal, Vol. 27 (1854), p. 550; Vol. 29 (1856), pp. 241–43; Vol. 33 (1860), p. 885.
Colburn's Railroad Advocate, Vol. 2 (Sept. 1, 1855), p. 3.
Engineering, London, Vol. 1 (1866), p. 396.
Engineering News, Vol. 4 (Feb. 10, 1877), p. 37.
Institution of Civil Engineers—Proceedings, London, Vol. 14 (1855), p. 460; Vol. 54 (1878), p. 182.
Journal of the Franklin Institute, Vol. 54 (1867), pp. 410–12.
Scientific American Supplement, Vol. 16 (1883), p. 6268.

CHAPTER XI: THE MIND AND THE HEART

Baker, R. P., "Roebling Family Linked with Rensselaer Institute," Brooklyn *Daily Eagle*, Brooklyn Bridge 50th Anniversary Section, May 24, 1933.
Engineering News, Vol. 8 (1881), pp. 83–87.
Greene, B. F., *The Rensselaer Polytechnic Institute. Its Reorganization in 1849–50; Its Condition at the Present Time; Its Plans and Hopes for the Future*, Troy, N. Y., 1855.
Information furnished by officials of Rensselaer Polytechnic Institute, 1941.
"Washington Augustus Roebling," in Scannell's *New Jersey's First Citizens and State Guide*," Vol. 1 (1917–18), pp. 431–34; Vol. 2 (1919–20), pp. 386–89.
"Washington Augustus Roebling," *Who's Who in America*, 1918–19, p. 2317, Marquis.

CHAPTER XII: "THE BRIDGE WILL BE BEAUTIFUL"

ALLEGHENY BRIDGE AT SIXTH STREET, PITTSBURGH

American Society of Civil Engineers—Transactions, Vol. 1 (1872), p. 32.
The Engineer, London, Vol. 56 (Dec. 7, 1883), p. 443.
Engineering, London, Vol. 4 (1867), pp. 98–99; Vol. 6 (1868), p. 1; Vol. 38 (1884), pp. 43–45.
Engineering News, Vol. 8 (1881), p. 280; Vol. 12 (1884), pp. 49–51.
Engineering News-Record, Vol. 106 (1931), p. 662.
"The Great Suspension Bridges of the United States," *Scientific American*, Vol. 40 (1879), pp. 335, 337.
Institution of Civil Engineers—Proceedings, London, Vol. 76 (1884), pp. 343–45.
Scientific American, Vol. 49 (1883), p. 320.
Scientific American Supplement, Vol. 16 (1883), p. 6269.
White, J., and von Bernewitz, M. W., *The Bridges of Pittsburgh*, Cramer Printing and Publishing Co., Pittsburgh, 1928, pp. 49, 84–85.
Wilkins, W. G., "The Reconstruction of the Sixth Street Bridge at Pittsburgh," *Proceedings of the Engineering Society of Western Pennsylvania*, Vol. 11 (1895), pp. 143–67.
Zeitschrift für Bauwesen, 1868, p. 499.

CHAPTER XIII: A RIVER DIVIDES

THE OHIO BRIDGE, CINCINNATI

American Railroad Journal, Vol. 20 (1847), p. 12; Vol. 29 (1856), pp. 584, 829.
Annual Report of the Covington and Cincinnati Bridge Company, Covington, Ky., March 7, 1859.
The Engineer, London, Vol. 11 (1861), p. 280.
Goss, C. F., *Cincinnati, the Queen City*, 1912.
Greve, C. T., *Centennial History of Cincinnati*, 1904.
Mechanics' Magazine, London, Vol. 47 (1847), p. 190.
Stevens, H. R., *The Ohio Bridge*, Ruter Press, Cincinnati, 1939.

CHAPTER XIV: THE SOLDIER

Archives of the Adjutant-General's Department, State House, Trenton.
Captains of the Civil War, Yale University Press, 1921.
Kaufmann, W., *Die Deutschen im amerikanischen Bürgerkriege*, Munich and Berlin, 1911.
Lyman, Col. Theodore, *Meade's Headquarters*, Little, Brown, 1922.
Original letters of Washington Roebling, 1861-65.
"Passing of Colonel Roebling Recalls his Distinguished Civil War Service," Trenton *Sunday Times-Advertiser*, Aug. 8, 1926.
Schuyler, Hamilton, *The Roeblings*, Princeton University Press, 1931.
Young, J. B., *The Battle of Gettysburg*, Harper, 1913.

CHAPTER XV: THE OHIO IS SPANNED

CINCINNATI BRIDGE

American Society of Civil Engineers, Transactions, Vol. 28 (1893), pp. 47-56, 358-71.
Annual Report of the President and Directors to the Stockholders of the Covington and Cincinnati Bridge Co. for year ending Feb. 28, 1867, Murphy & Bechtel, Trenton, 1867.
Engineering, London, Vol. 2 (1866), p. 307; Vol. 4 (1867), pp. 25, 346-47.
Engineering Record, Vol. 27 (1893), pp. 434-35; Vol. 38 (1898), pp. 314-16, 554-55.

Farrington, E. F., *A Full and Complete Description of the Covington and Cincinnati Suspension Bridge with Dimensions and Details of Construction*, J. P. Lindsay & Co., Cincinnati, 1867.
Harper's Weekly, Feb. 29, 1867.
Hildenbrand, W., "Cable Making for Suspension Bridges," *Institution of Civil Engineers—Proceedings*, London, Vol. 51 (1878), p. 300.
Institution of Civil Engineers—Proceedings, London, Vol. 131 (1897–98), pp. 400–01.
"Largest Bridge in the World," *The Engineer*, London, Vol. 20 (1865), p. 354.
Maw, W. H., and Dredge, James, *Modern Examples of Road and Railway Bridges*, pp. 159–65, London, 1872.
Railroad Gazette, Vol. 29 (1897), pp. 644–45.
Roebling, J. A., "The Cincinnati Suspension Bridge," *Engineering*, London, Vol. 4 (1867), pp. 22–23, 49, 74–76, 98–99, 140–41.
——— *Report to the President and Board of Directors of the Covington and Cincinnati Bridge Company*, April 1, 1867, Trenton, 1867.
Scientific American, Vol. 40 (1879), pp. 335, 337.
U. S. Engineers (Gov.), "Covington and Cincinnati Suspension Bridge," *Report of the Chief of Engineers* (U. S. Army), Vol. 2 (1871), pp. 414–19.

WIND ACTION ON SUSPENSION BRIDGES

am Ende, Max, "Suspension Bridges with Stiffening Girders," *Institution of Civil Engineers—Proceedings*, London, Vol. 137 (1898–99), pp. 306–42.
Annales des Ponts et Chaussées (Memoires et Documents), Vol. 2 (1859), pp. 249–329. (Roche-Bernard Bridge, France, built 1840, wrecked by a storm in 1852. Counter cables added in the restoration.)
Averill, W. A., "Collapse of the Tacoma Narrows Bridge," *Pacific Builder and Engineer*, Dec. 1940, pp. 20–27.
Barber, Frank, "Could the Collapse of the Tacoma Narrows Bridge Have Been Avoided?" *Roads and Bridges*, Toronto, June 1941.
Barlow, P. W., "On the Mechanical Effect of Combining Girders and Suspension Chains," *The Practical Mechanic's Journal*, Vol. 2 (1857–58), pp. 218–19.
Barlow, Peter (and Staff), "Observations on the Niagara Railway Suspension Bridge," *The Civil Engineer and Architect's Journal*, Vol. 24 (1861), pp. 53–55, 118.
Bender, C. B., "The Design of Structures to Resist Wind Pressure," *Institution of Civil Engineers—Proceedings*, London, Vol. 69 (1882), pp. 106–09. (See also pp. 125–26.)

BIBLIOGRAPHY

Bender, C. B., "Suspension Bridges of any Desired Degree of Stiffness," *Van Nostrand's Engineering Magazine*, Vol. 25 (1881), pp. 399–07.

Bowers, N. A., "Model Tests Showed Aerodynamic Instability of Tacoma Narrows Bridge," *Engineering News-Record*, Vol. 125 (Nov. 21, 1940), p. 44.

——— "Tacoma Narrows Bridge Wrecked by Wind," *Engineering News-Record*, Vol. 125 (Nov. 14, 1940), p. 1.

Brunel, I. K., and Cowper, E., "The Action of Roadway due to Vibration and Proposed Stiffening," *Institution of Civil Engineers—Proceedings*, London, Vol. 1 (1841), p. 77.

Cavarallo, M., "Note sur les conditions de stabilité des ponts suspendus," *Annales des Ponts et Chaussées (Memoires et Documents)*, 1852, Vol. 2, pp. 211–26.

Cissel, J. H., "Bibliography of Suspension Bridges, with Particular Regard to Wind and Dynamic Effects," *American Society of Civil Engineers—Proceedings*, Vol. 69 (Dec. 1943), pp. 1581–85.

——— "Stiffness as a Factor in Long Span Suspension Bridge Design," *Roads and Streets*, Vol. 84 (April 1941), pp. 64–72.

Cox, Homersham, "The Vibratory Strains of Suspension Bridges," *The Civil Engineer and Architect's Journal*, Vol. 12 (1849), pp. 107–10.

Dredge, James, "Principles of Bridge Building," *Mechanics' Magazine*, London, Vol. 34 (1841), pp. 472–73.

"Dynamic Wind Destruction" (editorial), *Engineering News-Record*, Vol. 125 (Nov. 21, 1940), p. 42.

Eldridge, C. H., "The Tacoma Narrows Bridge," *Civil Engineering*, Vol. 10 (May 1940), pp. 299–302.

The Failure of the Suspension Bridge over Tacoma Narrows, a report to Paul Carew, chairman, Narrows Bridge Loss Committee, June 1941.

The Failure of the Tacoma Narrows Bridge, a report to the Hon. John M. Carmody, Administrator, Federal Works Agency, Washington, D. C., March 28, 1941.

"Failure of the Tacoma Narrows Bridge," Report of the Special Committee of the Board of Direction, *American Society of Civil Engineers—Proceedings*, Vol. 69 (Dec. 1943), pp. 1555–86.

Farquharson, F. B., "Dynamic Tests on Bridge Models," *Engineering News-Record*, Vol. 127 (July 3, 1941), p. 73.

Finch, J. K., "Wind Failures of Suspension Bridges, or Evolution and Decay of the Stiffening Truss, *Engineering News-Record*, Vol. 126 (1941), pp. 402–07, 459.

The Gentleman's Magazine, Vol. 88 (1) (Mar. 1818), p. 268. (Dryburgh Abbey Bridge, Scotland, built 1817, completely destroyed by a gale on Jan. 15, 1818.)

Humphreys, A. A., "History of the Bridge Across the Ohio at Wheeling," *Report of the Chief of Engineers* (U. S. Army), 1878, Part 2, pp. 1029–33.
Laboratory Studies on the Tacoma Narrows Bridge at the University of Washington, July 1941.
Mills, B. D., "Aerodynamic Action on Wires and Bridges," *Engineering News-Record*, Vol. 126 (1941), pp. 167–68.
Moisseiff, L. S., "Growth in Suspension Bridge Knowledge," *Engineering News-Record*, Vol. 123 (Aug. 17, 1939), pp. 46–49.
"Montrose Suspension Bridge," *The Civil Engineer and Architect's Journal*, Vol. 1 (1838), pp. 381, 417. (Bridge in Scotland, built 1829, partially destroyed by a gale on Oct. 11, 1838.)
"Narrows Nightmare," *Time*, Nov. 18, 1940, p. 21.
Pigeaud, E., "Action du Vent sur un Tablier de Pont Suspendu," *Le Génie Civil*, Vol. 87 (1925), pp. 83–85.
Provis, W. A., "Observations on the Effect of the Wind on the Suspension Bridge over the Menai Straits," *Institution of Civil Engineers—Proceedings*, London, Vol. 1 (1841), pp. 74–81.
Rathbone, T. C., and Turner, C. A. P., "Tacoma Narrows Collapse," *Engineering News-Record*, Vol. 125 (Dec. 5, 1940), p. 40.
Raymond, C. W., Bixby, W. H., and Burr, Edward, "The Maximum Practical Length for Suspension Bridges," *Engineering News*, Vol. 32 (1894), pp. 423–25, 444–46, 463–65.
Report of the Board of Investigation, Tacoma Narrows Bridge, to James A. Davis, Acting Director of Highways, Olympia, Wash., 1941.
Robinson, S. W., "The Vibration of Bridges," *Institution of Civil Engineers—Proceedings*, London, Vol. 90 (1886–87), pp. 475–78.
Roebling, J. A., "The Cincinnati Bridge," *Engineering*, London, Vol. 7 (1869), p. 300.
——— "Niagara Suspension Bridge—Report to the Directors of the Niagara Falls International Railway Suspension Bridge Company," *Appleton's Mechanics' Magazine and Engineers' Journal*, Vol. 2 (1852), pp. 193–95.
——— "Some Remarks on Suspension Bridges, and on the Comparative Merits of Cable and Chain Bridges," *American Railroad Journal*, Vol. 12 (Apr. 1, 1841), pp. 193–96.
Royen, N., "Calculation of Extent to which Cable in Suspension Bridges Takes Up Wind Load," *Der Eisenbau*, Vol. 10 (1919), pp. 239–43.
Russell, J. S., "On the Vibration of Suspension Bridges and Other Structures; and the Means of Preventing Injury from this Cause," *Journal, Royal Scottish Society of Arts*, Vol. 1 (1839), pp. 305–07. (Describes destruction of span at Brighton, Nov. 30, 1836.)

"Stabilization of Suspension Bridges," *The Engineer*, London, Vol. 171 (1941), pp. 116–17, 400–01.

"Stays and Brakes Check Oscillation of Whitestone Bridge," *Engineering News-Record*, Vol. 125 (Dec. 5, 1940), p. 54.

Steinman, D. B., *Bridges and Aerodynamics*, reprint from *Proceedings, American Toll Bridge Association*, 1941, pp. 49–57.

——— "The Failure of the Tacoma Narrows Bridge," *Bulletin, American Society of Swedish Engineers*, Oct. 1941, pp. 10–11, 32–33.

——— "Rigidity and Aerodynamic Stability of Suspension Bridges," *Proceedings, American Society of Civil Engineers*, Vol. 69 (1943), pp. 1361–97; Discussions, Vol. 70 (1944), pp. 211–20, 410–12, 689–94, 1003–12, 1063–84, 1573–84, Vol. 71 (1945), pp. 487–535.

——— "Stay Systems for Suspension Bridges," *Engineering News-Record*, Vol. 126 (Feb. 27, 1941), pp. 36–37; May 8, 1941, pp. 60–61.

——— "Suspension Bridge Stability," *Engineering News-Record*, Vol. 127 (Dec. 18, 1941), p. 51.

——— "The Tacoma Bridge Report," *Engineering News-Record*, Vol. 127 (Aug. 14, 1941), pp. 59–61.

"Suspension Bridges," *Journal of the Franklin Institute*, Vol. 34 (1847), pp. 231–33.

"Suspension Bridges and Wind Resistance," by the staff of Modjeski and Masters, *Engineering News-Record*, Oct. 23, 1941, p. 97.

"Tacoma Narrows Bridge Crashes in Puget Sound," *Life*, Nov. 18, 1940.

Tyrrell, H. G., *History of Bridge Engineering*, Chicago, 1911, p. 209. (Union Bridge, over the Tweed at Berwick, the first highway suspension bridge in Great Britain, opened July 1820, blown down by a storm within six months.)

"Why the Tacoma Narrows Bridge Failed," *Engineering News-Record*, Vol. 126 (May 8, 1941), p. 75.

"The Wire Bridge over the Ohio" (Wheeling Bridge), *American Railroad Journal*, Vol. 22 (1849), pp. 626–27.

CHAPTER XVI: A CITY'S DREAM

American Railroad Journal, Vol. 4 (1835), pp. 4–5; Vol. 57 (1883), p. 115.

Brooklyn Bridge: 1883–1933, by the City of New York Department of Plant and Structures, 1933.

Brooklyn *Daily Eagle*, Semicentennial issues, Apr. 19–25, May 24, 1933.

Byrne, E. A., "The Brooklyn Bridge—A Half Century of Service," *Civil Engineering*, Vol. 3 (June 1933), pp. 299–303.

BIBLIOGRAPHY 439

The Engineer, London, Vol. 17 (Feb. 19, 1864), p. 107. (Roebling's proposal, in his own words, to build a bridge between New York and Brooklyn in one span of 1600 ft.)
Engineering News, Vol. 10 (May 26, 1883), pp. 241-42.
Howard, H. W. B., *History of the City of Brooklyn*, 1893.
Mechanics' Magazine, New York, Vol. 5 (Jan. 10, 1835), pp. 17-19.
Pope, Thomas, *A Treatise on Bridge Architecture*, Alexander Niven, New York, 1811.
Putnam, H., "Brooklyn," in L. P. Powell's *Historic Towns of the Middle States*, 1899.
van Wyck, F., *Keskachauge, or the First White Settlement on Long Island*, Putnam's, 1924.
Welch, M. S., *Vrouw Knickerbocker: The Romance of the Building of Brooklyn*, Dorrance, Philadelphia, 1926.

CHAPTER XVII: THE GOAL IN SIGHT

American Railroad Journal, Vol. 42 (1869), pp. 313-14, 477.
Engineering, London, Vol. 4 (1867), pp. 25, 71, 392-93, 423-24; Vol. 7 (1869), p. 223; Vol. 8 (1869), pp. 196, 409-10.
Engineering News, Vol. 8 (Apr. 30, 1881), pp. 171, 172.
Hildenbrand, Wilhelm, *Die Hängebrücken von der ältesten bis zur neuesten Zeit*, Lecture before the Technischen Verein von New York, New York, 1905.
In Memoriam: John A. Roebling, Trenton, 1869.
"John A. Roebling, Builder of Brooklyn Bridge, Died as Result of Injuries," Hartford *Courant*, Dec. 20, 1931.
Journal of the Franklin Institute, Vol. 54 (1867), pp. 242-51, 305-11; Vol. 55 (1868), p. 367; Vol. 58 (1869), pp. 147, 220, 361-63.
Last Will and Testament of John A. Roebling.
New York Bridge Company. An Act to Incorporate the Company, New York, 1867.
Roebling, J. A., *Long and Short Span Bridges* (published after his death by his son, Washington A. Roebling), Van Nostrand, 1869.
―― *Report of J. A. Roebling to the President and Directors of the New York Bridge Company on the Proposed East River Bridge*, 48 pages, Brooklyn, 1867.
Report of the Board of Consulting Engineers to the Directors of the New York Bridge Company, The Standard Press Print, Brooklyn, 1869.
Van Nostrand's Eclectic Engineering Magazine, Vol. 1 (1868), p. 370.

CHAPTER XVIII: DOWN IN THE CAISSONS

Annales des Ponts et Chaussées (*Memoires et Documents*), Vol. 1 (1874), pp. 352–400.
Annual Reports of the Chief Engineer and General Superintendent of the East River Bridge, June 12, 1870, Eagle Book and Job Printing Dept., Brooklyn, 1870.
Engineering, London, Vol. 8 (1869), pp. 409–10; Vol. 9 (1870), pp. 219, 275–76, 408; Vol. 10 (1870), pp. 31, 68, 275, 322, 484; Vol. 13 (1872), pp. 117, 135; Vol. 14 (1872), pp. 282, 297–98.
English Mechanic & Mirror of Science, Vol. 11 (1870), p. 465.
Green, S. W., *A Complete History of the New York and Brooklyn Bridge: From Its Conception in 1866 to Its Completion in 1883*, New York, 1883.
The Iron Age, Vol. 11 (May 22, 1873), p. 1.
Journal of the Franklin Institute, Vol. 59 (1870), pp. 223–24; Vol. 60 (1870), pp. 32, 217–19, 219–21; Vol. 61 (1871), pp. 75, 364–65; Vol. 62 (1871), pp. 147, 222–23; Vol. 65 (1873), pp. 149–50, 298; Vol. 66 (1873), pp. 223–26.
New York Bridge Company. *Report of the Chief Engineer of the East River Bridge on Prices of Materials and Estimated Cost of the Structure, June 28, 1872*, Brooklyn, 1872.
New York Bridge Company. *Report of the Executive Committee of the East River Bridge on the Cost of the Structure and Land, December 1873*, Eagle Print, Brooklyn, 1873.
New York Bridge Company. *Report of the General Superintendent of the New York Bridge Company*, Eagle Print, Brooklyn, 1873.
Report of the Chief Engineer and General Superintendent to the Board of Directors of the New York Bridge Company, June 5, 1871, Eagle Book and Job Printing Dept., Brooklyn, 1871.
Reports of the Executive Committee, General Superintendent and Treasurer, 1872–1873, 1873–1874.
Reports of the Executive Committee and Treasurer, Brooklyn, 1872.
Roebling, W. A., "The East River Bridge," *Engineering*, London, Vol. 12 (1871), pp. 15–16; Vol. 14 (1872), pp. 113, 181–82, 198; Vol. 15 (1873), p. 79.
——— "Second Annual Report of Chief Engineer," *Van Nostrand's Eclectic Engineering Magazine*, Vol. 5 (1871), pp. 381–88.
——— "East River Bridge—Chief Engineer's Report," *Van Nostrand's Eclectic Engineering Magazine*, Vol. 7 (1872), pp. 199–200.

Roebling, W. A., "East River Bridge, New York," *Engineering*, London, 1873, Vol. 15, pp. 117–18, 134, 165, 198, 248, 254, 278, 296, 381, 434; Vol. 16, pp. 18, 81–82, 216–17, 238, 306, 330, 349.
——— *Pneumatic Foundations of the East River Suspension Bridge, New York*, 92 pages, 7 plans, 5 plates, G. W. Averell, New York, 1872.
Scientific American, Vol. 27 (1872), pp. 24, 201.
Van Nostrand's Eclectic Engineering Magazine, Vol. 2 (1870), pp. 439, 629; Vol. 3 (1870), p. 439; Vol. 4 (1871), pp. 13, 168, 375, 478, 666; Vol. 6 (1872), p. 448; Vol. 7 (1872), pp. 399–407.

CHAPTER XIX: OF GRANITE AND STEEL

THE BROOKLYN BRIDGE

American Society of Civil Engineers, Transactions, Vol. 6 (1877), pp. 7–27; Vol. 9 (1880), pp. 162–72.
Annales des Ponts et Chaussées, 1874.
Articles and illustrations in *Leslie's Weekly, Harper's Weekly, Scientific American*, etc., 1869–83.
Barnes, A. C., *The New York and Brooklyn Bridge*, Brooklyn, 1883.
"The Brooklyn Bridge," *Harper's New Monthly Magazine*, Vol. 66 (May 1883), pp. 925–46.
Communication to the Board of Aldermen of New York by the President of the Trustees of the New York and Brooklyn Bridge, Made Feb. 3, 1876, Eagle Print, Brooklyn, 1876.
A Complete History of the New York and Brooklyn Bridge from its Conception in 1866 to its Completion in 1883, with Sketches of the Lives of J. A. Roebling, W. A. Roebling, and H. C. Murphy, published by S. W. Green, New York, 1883.
Conant, W. C., *A History of the Bridge* (reprinted from *Harper's Weekly*), No. 331 of *Harper's Franklin Square Library*, Harper, 1883.
Documents Printed in Pursuance of the Order of the Board of Trustees (of the New York and Brooklyn Bridge), July 12, 1877, Brooklyn, 1877.
The Engineer, London, 1874, 1877, 1881.
Engineering, London, Vol. 20 (1875), p. 385; Vol. 21 (1876), p. 139; Vol. 22 (1876), pp. 171, 320; Vol. 23 (1877), pp. 10, 262, 432, 509; Vol. 25 (1878), pp. 53–54, 121, 171; Vol. 27 (1879), pp. 27, 512; Vol. 28 (1879), p. 96; Vol. 31 (1881), pp. 216, 346, 356, 400, 580; Vol. 33 (1882), pp. 111, 219; Vol. 35 (1883), pp. 489–90.

BIBLIOGRAPHY

Engineering News, Vol. 3 (1876), pp. 230, 262, 350–51, 387–88; Vol. 4 (1877), p. 179; Vol. 5 (1878), pp. 86–88, 265–66, 270, 278; Vol. 6 (1879), pp. 5, 97, 182, 219, 382; Vol. 8 (1881), pp. 171–72, 181–83, 191–92, 201–02, 212–13, 223–24, 232–33, 262–63, 273, 283, 291–92, 301–02, 313–14, 334–35, 343–44, 353–56, 363–64, 419, 465–66, 497–98, 506; Vol. 9 (1882), pp. 12–13, 25, 48–49, 81, 97, 104, 300, 401; Vol. 10 (1883), pp. 20–21, 22, 25, 81, 132, 174, 237, 240, 241.

Farrington, E. F., *Concise Description of the East River Bridge, with Full Details of the Construction*, C. D. Wynkoop, New York, 1881.

Hildenbrand, Wilhelm, *Cable Making for Suspension Bridges*, Van Nostrand, 1877.

Hill, Albert, *An Analysis of the Specifications for Steel Cable Wire for the East River Suspension Bridge*, New York, 1877.

"How Times Have Changed" (Building the Towers of Brooklyn Bridge), Boston *Post*, Aug. 6, 1928.

An Illustrated Description of the New York and Brooklyn Bridge (pamphlet), John A. Roebling's Sons Co., Trenton, 1899.

"Inspection Proves Brooklyn Bridge Safe and Sound," New York *Tribune*, Oct. 31, 1920.

The Iron Age, Vol. 16, Sept. 23, 1875, p. 11; Vol. 17, Apr. 27, 1876, p. 14; Vol. 18, Aug. 17, p. 16; Nov. 30, p. 1; Dec. 7, p. 14; Vol. 19, Jan. 4, 1877, p. 5; Jan. 17, p. 14; Jan. 25, p. 14; Feb. 1, p. 1; Vol. 20, Oct. 11, 1877, p. 20; Dec. 6, p. 16; Vol. 21, Jan. 3, 1878, p. 14; June 6, p. 19; June 13, p. 1; Vol. 22, July 4, 1878, p. 26; Oct. 10, p. 18; Vol. 24, July 10, 1879, p. 15; Nov. 20, p. 15; Vol. 25, Mar. 11, 1880, p. 22; Vol. 27, Feb. 24, 1881, p. 13; Vol. 29, Apr. 6, 1882, p. 14; June 29, p. 14; Vol. 30, July 13, 1882, p. 15; July 27, p. 15; Aug. 24, p. 15; Aug. 31, pp. 7, 14; Sept. 21, p. 14; Vol. 31, May 3, 1883, p. 14; May 31, p. 15; Vol. 33, May 22, 1884, p. 23.

Johnson's Cyclopedia, "Bridges," 1896.

Mechanics, New York, Vol. 1 (1882), pp. 193, 242, 453; Vol. 2 (1882), pp. 33, 64, 143, 146, 264; Vol. 3 (1883), p. 318.

Mumford, J. K., *Outspinning the Spider. The Story of Wire and Wire Rope*, New York, 1921.

Mumford, Lewis, "The Brooklyn Bridge," *American Mercury*, Vol. 23 (Aug. 1931), pp. 447–50.

New York and Brooklyn Bridge Proceedings, 1867–84.

New York Evening Post, Apr. 18, 1883.

New York State Statutes. Acts of the State of New York and of the United States in Relation to the New York and Brooklyn Bridge, 1875.

Nouvelles Annales de la Construction, 1879, 1880.

Railway Age, Vol. 8 (1883), pp. 220, 277, 307.

The Railroad Gazette, Vol. 15 (1883), pp. 30, 253, 348, 482, 531.

BIBLIOGRAPHY

Report of the Officers of the New York Bridge Company to the Board of Directors, February 1875, Eagle Print, Brooklyn, 1875.

Roebling, W. A., *Communication from the Chief Engineer in Regard to the Method of Steam Transit over the East River Bridge. Made to the Board of Trustees, March 4, 1876*, Brooklyn, 1878.

—— *Report of the Chief Engineer of the New York and Brooklyn Bridge, January 1, 1877*, Eagle Print, Brooklyn, 1877.

—— *Report of the Chief Engineer on the Strength of the Cables and Suspended Superstructure of the (New York and Brooklyn) Bridge, Made to the Board of Trustees, January 9, 1882*, Eagle Book and Job Printing Dept., Brooklyn, 1882.

Schweizerische Bauzeitung, 1883.

Scientific American, Vol. 30 (1874), pp. 63–64, 408; Vol. 31 (1874), pp. 9, 289, 343; Vol. 32 (1875), pp. 7, 384; Vol. 33 (1875), p. 309; Vol. 34 (1876), pp. 15–16, 289, 377; Vol. 35 (1876), pp. 99, 151, 169, 195, 293; Vol. 36 (1877), pp. 127, 130, 143, 225; Vol. 37 (1877), pp. 63, 66; Vol. 38 (1878), pp. 207–08, 303, 306; Vol. 39 (1878), pp. 287–88; Vol. 40 (1879), pp. 209, 335, 337; Vol. 42 (1880), pp. 6, 281, 408; Vol. 43 (1880), p. 305; Vol. 44 (1881), pp. 3, 31–32, 319, 322, 352; Vol. 46 (1882), pp. 33, 97; Vol. 47 (1882), p. 84; Vol. 48 (1883), pp. 129, 256, 305, 319–20, 324–26.

Scientific American Supplement, Vol. 1 (1876), p. 289; Vol. 2 (1876), pp. 675, 755–56; Vol. 3 (1877), pp. 899, 932; Vol. 7 (1879), p. 3050; Vol. 15 (1883), p. 6044; Vol. 16 (1883), pp. 6265–68.

Steinman, D. B., *Suspension Bridges: Their Design, Construction and Erection*, John Wiley, 1929.

Testimony Taken by the Committee on Commerce and Navigation of the Assembly, in Relation to the New York and Brooklyn Bridge, Pursuant to the Resolution of the Assembly, passed February 5, 1879, 2 Vols. Albany, 1879.

Van Nostrand's Eclectic Engineering Magazine, Vol. 11 (1874), p. 91; Vol. 15 (1876), pp. 90, 380–81; Vol. 16 (1877), pp. 430–38; Vol. 17 (1877), pp. 171–84, 193–201, 289–99; Vol. 18 (1878), pp. 284, 380; Vol. 21 (1879), pp. 261–62, 394.

"Washington Roebling—Builder of Brooklyn Bridge," *Engineering News-Record*, Vol. 97 (1926), p. 232.

Watson, W. J., *Bridge Architecture*, William Helburn, Inc., New York, 1927.

"Wire and Space," *Fortune*, Vol. 3 (Jan. 1931), pp. 83–88.

For a more complete and detailed bibliography, see:

Jakkula, A. A., *A History of Suspension Bridges in Bibliographical Form*, published by the U. S. Public Roads Administration and the

A. and M. College of Texas, College Station, Texas, 1941. ("Brooklyn Bridge," pp. 198–236.)

CHAPTER XX: THE BRIDGE IS DEDICATED

Brooklyn *Daily Eagle*, Bridge Dedication Issue, May 24, 1883; Brooklyn Bridge 50th Anniversary Section, May 24, 1933.
"City Will Celebrate 50th Year of Brooklyn Bridge May 24, Once Daring Engineering Feat," New York *Herald Tribune*, May 14, 1933.
Douglass, Harvey, "The Story of the Brooklyn Bridge," Brooklyn *Daily Eagle*, April 19–25, 1933.
Hewitt, A. S., *Address on the Opening of the New York and Brooklyn Bridge, May 24, 1883*, J. Polhemus, New York, 1883.
J. S. T. Stranahan, W. C. Kingsley, S. Low, S. W. Green's Sons, New York, 1883.
Mumford, Lewis, *Sticks and Stones*, Harcourt, Brace, 1924.
Opening Ceremonies of the New York and Brooklyn Bridge, May 24, 1883, Eagle Job Printing Dept., Brooklyn, 1883.
Report of the Special Committee of the Common Council of the City of Brooklyn upon the Celebration of the Opening of the New York and Brooklyn Bridge, Brooklyn, 1883.
Schuyler, Montgomery, *The Bridge as a Monument* (Harper's Franklin Square Library, No. 331), Harper, 1883.

WASHINGTON A. ROEBLING

American Journal of Science, Sept. 1917.
"Brooklyn Bridge Engineer is Dead," Brooklyn *Daily Times*, July 22, 1926.
"A Builder of New York and His Bridge," New York *Times Magazine*, Dec. 29, 1929.
"Colonel Roebling at Eighty," Brooklyn *Eagle*, May 28, 1917.
"Colonel Roebling, Head of Great Mills, Tells How to Keep Young and on the Job at 84 Years," Trenton *Evening Times*, June 13, 1921.
"Colonel Washington A. Roebling," Trenton *Gazette*, July 22, 1926.
"Col. Washington Roebling, Brooklyn Bridge Builder, Graduate of R.P.I., Dead," Troy (N. Y.) *Record*, July 22, 1926.
"Keeping Active at 84," Kansas City *Star*, June 16, 1921.
"A Man Who Built a Bridge," Schenectady *Union Star*, July 23, 1926.
"New York Owes Much to Trentonian," Trenton *Sunday Times-Advertiser*, Jan. 12, 1930.

"One of Pennsylvania's Outstanding Engineers," *The Pennsylvania Engineer*, Nov. 1936.
"Roebling's Bridge Dream Came True," New York *Evening World*, July 22, 1926.
"Roebling's Penmanship Lauded by Senate President Cummins," Trenton *Sunday Times-Advertiser*, Jan. 18, 1925.
The Scientific Monthly, Vol. 25 (Oct. 1927), pp. 318–20.
"Washington A. Roebling," New York *Telegram*, July 22, 1926.

INDEX

INDEX

accidents in bridge construction, 95, 175, 282, 289, 340, 350, 358, 375, 376, 395, 406
Adams, Col. Julius W., 304, 316
Aderhold, hotel-keeper, 116, 117
Alexandria, Virginia, 253, 254
Allegheny Mountains, railroad surveys, 51–54; portage railway, 33, 61, 72, 73
Allegheny and Portage Railroad, 47, 50, 72, 111, 147
Allegheny River, 49, 50, 89; aqueduct, 80–89, 99; bridge, 205–216
Allen, Horatio, engineer, 316
American Railroad Journal, 58, 71, 86, 97, 124, 159
American Society of Civil Engineers, 316, 404
anchorages, 87, 92, 138, 146, 206, 274, 312, 378, 379, 391
Angelrodt, Carl, 40
Angers bridge, failure, 166
aqueducts, 79–89, 103–106, 145, 166
Arthur, Chester Alan, 410, 411, 415
Auener, Dr. Wilhelm, biographer, 155
August Eduard, the, 22, 26, 27, 139
Aurora, Oregon, colony, 36

Bach, Johann Sebastian, 5, 8
Baehr, Ferdinand, 30, 33, 35, 37, 110
Baltimore, 34, 125, 256
Barlow, Peter W., 185, 186
Bauer, Saxonburg miller, 116
Beethoven, Ludwig van, 16, 17
Belgium, revolution, 19
Berlin, 11, 31; University of, 11
Bernigau, cabinet maker, 116, 117
Bert, Dr. Paul, 362
Berwick bridge, failure, 166

Bethel, Missouri, colony, 36
Blücher, Marshal von, 112
Bollert, Saxonburg colonist, 118
Bowstead, James, 73
Bragg, Braxton, 268
Bramshall, Captain, 251
Bremen, 21, 22, 34
bridge failures, 12, 79, 95, 166, 170–173, 185, 189, 191, 198, 228, 396
Brighton Chain Pier bridge, 166
Britannia Bridge, 161, 166, 186
Brooklyn, 295–301, 313, 314, 319, 368, 403, 406, 411, 412, 415
Brooklyn Bridge, 301–418; a city's dream, 295–300; proposed by Roebling, 300; the enterprise launched, 301–304; Roebling selected as engineer, 304–305; Roebling makes surveys and plans, 306–313; he overcomes opposition, 313–318; the fatal accident, 319–321; Colonel Roebling carries the work forward, 323–324; down in the caissons, 324–367; the "Big Blowout," 339–341; the "Great Fire" in the caisson, 343–347; Colonel Roebling and men trapped in flooding caisson, 347–348; work in the New York caisson, 351–367; paralysis and deaths from caisson disease, 361–366; Colonel Roebling stricken, 366–367; continues to direct the work, 369–408; financial and political difficulties, 370–408; construction accidents, 374–377; anchorages built, 378–379; preparations for cable stringing, 380–391; the first aerial crossing, 382–384; the footbridge, 384–388; spinning the cables, 390–398; fraud by wire

449

INDEX

contractor discovered, 392-394; squeezing the cables, 394-395; more opposition and financial difficulties, 396-398; steel adopted for trusses and floorbeams, 398-400; increased weight and capacity attacked, 400-402; attempt to displace Colonel Roebling, 402-404; work completed, 404-408; the bridge is dedicated, 409-418
Brooklyn *Eagle*, 369, 514
Broughton bridge, failure, 166
Brown, George H., 250
Brown, Rev. John C., 320
Brown, Sir Samuel, bridgebuilder, 166
Brownsville, Pa., 90, 91, 217
Buck, Lefferts L., 198, 386, 387
Burnside, General A. E., 260
Butler County, Pa., 33, 45, 108
Butler County Historical Society, 108
Butler, John B., 67, 70
Buttendorf, Fritz, employee, 177

cables, hemp, 63, 64; replaced by wire rope, 67, 68, 69; parallel wire, construction, 75, 76, 79, 81, 82, 84, 93, 104, 181, 284; spinning, 81, 84, 93, 94, 169, 209, 279-281, 390-398; wrapping, 81, 94, 181, 281, 395; transatlantic, proposed by Roebling, 127-128, 145, 146
caisson disease, 338, 361-367
Campbell, William H., 147
canary breeding, 42, 109
Cassatt, A. J., 198
Central Park, New York, obelisk, 142
Chancellorsville, 257, 265
Chesapeake and Ohio Canal, 56
Chester and Holyhead Railway, 161
Chicago, 70, 71, 231
cholera epidemic, in Germany, 39; at Niagara Falls, 174-175
Cincinnati, 207, 218-224, 238, 245, 268, 269, 281, 286
Cincinnati-Covington Bridge, first projected, 219; early plans, 220-223; financial difficulties, 219, 229, 238, 239, 242, 267, 273; legislation, 223-230; Roebling becomes the engineer, 231; construction, 231-246, 269-290; W. A. Roebling's part in building, 275-290; opening, 286-288; service, 290-292
Civil Engineer, designation, 68
Civil Engineers, American Society of, 316, 404
Civil War, 243-246; John Roebling's activities in, 123, 247, 248, 249; W. A. Roebling's service in, 247-266
Clark, Colonel Emmons, 410
Clement, Saxonburg pastor, 118
Clermont, the, 295
Cleveland, Grover, 410, 411, 415
Coffee, Peter, pilot, 296
Cold Spring, N. Y., 265
Collingwood, Francis, 198, 381, 382
Columbus, Ohio, 113
Company A, National Guard, New Jersey, 248
Cooper, Peter, 71, 135, 136, 175
Cooper, Theodore, 198, 216
Corson, Dr., 262
Couper, William, sculptor, 321
Covington, Ky., 220-222, 228, 229, 239, 242-245, 264, 268, 269, 271, 281, 286
cradling of cables, 94
Craig, Neville B., editor, 91
Cynthiana, Kentucky, 268

Daniel Webster, the, 261-262
Danville, Kentucky, 168
Darmstadt, 21, 22
Dayton, Ohio, 238
Delaware and Hudson Canal, 69, 101-105, 145, 166
Delaware and Raritan Canal, 136
Delaware aqueduct, 103-105, 166
Delaware City, 28
Delaware River, 102, 103, 149
Dickens, Charles, 62
Dietelyn, J. F. W., professor, 11, 12
Dircksen, Cornelius, ferryman, 295
Dubail, Madam, innkeeper, 115

INDEX 451

Duke of Newcastle, the, 215
Dumba, Constantin, ambassador, 152
Dumfries, Virginia, 252
Durham bridge, failure, 166

Early, General Jubal A., 264
East River, 296, 297, 299, 303
East River Bridge. *See* Brooklyn Bridge
Edison, Thomas, 341
Edson, Franklin, mayor, 410
Eisenhardt, brother-in-law of Baehr, 110
Ellet, Charles, rival bridgebuilder, 55–60, 80, 86, 158–166, 170–173, 192, 221, 227
Emerson, Ralph Waldo, 130, 204
Engineering (London), 289, 306, 307
engineering as a profession, 111, 226
engineering education, 11, 12, 196, 197
Engineering News, 215, 401
Erfurt, 3, 10, 48
Erie Canal, 29, 47
Eschwege, 9, 23
Estabrook, Henry D., speaks at unveiling of Roebling monument, 321, 417
Etzler, Y. A., 18, 19, 20, 22, 34
Everson, W. H., ironmaster, 92, 93
Eytelwein, J. A., professor, 12

Fairmount Bridge, built by White and Hazard, 12, 78, 79; built by Ellet, 57, 86, 221
Farrington, E. F., master carpenter, 274, 278, 289, 332, 382–384
Field, Cyrus W., 128, 412
Finley, James, bridgebuilder, 12
Finnell, John W., 244, 245, 267
Fisher, Otis, 264
Fort Sumter, 245, 247
Frankfort, Kentucky, 268
Fredericksburg, Virginia, 253, 254, 256, 257, 262, 265
Freeport, Pennsylvania, 50
Frémont, General John C., 123

French Revolution, 13; 1830 July Revolution, 18
Front Royal, Virginia, 253
Fulton Ferry, the, 295, 296, 319, 327
Fulton, Robert, 295, 297

Gerstner family, Saxonburg, 119
Gettysburg, Battle of, 258–261, 265
Giesy, Helena, romance, 36
Gill, E. H., engineer, 48, 49, 100
Glasgow, Kentucky, 267
Golden Gate Bridge, 149, 180
Goethe, Johann Wolfgang von, 3, 7, 8, 16
Gosewisch, Saxonburg artist, 118
Gotha, 3, 118
Grabe, August, Saxonburg colonist, 34
Grant, Ulysses S., 318
Great Western Railway of Canada, 158, 179
Greeley, Horace, editor, 313
Greenwood, Mr., bridge director, 230, 232

Haigh, J. Lloyd, contractor, 388, 392
Halleck, General H. W., 260
Harper, John, 215
Harpers Ferry, 253, 255, 256, 265
Harrisburg, Pa., 33, 51, 74, 75, 84
Harrison, William Henry, 219
Hegel, Georg Wilhelm Friedrich, 13, 14, 16, 39, 128
Heine, Heinrich, 16
Helmbold, Ernest, Saxonburg colonist, 115, 119
Henne, German teacher, 113, 194
Henry Barclay, the, 21, 22, 34
Herting, Ernst, Saxonburg colonist, 41, 109, 110, 112, 114, 140
Herting, Johanna, 41. *See* Roebling, Mrs. John A.
Herting, Leonora, 114
Hewitt, Abram S., 135, 136, 300, 387, 413–415
High Falls aqueduct, 105, 119, 137, 165, 195
Hollidaysburg, Pa., 50, 61, 62, 72

INDEX

"Honeymoon Bridge," 190–192
Hood, General John Bell, 265
Hooker, General Joseph, 252
Horn, Saxonburg colonist, 117
Humphries, General A. A., 318
Huntingdon, Pennsylvania, 33
hydropathy, Roebling's interest in, 120, 174, 275

inclined planes, 61–63, 72, 73, 141, 168
inclined stays, 94, 171, 172, 180, 282, 283, 285, 310

Jackson, President Andrew, 29
Jay, John, 412
Jena, 3, 7, 13, 17
Johnson, Jeremiah, historian, 297
Johnstown, Pa., 50, 61, 62, 66
Journal of Commerce, Roebling's letters in, 127, 145, 146, 300

Kalbfleisch, Martin, mayor, 313, 315
Kant, Immanuel, 16
Keefer, Samuel, 158, 190, 192
Keil, Dr. William, 36
Kentucky River Bridge, 168, 169, 213, 260
Kenzie, Mr., carpenter, 175
Kingsley, William C., 301, 302, 303, 314, 371, 406, 414
Kinsella, Thomas, editor, 369
Kirkwood, John J., engineer, 316
Kittanning, Pennsylvania, 51
Kleber family, Pittsburgh, 118
Knoch, Amelia Roebling, 107, 117
Knoch family, blacksmiths, 117
"Know Nothing" Movement, 74
Koch, Dr., Saxonburg doctor, 120
Kunze, and son, mechanics, 140

Lackawaxen, Pa., 102, 103, 165
Lackawaxen aqueduct, 104, 137
Lafayette, General, 56
Lamb, Saxonburg colonist, 117, 118
Lange, Carl, 138, 139
Latrobe, Benjamin H., 316
Laurel Hill, 52, 53

Lee, General Robert E., 258
Lehigh Water Gap, 63
Lentz, Major Carl, 266
Lewiston-Queenston bridge, 176, 189, 192
Lexington, Ky., 220, 267, 268, 292
Lexington and Southern Kentucky Railroad, 168
Licking River, 222; bridge, 227, 228
Lies, Ludwig, 9
Lincoln, Abraham, 245, 247, 248, 263, 276
Lincoln, Nebraska, 163
Lindenthal, Gustav, 88
Long, Colonel, Kentucky, 222, 223
Louisville, Kentucky, 219, 292
Low, Seth, mayor, 403
Luther, Martin, 3, 8
Lyman, Colonel Theodore, 261

Macdonald, Charles, bridgebuilder, 198
Mahan, D. H., professor, 98
"Manifest Destiny," 222
Martin, Charles C., engineer, 198, 381, 382, 404
Mauch Chunk, Pennsylvania, 62, 63
Mayerstedt, Saxonburg weaver, 117
McAlpine, William J., engineer, 316
McClellan, General George B., 252
McCue, Alexander, 301, 302, 315
McDowell, Dr., Pittsburgh, 98
McDowell, General Irvin, 252
McNulty, George W., engineer, 381
Meade, General George Gordon, 258, 259, 261
Meigs, General Montgomery C., 253
Meissner, Carl, 40, 168
Meissner, Mrs. Carl (Amalia Roebling), 39, 40, 168
Meissner, Fritz, 39, 168, 211, 241
Memphis, 219, 267, 292
Menai Straits Bridge, 186, 307
Methfessel, Mrs. Anton Gottlieb (Laura Roebling), 120, 202, 212
Metternich, Prince von, 16
Minor, D. K., editor, 71, 86, 97, 159, 160, 162

INDEX

Monongahela Bridge, 89–100, 160
Monongahela River, 49, 89, 90, 96
Montrose bridge, 166
Monument to Roebling, 321, 417
Morgan, Colonel John, 267, 268, 271, 272, 273
Morgan, John Pierpont, 258
Morris Canal, 69, 168
Morris, Robert, patriot, 33
Mount Pisgah, inclined plane, 62, 63
Mühlhausen, Thuringia, 3, 4, 7, 18, 19, 20, 21, 37, 40, 41
Münzer, Thomas, reformer, 8
Murphy, Senator Henry C., 301, 302, 303, 304, 315
Murphy, Colonel William A., 251

Napoleon, 7, 13, 16, 17, 112
Napoleonic wars, 7, 8, 13, 17, 110, 112, 218
Nashville, 267
Nassau, the, 295, 296
National Guard, New Jersey, 248
National Guard, New York, 248, 410, 412
National Road, the, 90
Neher, cornetist, 116, 117, 118
Neversink aqueduct, 105, 195
New Brighton, Pa., 119, 213
New Castle, Pennsylvania, 48
New Harmony, Indiana, 34
New Orleans, 50, 219, 229, 267
Newport, Ky., 227, 228, 245, 281, 286
New York Bridge Company, 303
New York Central Railroad, 158, 179
New York City, 48, 125, 138, 299, 315, 378, 379, 397
Niagara-Clifton Bridge, 190–192
Niagara footbridge, 163–165
Niagara Railway Suspension Bridge, 145, 157, 158, 167–189
Ninth Regiment, N. Y. State Militia, 248
North Sea Mine Barrage, 153, 154

Obelisk in Central Park, 142
Ohio River, 49, 170, 217, 225, 235, 236, 237, 290–292

Ohio River Bridge. *See* Cincinnati Bridge, *also* Wheeling Bridge
Oil City Bridge, 142
O'Neill, John, 73, 74
Oregon, 36
Oregon Statesman, editorials, 36
Owen, Robert, colony, 34
Owl's Head, meeting at, 301–303

Paine, Colonel William H., 319, 398
Panic of 1837, 222
Panic of 1857, 239–241
Paris, July Revolution (1830), 18
Paris, Kentucky, 268
patents, by Roebling, 46, 47, 66, 75, 146, 213
Pennsylvania, 32–34, 67, 69, 74, 75
Pennsylvania Canal, 32, 33, 47, 50, 79, 80, 85, 240; Commission, 67, 84
Pennsylvania Railroad, 51, 126; Roebling forecasts, 124–127
Philadelphia, 57, 59, 60, 62, 74, 78, 101, 124–127; Roebling's first impressions, 30, 31
Pittsburgh, 32–34, 51, 62, 74, 80, 88, 89, 99, 113, 114, 207, 212, 217; bridges built by Roebling, 79–89, 89–100, 205–216; W. A. Roebling attends school in, 114, 194; his impressions of, 207, 210
Pittsburgh-Allegheny Bridge, 205–216
Polycarp, Saint, Bishop of Smyrna, 4
Poolesville, Maryland, 249
Pope, General John, 255
Pope, Thomas, 297
Port Jervis, New York, 105
portage railways, 33, 61, 72, 73
Priessnitz, Vincenz, 174
Prime, New York historian, 298
Prince of Wales, 185, 215
Provis, John, English engineer, 186

Quantico, Virginia, 251
Quebec Bridge, 198
Queenston, Ontario. *See* Lewiston-Queenston bridge

454 INDEX

Race Street Bridge, Cincinnati, 221
Ransom, Richard, 230, 231, 238, 239, 242
Rapp, George, colony, 34
Rappahannock River, 254, 255, 257
Reading, Pennsylvania, 32
Regnitz River bridge, 14
Rensselaer Polytechnic Institute, 195–199, 203
Rhule, David, 255, 256
Richmond, Virginia, 257, 264
Ricketts, Palmer C., Director, 199
Riedel, Edmund, 114, 241, 320
Riedel, Julius, 112–114, 116, 194
Riedel, Mrs. Julius (Leonora Herting), 114
Robinson, William, river pilot, 99
Roebling, Amelia. *See* Knoch, Amelia Roebling
Roebling, Charles Gustavus, 141, 142, 199, 200–204, 324
Roebling, Christoph Polycarpus, 4, 6, 11, 16, 20, 40
Roebling, Frau Christoph Polycarpus (Friederike Dorothea Mueller), 4, 5, 6, 16, 22, 39, 131
Roebling, Edmund, 170
Roebling, Elvira, 120, 279
Roebling, Emily, 120
Roebling, Ernst Wilhelm, publisher, 23
Roebling, Ferdinand William, 120, 200, 203, 207, 241, 324
Roebling, Friederike Amalia, 11, 39, 40. *See also* Meissner, Mrs. Carl
Roebling, Hermann Christian ("Christel"), 11, 30, 33, 40
Roebling's, John A., Sons Company, 380, 388
Roebling, Mrs. John A. (Johanna Herting), wife of Roebling, 41, 140, 144, 168, 208, 211, 212, 273, 275; children, 41, 120, 141, 147, 170, 200, 210, 212
Roebling, John August, birth, 3, 5, 17; childhood scene, 4–8; education, 8–14, 36; influence of Hegel, 12, 128; inspired by "miracle" bridge, 14; builds roads in Westphalia, 15, 16, 19; a rebellious liberal, 17; decides to emigrate, 18, 19; organizes party of colonists, 19–21, 34, 37; farewell to Mühlhausen, 21; voyage to America, 22–28, 139, 175; impressions of Philadelphia, 30; comments on American life, 31, 32; attitude toward Slavery, 32; as a pioneer farmer, 33–38, 42, 108; founds village of Saxonburg, 38; writes friends to join him, 35, 37, 40; death of mother, 39; marriage, 41; children, 41, 120, 141, 147, 170, 200, 210, 212; death of brother, 43–45, 109; becomes a naturalized citizen, 45; builds dams and locks on canal, 49, 50; surveys railroad route over Allegheny Mtns., 51–54, 66; seeks work on bridgebuilding, 55–60; invents wire rope, 63–65; makes first wire rope, 65, 66, 111; secures acceptance of wire rope, 67–77; builds suspension aqueduct over Allegheny River, 79–86; builds his first suspension bridge (over Monongahela River), 89–100; builds suspension aqueducts on Delaware and Hudson Canal, 103–106; forecasts future of railroads, 124–127; forecasts transatlantic cable, 127–128, 145–146; writes on "Matter and Spirit," 128–130; moves to Trenton, 135–136; places Charles Swan in charge, 136, 142; letters to Swan, 137–139, 143, 144, 167; proposes span over Niagara, 145, 157; competition with Ellet, 158–163; builds Niagara Railroad Suspension Bridge, 165–189; rebuilds Wheeling Bridge, 173; cholera epidemic at Niagara, 174, 175; letters in Panic of 1857, 240–241; starts Kentucky River Bridge, 168, 169; Niagara bridge completed, 177–179; report to directors, 179–183; builds Pittsburgh-Allegheny

INDEX

bridge, 205–216; early plans for Cincinnati Bridge, 223–227; engaged to build Cincinnati Bridge, 231; starts foundations, 232, 233; difficulties and interruptions, 234–246; activities during Civil War, 123, 247, 248, 249; resumes work on Cincinnati Bridge, 270–290; death of wife, 275; feels strain of years, 279; completes Cincinnati Bridge, 270–290; proposes Brooklyn Bridge, 300; appointed chief engineer, 304, 305; makes surveys and plans, 307, 308; reports on plans, 308–313; plans approved by board of consulting engineers, 316–318; accident on work, 319; death, 320; tributes to his genius, 413–418; inventions, 45–50, 54, 64–66, 87, 88, 123, 146, 147, 213, 284, 285; writings, 14, 23, 58, 71, 86–89, 97, 98, 124–129, 145–147, 161, 162, 179–184, 224–226, 230, 270, 284, 285, 300, 308–313; personality and character, 6–11, 14–22, 24–27, 32, 36, 37, 39, 43, 45, 60, 74, 76, 77, 82, 83, 121–124, 131, 139, 140, 144–148, 167, 169–170, 174–177, 181, 194, 203–208, 214, 226, 240, 247–249, 258, 262, 270, 273, 275, 276, 279, 313, 314, 318, 320–322, 418

Roebling, John Augustus, II, 199

Roebling, Carl (Karl Friedrich), 11, 16, 22, 24, 30, 33, 37, 40, 43, 107

Roebling, Laura (Mrs. Anton Gottlieb Methfessel), 120, 202, 212

Roebling monument at Trenton, 321, 417

Roebling, town of, New Jersey, 141, 149

Roebling, Mrs. Washington A., 261, 265, 369, 399, 404, 413, 414

Roebling, Washington Augustus, birth, 41, 44; characteristics, 107, 200, 201, 203; childhood, 68, 69, 107, 109, 120, 121, 122, 202; reminiscences, 65, 68, 69, 108–122, 140, 142, 194; education, 112, 113, 114, 194–198, 205; benefactions, 198–199; family influence on character, 200–204; work on Allegheny Bridge, 205–216; Civil War experiences, 247–266; appearance and manner, 261, 263; marriage, 265; work on Cincinnati Bridge, 275–290; trip to Europe, 306, 323; succeeds father on Brooklyn Bridge, 323, 324; down in the caissons, 324–367; stricken by caisson disease, 366, 367; completes the Brooklyn Bridge, 368–408; tributes at dedication, 412–415

Roebling, Washington Augustus, III, 142

Roosevelt, Robert B., 400, 401

Royal Polytechnic Institute, Berlin, 11–14, 49

Sandy and Beaver Canal, 48, 49

Saxonburg, Pennsylvania, 33, 34, 38–41, 65–69, 76, 81, 83, 107, 108–120, 140, 141, 149

Schelling, Friedrich Wilhelm J. von, 13

Schiller, Johann Christoph Friedrich von, 3

Schlatter, Charles L., 51, 53, 66, 80

Schubert, Franz, 16

Schuylkill Falls, bridge at, 12, 78, 79

Scott, General Winfield, 226

Seguin, French engineer, 161

Seibert, Conrad, 140

Serrell, Edward W., 158, 176, 189, 192

Serrell, John J., 316

Seventh Regiment, National Guard, 410

Shinkle, Amos, 229, 230, 232, 234, 245, 246, 267, 286, 288, 289, 290

Sickles, General D. E., 250, 259

Sixth Street Bridge, Pittsburgh, 205–216

Smith, Dr. Andrew H., 362

Smith, Kirby, 268, 269

INDEX

Smithfield Street bridge, Pittsburgh, 89–100
Snodgrass, John, 72–74
South Carolina, 34, 141
St. Louis, Missouri, 40, 56, 124–127
stays, inclined, 94, 171, 180, 282, 283, 285, 310
steel, for bridges, 309, 398, 399; for wire cables, 309, 385–387
Steele, J. Sutton, engineer, 316
Stephenson, Robert, 161, 166, 186
stiffening trusses, 95, 99, 180, 183, 310
Stockton, Mrs., neighbor, 142
Stone, Ethan, 219
Stranahan, James S. T., 400, 402
Stuart, Charles B., 157, 158, 159, 160, 161, 163, 192
Stuebgen, Saxonburg blacksmith, 116
Stueler, Friederich August, 10
Summit, Pennsylvania, 147
suspension aqueducts, 79–88, 99, 103–106, 145, 166
suspension bridges, Roebling's early interest in, 12, 14; early history of, 12, 14, 78, 79; failures of, 12, 79, 95, 166, 170–173, 185, 189, 191, 228, 396
Suspension Bridge, New York, 168
Swan, Charles, 136, 137, 142, 143, 167, 195, 196, 205, 273

Tacoma Narrows Bridge, 171, 185
Telford, Thomas, 161, 186
Thaw, John, 96
Thierry, Edward, 48
Thuringia, 3, 17, 35, 42
Times, New York, 171, 400, 401
Titanic disaster, 142
Tod, Governor, 269
Tolly, Saxonburg colonist, 118, 119
"Tom Thumb," first locomotive, 135
Townsend, R., & Co., 80
Trenton Academy, 194, 195
Trenton Iron Company, 135, 136, 176
Trenton, New Jersey, 135, 136, 145, 148, 247; Roebling monument in, 321, 417

Trinity Church, New York, 374, 411
Troy, New York, 195–199, 203
Trujillo, Francisco, engineer, 198
Tweed, William M. ("Boss"), 315, 370
Twenty-Third Regiment, National Guard, 412

Unger, Dr. Ephraim Solomon, 10, 13, 48

Vienna, Congress of, 17

Wallace, General Lew, 269
Walsh, Homan, kite exploit, 163
Warren, Emily, 261, 265. *See also* Roebling, Mrs. Washington A.
Warren, General Gouverneur K., 258, 259, 261, 262, 265
Washburn, J. & C., 71, 72, 168
Washington, George, 41, 378
Waterloo, battle of, 110
Waterloo, Iowa, 231
Weimar, 3
Weldon Railroad, 265
Wernwag, Louis, bridgebuilder, 89
Westphalia, 15, 19, 42
Wheeling *Intelligencer*, 171
Wheeling, West Virginia, bridge at, 170–173, 181–183, 227
White and Hazard, first wire bridge, 12, 78, 79
Wickenhagen, Saxonburg colonist, 116
Wickersham, Colonel S. M., 97
Wilderness, Battle of the, 261, 263
Wilkins, William, 91
Williamsburg Bridge, over East River, 142, 198, 308, 387
Winkler, J. K., *Morgan the Magnificent*, 258
wire, 93, 105, 136, 141, 188, 189, 309, 385, 386
wire cables, parallel, 75, 76, 79, 81, 82, 84; Roebling's contributions, 80, 81, 284; spinning, 81, 84, 93, 169, 280, 390–392; squeezing, 81; wrapping, 81, 281

wire rope, conceived by Roebling, 64, 65; replaces hemp cables on portage railways, 66, 69; first wire rope factory, in Saxonburg, 65; increased demand for, 69, 70; Roebling industry removed to Trenton, 135, 136; patents, 65, 146; progress of industry, 147, 148, 149, 150, 151; demand for, during World War I, 152–154
Worcester, Massachusetts, 71, 168
World War I, 152–156, 173
World War II, 173
Wurts, William, Charles, and Maurice, 101, 102

Young, Andrew, contractor, 59, 60
Young, Jesse Bowman, the *Battle of Gettysburg*, 259

Zschokkes, Heinrich, *Hours of Devotion*, 113

TECHNOLOGY AND SOCIETY

An Arno Press Collection

Ardrey, R[obert] L. **American Agricultural Implements.** In two parts. 1894

Arnold, Horace Lucien and Fay Leone Faurote. **Ford Methods and the Ford Shops.** 1915

Baron, Stanley [Wade]. **Brewed in America:** A History of Beer and Ale in the United States. 1962

Bathe, Greville and Dorothy. **Oliver Evans:** A Chronicle of Early American Engineering. 1935

Bendure, Zelma and Gladys Pfeiffer. **America's Fabrics:** Origin and History, Manufacture, Characteristics and Uses. 1946

Bichowsky, F. Russell. **Industrial Research.** 1942

Bigelow, Jacob. **The Useful Arts:** Considered in Connexion with the Applications of Science. 1840. Two volumes in one

Birkmire, William H. **Skeleton Construction in Buildings.** 1894

Boyd, T[homas] A[lvin]. **Professional Amateur:** The Biography of Charles Franklin Kettering. 1957

Bright, Arthur A[aron], Jr. **The Electric-Lamp Industry:** Technological Change and Economic Development from 1800 to 1947. 1949

Bruce, Alfred and Harold Sandbank. **The History of Prefabrication.** 1943

Carr, Charles C[arl]. **Alcoa, An American Enterprise.** 1952

Cooley, Mortimer E. **Scientific Blacksmith.** 1947

Davis, Charles Thomas. **The Manufacture of Paper.** 1886

Deane, Samuel. **The New-England Farmer,** or Georgical Dictionary. 1822

Dyer, Henry. **The Evolution of Industry.** 1895

Epstein, Ralph C. **The Automobile Industry:** Its Economic and Commercial Development. 1928

Ericsson, Henry. **Sixty Years a Builder:** The Autobiography of Henry Ericsson. 1942

Evans, Oliver. **The Young Mill-Wright and Miller's Guide.** 1850

Ewbank, Thomas. **A Descriptive and Historical Account of Hydraulic and Other Machines for Raising Water,** Ancient and Modern. 1842

Field, Henry M. **The Story of the Atlantic Telegraph.** 1893

Fleming, A. P. M. **Industrial Research in the United States of America.** 1917

Van Gelder, Arthur Pine and Hugo Schlatter. **History of the Explosives Industry in America.** 1927

Hall, Courtney Robert. **History of American Industrial Science.** 1954

Hungerford, Edward. **The Story of Public Utilities.** 1928

Hungerford, Edward. **The Story of the Baltimore and Ohio Railroad, 1827-1927.** 1928

Husband, Joseph. **The Story of the Pullman Car.** 1917

Ingels, Margaret. **Willis Haviland Carrier, Father of Air Conditioning.** 1952

Kingsbury, J[ohn] E. **The Telephone and Telephone Exchanges:** Their Invention and Development. 1915

Labatut, Jean and Wheaton J. Lane, eds. **Highways in Our National Life:** A Symposium. 1950

Lathrop, William G[ilbert]. **The Brass Industry in the United States.** 1926

Lesley, Robert W., John B. Lober and George S. Bartlett. **History of the Portland Cement Industry in the United States.** 1924

Marcosson, Isaac F. **Wherever Men Trade:** The Romance of the Cash Register. 1945

Miles, Henry A[dolphus]. **Lowell, As It Was, and As It Is.** 1845

Morison, George S. **The New Epoch:** As Developed by the Manufacture of Power. 1903

Olmsted, Denison. **Memoir of Eli Whitney, Esq.** 1846

Passer, Harold C. **The Electrical Manufacturers, 1875-1900.** 1953

Prescott, George B[artlett]. **Bell's Electric Speaking Telephone.** 1884

Prout, Henry G. **A Life of George Westinghouse.** 1921

Randall, Frank A. **History of the Development of Building Construction in Chicago.** 1949

Riley, John J. **A History of the American Soft Drink Industry:** Bottled Carbonated Beverages, 1807-1957. 1958

Salem, F[rederick] W[illiam]. **Beer, Its History and Its Economic Value as a National Beverage.** 1880

Smith, Edgar F. **Chemistry in America.** 1914

Steinman, D[avid] B[arnard]. **The Builders of the Bridge:** The Story of John Roebling and His Son. 1950

Taylor, F[rank] Sherwood. **A History of Industrial Chemistry.** 1957

Technological Trends and National Policy, Including the Social Implications of New Inventions. Report of the Subcommittee on Technology to the National Resources Committee. 1937

Thompson, John S. **History of Composing Machines.** 1904

Thompson, Robert Luther. **Wiring a Continent:** The History of the Telegraph Industry in the United States, 1832-1866. 1947

Tilley, Nannie May. **The Bright-Tobacco Industry, 1860-1929.** 1948

Tooker, Elva. **Nathan Trotter:** Philadelphia Merchant, 1787-1853. 1955

Turck, J. A. V. **Origin of Modern Calculating Machines.** 1921

Tyler, David Budlong. **Steam Conquers the Atlantic.** 1939

Wheeler, Gervase. **Homes for the People,** In Suburb and Country. 1855